C0-AJT-123

Evolutionary Biology

VOLUME 19

A Continuation Order Plan in available for this series. A continuation order will bring delivery of each new volume immediately upon publication. Volumes are billed only upon actual shipment. For further information please contact the publisher.

Evolutionary Biology

VOLUME 19

Edited by

MAX K. HECHT
Queens College of the City University of New York
Flushing, New York

BRUCE WALLACE
Virginia Polytechnic Institute and State University
Blacksburg, Virginia

and

GHILLEAN T. PRANCE
New York Botanical Garden
Bronx, New York

PLENUM PRESS • NEW YORK AND LONDON

The Library of Congress cataloged the first volume of this title as follows:

Evolutionary biology. v. 1– 1967–
New York, Appleton-Century-Crofts.
v. illus. 24 cm annual.
Editors: 1967– T. Dobzhansky and others.
1. Evolution—Period. 2. Biology—Period. I. Dobzhansky, Theodosius Grigorievich, 1900–
QH366.A1E9 575′.005 67-11961

ISBN 0-306-42134-8

A Division of Plenum Publishing Corporation
233 Spring Street, New York, N.Y. 10013

Printed in the United States of America

Contributors

Marc Bekoff • *Department of Environmental, Population, and Organismic Biology, University of Colorado, Boulder, Colorado 80309*

Charles E. Blohowiak • *Computing Center, Virginia Polytechnic Institute and State University, Blacksburg, Virginia 24061*

John A. Byers • *Department of Biological Sciences, University of Idaho, Moscow, Idaho 83843*

Robert R. Capranica • *Section of Neurobiology and Behavior, Division of Biological Sciences, Cornell University, Ithaca, New York 14853*

Dewey M. McLean • *Department of Geological Sciences, Virginia Polytechnic Institute and State University, Blacksburg, Virginia 24061*

Laurence D. Mueller • *Department of Zoology and Program in Genetics, Washington State University, Pullman, Washington 99164*

Eviatar Nevo • *Institute of Evolution, University of Haifa, Haifa, Israel*

Monica Riley • *Biochemistry Department, State University of New York at Stony Brook, Stony Brook, New York 11794*

Bruce Wallace • *Department of Biology, Virginia Polytechnic Institute and State University, Blacksburg, Virginia 24061*

Preface

Evolutionary Biology, of which this is the nineteenth volume, continues to offer its readers a wide range of original articles, reviews, and commentaries on evolution, in the broadest sense of that term. The topics of the reviews range from anthropology and behavior to molecular biology and systematics.

In recent volumes, a broad spectrum of articles have appeared on such subjects as natural selection among replicating molecules *in vitro*, mate recognition and the reproductive behavior in *Drosophila*, evolution of the monocotyledons, species selection, and the communication network made possible among even distantly related genera of bacteria by plasmids and other transposable elements. Articles such as these, often too long for standard journals, are the stuff of *Evolutionary Biology*.

The editors continue to solicit manuscripts on an international scale in an effort to see that every one of the many facets of biological evolution is covered. Manuscripts should be sent to any one of the following: Max K. Hecht, Department of Biology, Queens College of the City University of New York, Flushing, New York 11367; Bruce Wallace, Department of Biology, Virginia Polytechnic Institute and State University, Blacksburg, Virginia 24061; Ghillian T. Prance, New York Botanical Garden, Bronx, New York 10458.

The Editors

Contents

1

Discontinuous Processes in the Evolution of the Bacterial Genome

MONICA RILEY

INTRODUCTION

Much of the change that takes place in genomic DNA in the course of evolution appears to be gradual in the sense that it is brought about by the sequential accumulation of individual single-base-pair replacements. However, changes of a more drastic type also take place, changes that alter genomic DNA by affecting more than one base pair as a result of a single genetic event. Rearrangements occur such as inversions, transpositions, substitutions, additions, and deletions. These changes, more drastic than point mutations, have the potential of changing more than one phenotypic function as a consequence of a single rearrangement event. They also have the potential of introducing discontinuities into the processes of more gradual evolutionary change, possibly enabling leaps of multiple changes of phenotypes against a background of steady, gradual change. We need to know more about the mechanisms, consequences, and frequencies of discontinuous genomic alterations, and we need to take such discontinuous changes into account when we undertake comparative analyses of amino acid and nucleotide sequences of evolutionarily related genes.

Studies on the evolution of bacterial DNA can contribute to the understanding of molecular mechanisms of evolutionary change. Bacteria are favorable materials for work on the molecular level, and for some

MONICA RILEY • Biochemistry Department, State University of New York at Stony Brook, Stony Brook, New York 11794.

bacteria a body of genetic data has been assembled as well. Information that bears on mechanistic questions is beginning to emerge from some of the studies on the molecular bases of evolutionary and mutational changes in the genomic DNA of bacteria. Studies on both naturally occurring and mutant genomic rearrangements at the molecular level are providing information on the types of discontinuous changes that can and do occur at the DNA level in bacteria, the molecular mechanisms by which they arise, and the phenotypic consequences of the rearrangements. In addition, with the recent improvements in methodology, more and more nucleotide sequences of evolutionarily related genes are accumulating, allowing detailed analysis of the relationships between the genes, base by base and codon by codon, providing ever more detailed information on the complexities of the processes by which divergence took place. A summary of selected studies bearing on discontinuous change in bacterial genomic DNA is presented here, with emphasis on work done with the enteric bacteria.

INTERNAL REARRANGEMENTS

A distinction can be made between the kinds of changes that occur within a single genome, to be dealt with in this section, and changes that involve interaction between all or parts of two different genomes, to be dealt with in the next section.

Large-Scale Internal Rearrangements

Molecular Characteristics

Internal rearrangements in bacterial DNA come in all sizes, from massive changes that affect the configuration of the entire genome down to changes involving only a few bases within a gene. They all share the property that they introduce a discontinuity in the DNA. New join points are created where DNA sequences are joined together that were not joined before. Transpositions or inversions create two new join points or junctions, one at each end of the moved segment. Deletions and tandem duplications create one new join point.

Inversions, duplications, and transpositions can be brought about by illegitimate recombination, site-specific recombination, replication error, or by homologous recombination between nonidentical repeated sequences in the genome. Hill and co-workers delineated one category of rearrangement. They showed that the genes for ribosomal RNA, which are present in multiple copies in the enterobacterial genome, serve as foci for unequal crossover events or errors in replication that result in rearrangements. Deletion or duplication of the DNA that lies between the *rrn* loci occurs when the *rrn* sequences are in direct repeat configuration; inversion occurs if the sequences are in inverted relationship (Hill *et al.*, 1977, Capage and Hill, 1979; Hill and Harnish, 1981, 1982). In addition to the *rrn* loci, other repeated sequences exist in the bacterial genome that can mediate internal recombination, serving as homologous sequences at end points of internal rearrangements (Lin *et al.*, 1984; Schmid and Roth, 1983*b*; Sammons and Anagnostopoulos, 1982). By experimental intervention, rearrangements can be engineered by introducing portable regions of homology at the positions desired as end points of the rearrangement. Transposons are convenient tools for this type of genetic manipulation. A transposon can be introduced at two desired locations in the genome, thus setting the stage for internal homologous recombination and rearrangement (Chumley and Roth, 1980).

Internal rearrangements of each of the three main types, duplications, transpositions, and inversions, arise spontaneously in nature and some of these can be selected and isolated in the laboratory. Duplication mutants arise readily in bacterial populations, but they are not stable in the absence of a selective pressure for their maintenance [for review, see Anderson and Roth (1977)]. Tandem duplications are formed and deleted at relatively high rates in bacteria, about once in 10^3 cell generations. In wild-type populations back mutations are produced at high frequency by excising duplicated material through homologous recombination between duplicate sequences. Thus, duplications both arise and are reversed at high frequencies. To freeze duplications, selections have been devised that require the continuous presence of a duplicated region, preventing excision of the redundant sequences. In this way, duplications of parts of the bacterial genome have been selected and stably maintained that range in size up to as much as one-third of the genome (e.g., Straus and Hoffman, 1975; Straus and Straus, 1976).

Some duplications are flanked by repeated sequences, as is the case, for instance, for duplicated segments that are flanked with *rrn* genes (Capage and Hill, 1979). Other mutant duplications do not have repeated sequences at their join points, such as a duplication of the *arg*E gene in *Escherichia coli* (Charlier *et al.*, 1979) and a duplication within the *lac*I gene (Calos *et al.*, 1978). Duplications occasionally arise naturally and become stabilized in wild-type bacterial populations without any deliberate application of selection. An ordinary laboratory strain of *E. coli*, strain KL399, was found recently to harbor a large duplication that is not present in the wild type, apparently suffering no ill effects from carrying the extra DNA (Dykstra *et al.*, 1984).

Another major kind of rearrangement is transposition. Transposition mutants have been isolated by selecting for uptake of a gene into a mutant genome that lacks the wild-type allele, preventing incorporation of the gene at its normal location. In this way, the *lac* genes of *E. coli* have been positioned at several novel locations around the genome (Casadaban, 1976). Transposition also can occur by excision of circles of DNA from the genome by homologous recombination between repeated sequences, followed by reintegration at a third copy of the repeated sequence located elsewhere in the genome. This mechanism was first documented in a system that utilized replicate *rrn* genes (Hill and Harnish, 1982). Transpositions also have been artificially engineered by introducing multiple copies of the transposon Tn10 into the genome by genetic manipulation in order to serve as the necessary repeated sequences (Chumley and Roth, 1980; Schmid and Roth, 1980). Insertion sequences (IS's) are self-transposable genetic elements. The effects of movement of these special genetic elements will be discussed in a later section.

Mutants have been isolated in which segments of *E. coli* chromosomal DNA are inverted. Some of these were found to be flanked with *rrn* loci in inverted configuration (Hill and Harnish, 1981). One such mutation was found to have occurred spontaneously during culture maintenance in the laboratory. Hill and Harnish (1981) discovered that a seemingly ordinary and much used laboratory strain of *E. coli*, strain W3110, had undergone an inversion of 18% of its total genomic DNA. The progenitor of strain W3110, strain W2637, apparently experienced inversion at some time during its cultivation in the laboratory during the last 30 to 40 years, suffering no apparent deleterious effects as a result. Another type of inversion was isolated by selecting for any rearrangement that would bring a promoter sequence up to a promoterless *his* gene in such a way as to restore gene expression (Schmid and Roth, 1983*a*,*b*). This selection resulted in isolation of mutants with many types of rearrangements, inversions among them. The inversion mutants were rare compared with other kinds of rearrangements. I will return to this point later.

Some rearrangements affect the configuration of regulation genes. Mutants were isolated with new configurations of the *arg*E gene relative to the *arg*CBH genes (Charlier *et al.*, 1983). A set of small rearrangements involving inversion and/or duplication created alternate versions of the cluster of *arg* genes. The wild type version transcribes these genes in one direction, using two promoters. Mutants were isolated that transcribed the *arg*E gene in opposite direction from the *arg* CBH genes and carried either duplications or triplications of regulatory sequences. These experiments show that rearrangements that create new configurations of regulatory units can and do arise.

Rearrangements during Divergence of Enterobacterial Genomes

Two rearrangements have been found in spontaneously arising variants of *E. coli* as mentioned above, an inversion in strain W2637 and a duplication in strain KL399. Chromosomal rearrangements have also occurred in the course of divergence of enterobacteria from one another. A comparison of the genetic maps of *Salmonella typhimurium* and *E. coli* showed that an inversion of about one-eighth of the genome occurred in one or the other of these genomes, as illustrated in Fig. 1 (Casse *et al.*, 1973). There is no known effect of this inversion on the function of any gene within the inverted segment. Another example involves the *lac* operon. A comparison of the nucleotide sequences of the *lac* operons of *Klebsiella pneumoniae* and *E. coli* shows that the orientation of the *lac* I gene of one operon is inverted with respect to the other *lac* operon (Fig. 2) (Buvinger and Riley, 1985). Both *lac*I genes function by exerting negative control on the *lac* operon genes, and both repressors are inactivated by the inducers lactose and isopropylthio-B-D-galactopyranoside (IPTG) (Reeve and Braithwaite, 1974), therefore the gross aspects of function of the two *lac*I genes have not been affected by this inversion.

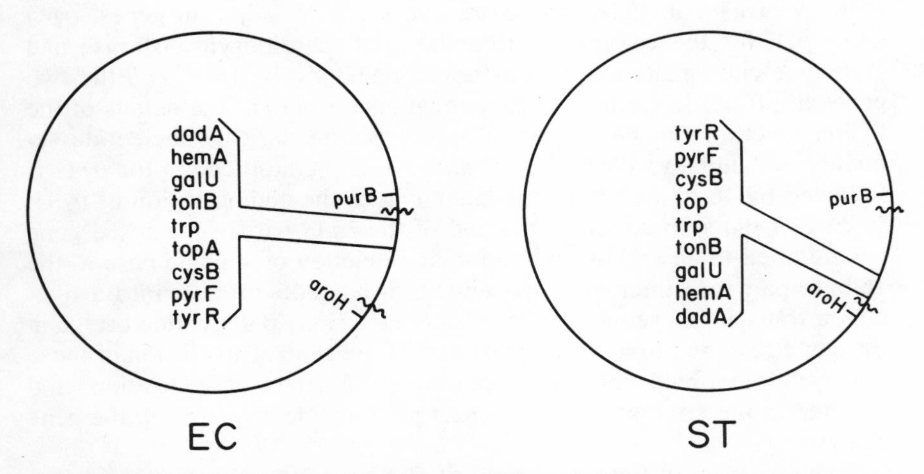

FIG. 1. Inversion in the genomes of *Escherichia coli* and *Salmonella typhimurium*. The locations of selected genes in the respective genetic maps are shown. The jagged lines indicate the end points of the inverted segment.

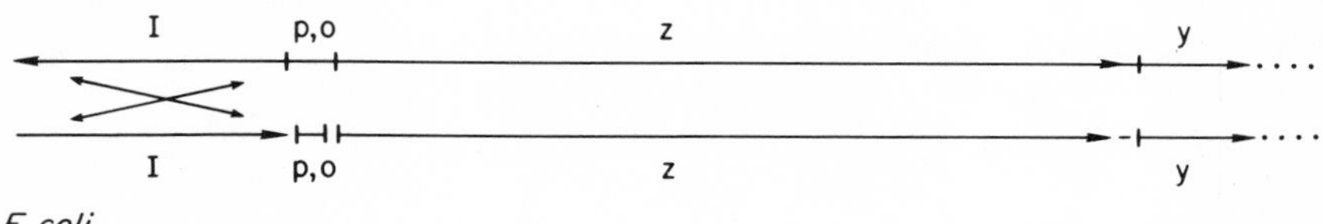

FIG. 2. Inversion of the *lac*I genes in the *lac* operon of *Escherichia coli* with respect to *Klebsiella pneumoniae*. Horizontal lines represent the genes for *lac*I (*lac* repressor), *lac*Z (-galactosidase), and P, O (promoter–operator regulatory region). Arrows above and below the lines show directions of transcription of the genes. Crossed arrows between the lines symbolize the inversion.

The genes of tryptophan biosynthesis are arranged differently in various bacteria, fungi and eukaryotes (Yanofsky, 1984). The genes are clustered together in *E. coli* and some other bacteria, but are separated into clusters of varied composition in still other bacteria and in *Neurospora crassa* and yeast. In some organisms there are two separate genes for two enzymes that are adjacent in the metabolic pathway, but in other organisms the genetic information for the two enzymatic activities is fused into one gene that codes for a multifunctional protein. The functional advantages of one arrangement versus another have not been fathomed. The molecular events by which fusion of two genes has occurred have been inferred by analysis of nucleotide and amino acid sequences. In *Serratia marcescens*, for instance, there are two separate, adjacent genes, *trp*G and *trp*D, for the enzymes anthranilate phosphoribosyltransferase and glutamine-chorismate amidotransferase, respectively. In *E. coli* the two genes are fused to form the bifunctional gene *trp*G,D. The details of the fusion junction can be examined by comparing the two nucleotide sequences (Yanofsky, 1984). In *S. marcescens* the stop codon for *trp*G is followed by 16 bases before the beginning of the coding region of *trp*D. In *E. coli* the stop codon at the end of the *trp*G part of the fused gene has mutated to an amino acid codon and deletion of a single base in the 16-base-pair (bp) intergenic spacer region has converted it into a five-codon translatable sequence that generates a short polypeptide bridge in the same reading frame as the *trp*D part of the gene. Thus fusion of these two genes involved one base replacement and one base deletion, and resulted in the creation of a new short polypeptide sequence in the gene product.

Comparison of the amino acid sequences of the *trp* polypeptides of *E. coli* and *N. crassa* reveals regions of homology that are interrupted in a few places by the appearance of short polypeptide segments that are present in one protein and not the other. These extra polypeptide segments might have arisen from past gene fusion events. Conversion of spacer nucleotide sequences into translated polypeptides introduces discontinuities in the evolutionary histories of parts of the amino acid sequences of such proteins, discontinuities in history that need to be taken into account when the sequences are analyzed from the evolutionary point of view.

Freedom of Genome Rearrangement

We can ask whether all genome rearrangements are equally free to form, or whether some rearrangements have deleterious effects, interfering with one or more vital functions of the cell. Is there freedom to form all possible rearrangements, or are there constraints? One possibility is that genes needing a higher copy number must be located near the origin of replication (Bachmann *et al.*, 1976). Part of the consequence of a rearrangement lies at the new join points. When a gene or a regulatory element is split by the rearrangement event and the two moieties of the split gene are joined to two other DNA sequences, then the function of the split DNA sequence is abolished. Thus, essential genes cannot lie at join points. If, however, an essential gene does not reside at an edge of the rearranged DNA, but lies totally within a segment of the DNA that is rearranged, the question arises of whether the function of that gene is affected by the rearrangement. The nucleotide sequence and genetic information of the gene itself is not altered in any way, but the location and orientation of the gene in the genome can be affected by the rearrangement event. Does gross location of a gene in the global map of the organism have any effect on function? Is there any information from studies on bacterial mutants about the effects of rearrangements on the function of moved genes?

There are mutants of *Bacillus subtilis* that have undergone massive rearrangements in the genome involving inversion and transposition, and there are merodiploid recombinants derived from these mutants that have undergone further rearrangement and partial duplication (Anagnostopoulos, 1976; Trowsdale and Anagnostopoulos, 1976; Schneider and Anagnostopoulos, 1983). In the so-called *trp*E26 rearranged strain, the *trp*E gene is split at a join point created by one of the rearrangement events, inactivating the gene. However, with respect to the many genes that lie within the rearranged segments, there is no known instance of loss of function, although there is a hint that an amino acid biosynthetic enzyme may be functioning poorly in one class of rearranged mutants.

There are numerous examples in *E. coli* of changes in the positions of genes that apparently have had no effect on the function of those genes. Examples were mentioned above of naturally occurring inversions, duplications, and transpositions that appear not to have suffered as a result of rearrangement. To examine this question further, the growth rates of some rearranged mutants have been determined relative to wild type in mixed culture experiments (Hill and Harnish, 1981, 1982). Although growth rate may not be a very good measure of relative fitness, a small competitive edge enjoyed by some rearranged mutants is worthy of note. In some cases, the mutants had growth rates that were indistinguishable from wild type, indicating that there was no measurable effect of changing the location or orientation of the rearranged genes under the conditions of the experiments; on the other hand, most of the rearranged mutants grew slower in mixed cultures, up to 4-5% slower relative to wild type, suggesting that some of these mutants had suffered mild deleterious effects from the rearrangements. Growth rate depressions of this magnitude can cause elimination of the slower growing strains from mixed cultures with wild type in a relatively short time.

There are indications that some rearrangements are lethal. In spite of using a sensitive assay, no inversions were detected in the *lac-att80* region of the *E. coli* chromosome (Konrad, 1977). In this experiment, two alleles of *lac*Z each containing a non-identical partial deletion within *lac*Z were placed at two distant loci in the chromosome of *E. coli*. The orientations of the partially deleted *lac*Z genes were such that two crossover events could generate a recombinant *lac*+ gene (and a doubly mutant *lac*− gene). In *E. coli* strains that carried a mutationally elevated recombination system, *lac*+ recombinants of this double crossover type were found. Another path for generation of *lac*+ recombinants could have been a properly placed single crossover which would generate a large inversion between the two *lac* alleles. Such *lac*+ recombinants would be characterized by carrying a certain defective phage attachment gene, since this gene would be disrupted by the single crossover recombination event. No *lac*+ recombinants of this type were observed, none with a single crossover and an inversion in the chromosomal DNA. Only double crossover recombinants with no genomic rearrangement were found.

In another experiment, as mentioned above, when a selection was applied that required repositioning of a promoter sequence without regard to the kind of rearrangement that accomplished the repositioning, duplications and transpositions were isolated, but very few inversions were found (Schmid and Roth, 1983a). Analysis of the inversions that *did* arise showed that the end points were not random, as if only a limited class of inversion events are viable (Schmid and Roth, 1983b).

Constraints on Genome Rearrangement

Since genome rearrangements are not often observed, the incidence of rearrangement phenomena may be low. There may be powerful constraints on gene movement in bacteria since closely related bacteria are observed to have similar genetic configurations, similarities that would be erased if a significant level of rearrangement could take place. One observes that there are only minor differences in the genetic maps of different *E. coli* strains. Relatively few rearrangement events have occurred in the genomes of strains of *E. coli* that were originally isolated at widely different places around the globe. Neither have there been very many rearrangements in the genomes of the evolutionarily related enterobacteria *E. coli*, *Shigella dysenteriae*, *Salmonella typhimurium*, and *Klebsiella pneumoniae* [see reviews by Sanderson (1976) and Riley and Anilionis (1978)]. However, more distantly related enterobacteria such as *Proteus mirabilis* do exhibit differences in genetic maps (Coetzee, 1979).

Congruence of maps extends different taxonomic distances in different areas of bacterial taxonomy. The maps of the two (*Bacillus*) species, *subtilis* and *lichenoformis*, are unlike the enteric map, but like each other (Rogolsky, 1970); similarly the maps of various (*Rhizobia*) and (*Agrobacterium*) species are like one another (Hooykaas *et al.*, 1982), but unlike those of other species. The locations of the genes that have been mapped in two *Streptomyces* species and a member of the *Nocardia* genus are similar to one another (Schupp *et al.*, 1975), suggesting that these bacteria may not be as widely separated as their taxonomic designations imply. At the other end of the sensitivity scale, some strains of *Pseudomonas* species have congruent maps, while others do not. The strains of *Pseudomonas aeruginosa* PAO and PAT, which came from Australia and South Africa, respectively, have very similar genetic maps (Holloway *et al.*, 1979), whereas the map of *Pseudomonas putida* differs (Dean and Morgan, 1983). *Pseudomonas aeruginosa* seems to be as far away from *Pseudomonas putida* with respect to genetic maps as *E. coli* is from *Proteus mirabilis*.

It is clear that more closely related bacteria have maintained closely similar genetic maps. *E. coli* and *S. typhimurium* are estimated to have diverged about 50 million years ago (Hori, 1976), a long time during which changes could have been absorbed, yet the two maps are similar. It appears either that the frequency of intragenomic recombination is low or that there have been constraints on the genetic activities of the bacterial genomes such that there are limitations on the amount of rearrangement that can be sustained.

The constraints on rearrangement might derive from the importance of maintaining the same gene order so as to permit synapsis and homologous genetic recombination. However, the recombination frequencies between *E. coli* and *S. typhimurium* are extremely poor, 10^{-4}–10^{-5} that of *E. coli* × *E. coli* matings. Thus, it seems unlikely that effects on the process of interspecific recombination exert much leverage in defining and limiting evolutionary mechanisms that were operative among enterobacteria. Alternatively, as suggested by Schmid and Roth (1983*b*), perhaps the constraints on inversion events relate to requirements concerning physical relationships between the origin and terminus of chromosomal replication, and perhaps there are also requirements concerning spatial relationships between genes that are required for genome replication. Beyond these considerations, there also may be effects of DNA configuration on gene expression that have not yet been recognized or analyzed. Whatever the nature of the constraint on gene arrangement, it is clear that some sort of control is maintained over bacterial genomes such that some kinds of rearrangements can and do occur, but at the same time the basic configurations of the genomes of related bacteria are maintained.

Genetic Maps as Phyletic Characters

Perhaps groupings should be made of bacteria that have similar genetic maps, distinguishing them from bacteria that have dissimilar gene arrangements. It might be that congruence of genetic maps is a taxonomically and evolutionarily meaningful criterion to apply in working out hierarchically comparable groupings of related organisms. For instance, it might be useful to have an indication in classification schemes that the enteric bacteria that have congruent maps are more closely related than the enteric bacteria that do not, and that *Nocardia* and *Streptomyces* species are so closely related that their maps seem to have the same basic features.

Genetic Rearrangement As a Mechanism of Phenotypic Alternation

There is a special kind of inversion that occurs at relatively high frequency in the genome of *S. typhimurium*. This is the site-specific inversion of a section of 970 base pairs of DNA at the locus of the genes for flagellar antigens. The facile inversion of this segment of DNA is the underlying molecular mechanism of a phenomenon of phenotypic alternation in *S. typhimurium* called "phase variation." The system has been worked out by Simon and co-workers (Simon *et al.*, 1980; Szekely and Simon, 1983) and is closely related to very similar systems in phage Mu (Kamp and Kahmann, 1981), phage P1 (Hiestand-Nauer and Iida, 1983), and prophage e14 in the *E. coli* chromosome (van de Putte *et al.*, 1984). In *S. typhimurium*, an enzyme coded by the *hin* gene recognizes a specific 14-bp sequence that is present in inverted repeat configuration on either side of the 970-bp segment of DNA. The enzyme catalyzes site-specific recombination between the inverted repeat sequences, resulting in inversion of the segment between them. By changing the orientation of a promoter element, changing its relationship to one of the flagellar genes and to a regulation gene, the inversion causes either activation or inactivation of genes, determining which of two flagellar antigen genes is expressed at any given time.

Another case of rapid alternation in phenotypes is also being worked out at the chromosomal level. In *Neisseria gonorrhaeae*, pilus anatigens undergo rapid alternations in state, and the alternations have been shown to be associated with events entailing rearrangements of genomic DNA (Meyer *et al.*, 1982; Meyer *et al.*, 1984). It seems reasonable to suppose that one type of genetic event that could result in the emergence of differentiated genomes for new lines of phyletic descent would be the loss of a mechanism for moving back and forth between alternate genomic states, thus freezing some genomes in one configuration, some in the other configuration.

Small-Scale Rearrangements: Divergence of Duplicate Genes

Some pairs of bacterial genes appear to have arisen by duplication and divergence. Analysis of the relationships between the nucleotide and amino acid sequences of such pairs of genes can tell us something about the events that took place in the course of their divergence. In enteric bacteria, there are duplicate gene pairs that are very closely related, in fact, almost identical, and there are gene pairs that have separated so far that homology in some parts of the genes can no longer be found. The most closely related gene pairs are related to each other by simple base replacements only, and appear to have diverged by sequential point mutations only. More distantly related gene pairs have more complex relationships that suggest that processes other than base replacement participated in the divergence within the coding regions, processes like microscale genetic rearrangements within the genes. Examples of these kinds of relationships will be described in the sections below.

In *E. coli*, two genes, *tuf*A and *tuf*B, that code for a protein synthesis factor have diverged very little, differing in only 13 codons, 12 of which are synonymous, one coding for a replacement amino acid (An and Frisen, 1980; Yokota *et al.*, 1980). In *S. typhimurium*, nucleotide sequences of the closely related transport genes *his*J and *arg*T have been determined (Higgins and Ames, 1981). The two genes are about 70% identical in both nucleotide sequence and amino acid sequence. The genes are related by simple base replacements only, about half of which change the amino acid that is coded, half do not and thus are silent mutations. The duplicate genes for ornithine transcarbamoylase in *E. coli* K12, *arg*I and *arg*F, apparently arose at some time after the divergence of strain K12 from other *E. coli* strains such as *E. coli* B or C since these latter strains do not have *arg*F genes. The nucleotide sequences and amino acid sequences of these two genes have been compared (Van Vliet *et al.*, 1984). Like the two gene pairs cited above, the *arg*I and *arg*F genes differ by base substitutions only. In this case, about twice as many of the base replacements are silent, producing synonymous codons, as the number that cause amino acid replacements. The secondary structures of the two proteins as deduced by computer program analysis are nearly identical.

Synonymous codon usage is not the same for all genes of a bacterium. Codon usage bias will be dictated in part by the level of expressivity that is characteristic of the gene (Grantham *et al.*, 1981). In the case of the *tuf* genes, a high level of expressivity is required, similar to the high level required of the genes for ribosomal proteins (Zengel *et al.*, 1984). These gene products are needed in large amounts in order to fulfill their cellular functions. A characteristic feature of high-expressivity genes is that, in contrast with genes that are not expressed at such high levels, the amino acid codons in the genes are the codons that correspond to the anticodons of the most abundant tRNA species in the cell, thus making it unlikely that expression of the gene would be limited by reason of scarcity of the necessary charged amino acids. One consequence of this codon usage control device is that high-expressivity genes have a lower capacity for absorbing silent mutations than do genes of low or moderate expressivity. In order to maintain the tRNA-related codon bias, some so-called synonymous codons would need to be avoided. Codons that correspond to rare tRNAs could confer a selective disadvantage to this group of intensively used genes.

Returning to the exercise of sequence comparisons, one can analyze the sequences of three closely related outer membrane protein genes in *E. coli*. The nucleotide sequences have been determined for the three genes , *omp*C (Mizuno *et al.*, 1983), *omp*F (Inokuchi *et al.*, 1982), and *pho*E (Overbeeke *et al.*, 1983). In all three pairwise comparisons, one finds evidence of base replacements, just as in the gene pairs cited above. However, in addition, these genes have numerous gaps in relation to each other. When the three nucleotide sequences are aligned in the three pairwise combinations, there are locations at intervals in the genes where

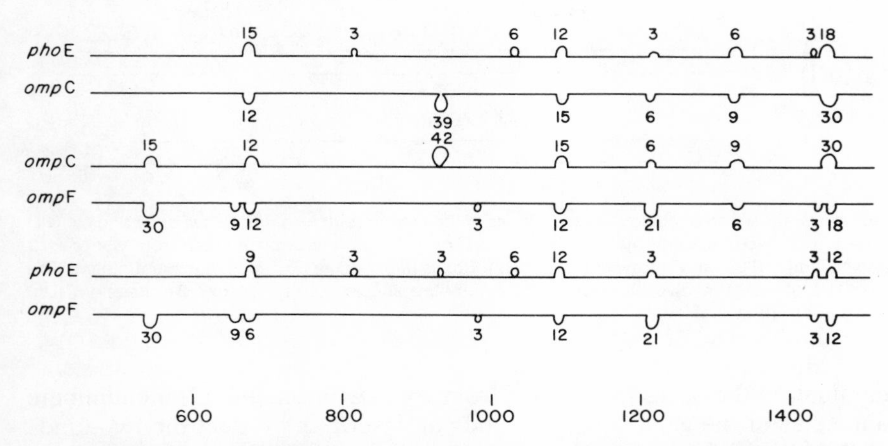

FIG. 3. Topography of comparisons of the *omp*C, *omp*F and *pho*E genes. Nucleotide sequence similarities in pairwise comparisons were maximized by introduction of gaps. The extra nucleotides are shown in the form of loops. Numbers by the loops designate the number of nucleotides in each discontinuity.

gaps must be introduced in order to maximize sequence identity. The numbers and sizes of the gaps are shown schematically in Fig. 3. The extra, unpaired bases are shown in the form of loops. Each loop contains a multiple of three bases, therefore there is no disturbance of the reading frame with respect to the amino acid sequence downstream of the loop (gap).

What was the molecular nature of the gaps? Mizuno *et al.* (1983) noted that some of the discontinuities in base sequence in the *omp* and *pho*E genes are flanked with short repeated sequences. Possibly errors were made during replication, either slippage that resulted in skipping over bases on one strand or stalled reiterative copying that inserted extra bases on the other strand. The ultimate effect of these extra bases or deleted bases is to insert extra amino acids or to delete amino acids from the outer membrane proteins encoded by the genes. Thus, a single event such as an error in replication results in an effect on several amino acids in the end product. Depending on the structure and function of the polypeptide chain at the location of the insertion or deletion, there is the potential for decisive functional consequences as a result of the multiple base changes.

The relationships among the three genes involve two important components, base pair replacements and gaps. To describe the relationships between the *omp* genes two kinds of parameters need to be specified: (1) the percent identity of nucleotides, determined for the paired regions only,

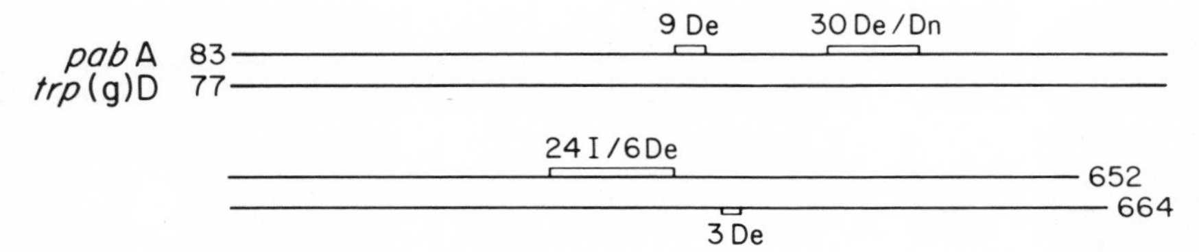

FIG. 4. Comparison of nucleotide sequences of the coding regions of the *E. coli pab*A and *trp*(G)D genes. Sequence data and nucleotide numbers for each gene have been taken from Goncharoff and Nicholls (1984) and Horowitz et al. (1982). Boxes indicate the positions of discontinuities. The numbers of bases involved in each discontinuity are shown above the box. (De) deletion; (Dn) duplication; (I) inversion.

eliminating the bases in the gaps from consideration; and (2) the numbers and sizes of the gaps. The two kinds of descriptors reflect the two kinds of processes that have contributed to the divergence of these genes—base replacement and insertion/deletion.

Another example of pairs of genes in *E. coli* that appear to have evolved by duplication and divergence are two pairs of genes in the pathways of biosynthesis of *p*-aminobenzoic acid and tryptophan. The gene pair *pab*A and *trp*(G)D, coding for biochemically similar glutamine amidotransferases, and the gene pair *pab*B and *trp*E, coding for *p*-aminobenzoate synthetase component I and anthranilate synthetase component I, respectively, evidently each diverged from common ancestral genes. The nucleotide sequences of the four genes have been determined (Nichols *et al.*, 1981; Horowitz *et al.*, 1982; Kaplan and Nichols, 1983; Goncharoff and Nichols, 1984). The relationship of the *pab*B gene and the *trp*E gene is straightforward, and seems to involve simple base replacements and four gaps, each of which entails insertion or deletion of a multiple of three bases, similar to the case of the *omp* genes (Goncharoff and Nichols, 1984). The relationship of the *pab*A gene and the *trp*(G)D gene is a little more complex (Kaplan and Nichols, 1983), and is shown schematically in Fig. 4. There are four discontinuities in the nucleotide sequences. Two are gaps, one of 9 bp, one of 3 bp. A complicated discontinuity is a 30-bp duplication in the *pab*A gene that appears in place of a different 30 base pair sequence that was apparently deleted. The fourth discontinuity is also complex and involves a 24-bp inversion and a 6-bp deletion in the *trp*(G)D gene. Altogether in this gene pair, 66 base pairs, or 10% of the total, were altered by mechanisms other than point mutation. The creation of small, internal rearrangements appears to be a powerful avenue for introducing rapid change in diverging gene pairs. The agent for such change might be the replication mechanism. Kaplan and Nichols (1983) have pointed out that short repeated sequences are present on either side of the sequences that were either duplicated or inverted. Perhaps the juxtaposition of the copies of the same base sequence in close proximity increases the chance of mistakes being made by the replication mechanism. On the other hand, some kind of recombination mechanism might be responsible for deletions or additions.

Another gene pair from *E. coli* appears to be evolutionarily related, the genes for ornithine transcarbamoylase, *arg*I, and aspartate transcarbamoylase, *pyr*B. The *arg*I gene was mentioned above in connection with its duplicate gene in strain K12, *arg*F. The *arg*I–*arg*F duplication evidently occurred relatively recently since only strain K12 is known to have the *arg*F gene. At some earlier time in an ancestral enterobacterium apparently the ancestor of the *arg*I and *pyr*B genes duplicated and diverged to give rise to the genes for two transcarbamoylase enzymes. Ornithine transcarbamoylase and aspartate transcarbamoylase differ in substrate specificity and serve two separate biosynthetic pathways, one for arginine synthesis, the other for pyrimidine synthesis. The *arg*I and *pyr*B genes map very close together, but are not contiguous, being separated by 4 kilobases. The nucleotide sequences of both the *arg*I and *pyr*B genes have been determined (Hoover *et al.*, 1983; Bencini *et al.*, 1983). Analysis of the relationship between the two sequences reveals complexities. Substantial portions of the two genes exhibit no discernible base sequence homology. The portions of the genes that are recognizably related embody many discontinuities, and the discontinuities are of a kind that create change in the gene product that extends beyond the site of the alteration.

Figure 5 shows the sequence of *arg*I from the beginning of the coding region [base 31 in the numbering scheme of Bencini *et al.* (1983) through base 216] and its relationship to the corresponding sequence at the beginning of the coding region of the *pyr*B gene. There are many gaps or loops, places where there are more bases in one sequence than in the other. These are pictured in the figure as if they were insertions in one gene, but they can as well be thought of as deletions in the other gene. Most of the groups of bases that were inserted or deleted are not multiples of three. This has the result of changing the reading frame and altering the amino acid sequence of the enzyme until another compensating change occurs downstream that restores the original reading frame. The shaded bars in Fig. 5 show the locations and lengths of the segments that underwent reading frame changes between bases 31 and 216. Figure 6 shows a similar analysis from base 385 of the *arg*I gene to base 504. Again, in this region many of the insertions or deletions did not occur in multiples of three, and thus the effect of the change in the base sequence on the amino acid sequence of the gene product extends beyond the actual site of the base sequence alteration. Figure 7 shows the amino acid sequences

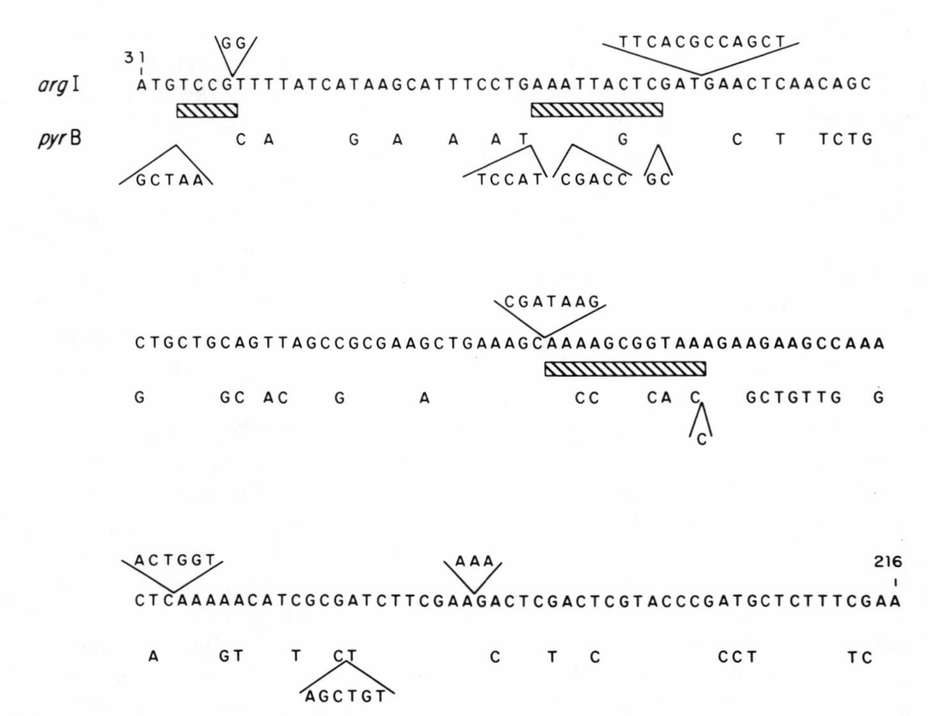

FIG. 5. Comparison of a section of the nucleotide sequences of the *arg*I and *pyr*B genes of *E. coli*. Nucleotide numbers ae those assigned to the *arg*I gene. Nucleotides that differ are shown either as replacements on the *pyr*B line or as inserts of bases above the *arg*I line or below the *pyr*B line. Sequence data has been taken from Hoover et al., 1983 and Bencini et al., 1983. Hatched boxes between the two lines show regions of frameshift.

of the two enzymes for these two regions. Where the amino acids are the same in the two enzymes, the symbols are boxed. Where there are gaps in one amino acid sequence relative to the other, a blank space is used. The bars between the two sequences locate all of the base sequence discontinuities, both the kind that produced frame shifts and those that did not. Evidently the discontinuities introduced by insertion/deletion events played at least as important a role as did simple base replacements in the process of divergent evolution of these two genes.

Figure 8 summarizes diagramatically the locations and distributions of the insertions/deletions in the two regions that have been analyzed here. For the first region, the overall percent identity of nucleotide sequence is 58%, while that of amino acid sequence is 43%, a considerably lower figure. Similarly, for the second region, the overall percent identity of

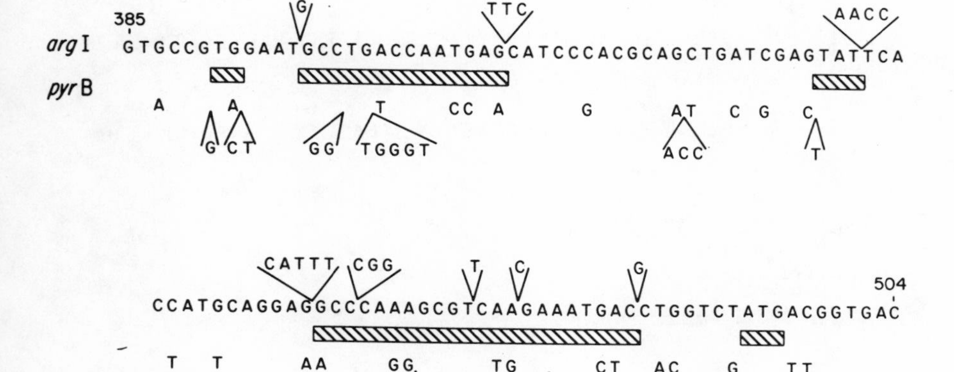

FIG. 6. Comparison of a section of the nucleotide sequences of the *arg*I and *pyr*B genes of *E. coli*. Nucleotide numbers are those of the *arg*I gene. The scheme is the same as for Fig. 5.

nucleotide sequence is 61% while that of amino acid sequence is 37%. For evolutionarily related genes that have undergone only nucleotide replacements, one might expect a higher identity in amino acid sequence than in nucleotide sequence because many nucleotide changes such as third position changes can be silent, generating synonymous codons. The reverse relationship for the *arg*I and *pyr*B genes arises because many nucleotide changes have exerted drastic effects on amino acid sequence as a consequence of insertions/deletions and frameshifts. In fact the great disparity between the degrees of relatedness of the nucleotide sequences and the amino acid sequences is accentuated if one examines the relatedness of only those portions of the nucleotide sequences that can be paired, that is the sequences between gaps. If one removes the insertion/deletion sequences from consideration and determines the percent identity of the nucleotide sequences in the remaining paired regions, one finds that the nucleotide sequence homology is 73% in the first region and 69% in the second region, much higher than the values for the corresponding amino acid sequences of 43 and 37%, respectively. The disparity is a measure of the participation of mechanisms other than single base replacements in the process of divergence, and could conceivably be used as a measure of the amount of discontinuous change that has occurred during evolution of any pair of genes.

After preparation of this manuscript, an independent analysis of the *arg*I and *pyr*B genes appeared (Houghten *et al.*, 1984). The amino acid and nucleotide sequence analyses of the enzymes and the genes are nearly

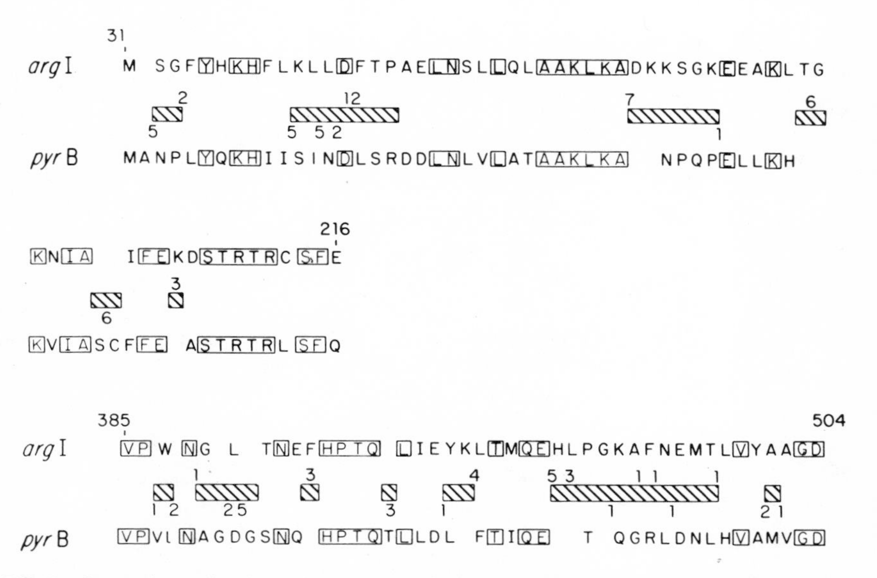

FIG. 7. Comparison of amino acid sequences for two sections of the *arg*I and *pyr*B genes. Nucleotide numbers are those of the *arg*I gene. Amino acids are shown by the single letter code. Boxes show amino acids that are identical, blanks signify gaps that were introduced to maximize identity of the two sequences. Open boxes between the sequences represent positions of discontinuity in the nucleotide sequences as shown in Figs. 6 and 7.

the same as the analysis presented here. In addition, these authors show that there is substantial conservation of structural conformation of the aspartate transcarbamoylase and the two ornithine transcarbamoylases, even in regions where conservation of primary amino acid sequence is poor.

The sequence comparison of the entirety of the *arg*I and *pyr*B genes shows that there are sharp discontinuities in the degree of similarity of the sequences in different parts of the genes. In some regions, no relationship can be seen at all. Other pairs of *E. coli* genes also show heterogeneity in degree of similarity within the genes. For instance, the *fus* gene of *E. coli* which codes for the EF-G protein synthesis elongation factor has amino acid sequence and nucleotide sequence similarities with the *tuf*A and *tuf*B genes which code for the EF-Tu elongation factor (Zengel *et al.*, 1984), but these sequence similarities are confined to the amino-terminal end of the protein. Three genes for sensory transducers in *E. coli* are related, the most closely related pair being the *tar* and *tsr* genes.

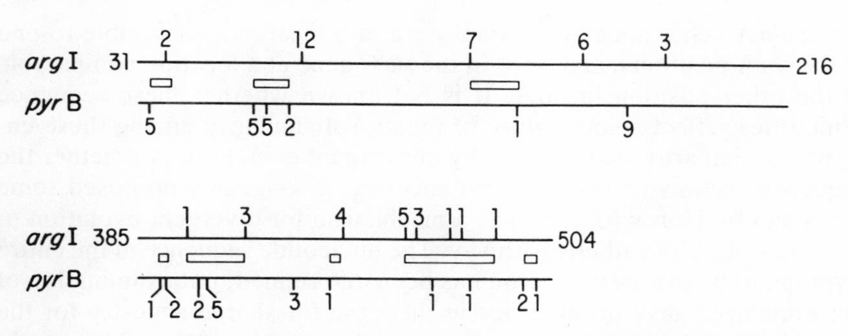

FIG. 8. Schematic summary of two sections of discontinuity within the sequences of the *argI* and *pyrB* genes. Horizontal lines represent nucleotide sequences. Numbers at the start and end of each sequence are the numbers assigned to the *argI* gene. Numbers at the tic marks indicate the positions and the numbers of unmatched nucleotides of each discontinuity. Open boxes in the center represent the extent of regions of discontinuity in amino acid sequence introduced by the insertion/deletion of nucleotides.

The genes divide into well-defined regions of good, fair and poor homology that seem to coincide with separate functional domains of the proteins (Krikos *et al.*, 1983).

Whenever an abrupt transition occurs between a region of high percentage of identity of amino acid or nucleotide sequences and a region of low identity, or a region with no discernable relationsip, such an abrupt transition could reflect division of the protein into separate functional domains. The functional or structural domains could exhibit different homology relationships if they had diverged at different rates by virtue of having different requirements for stringency of sequence conservation. Alternatively, sharp boundaries in degree of sequence identity could result from recombinational events that generated a recombinant gene having component segments derived from different ancestral sources. The possibility of a checkered past exists for any gene that has strong discontinuities in sequence relationships. Recombination events will be discussed further in the next section.

When comparing the nucleotide sequences of bacterial genes that are even more tenuously related, it becomes increasingly difficult to discern the molecular events that took place during their evolution. The nucleotide sequences of the genes for the enzymes of biosynthesis of threonine in *E. coli* have been determined (Parsot *et al.*, 1983). The three contiguous genes form an operon. The great majority of the three sequences appear not to be related, but there is a 35-amino acid region that appears twice in the *thr*A gene, once in the *thr*B gene at a location comparable to one of the *thr*A positions, and once in the *thr*C gene at a location comparable to the other position in *thr*A. It is not known whether these sequence similarities reflect some feature of functional similarity among these enzymes, a similarity that evolved by convergent evolution, or whether the sequences echo an ancient shared ancestry, as originally proposed some years ago by Horowitz (1965) as a mechanism for divergent evolution of the genes of a biosynthetic pathway. The nucleotide sequence of the entire tryptophan biosynthesis operon has been determined, but examination of the sequences gave no evidence in this case for shared ancestry for the several genes involved in biosynthesis of tryptophan (Yanofsky *et al.*, 1981).

INTERACTIONS BETWEEN TWO GENOMES

Although mutation is clearly an important source of genetic diversity in bacteria, it is not the only source. Some of the generation of genetic diversity and production of evolutionary change in bacteria undoubtedly results from lateral exchanges involving genetic recombination activities, the standard sexual recombination between closely related homologous DNAs, as well as site-specific recombination, and the kind of nonhomologous or illegitimate recombination that does not require homology or relatedness between the participating DNAs (reviewed by Low and Porter, 1978). Incorporation of a segment of DNA into the bacterial genome from another source introduces discontinuous change. New DNA is acquired, new join points are created, and an old sequence is interrupted. The acquired DNA can have a different evolutionary history from the genomic DNA of which it is now a part. Complexities are created by joining DNAs of separate origins and lineages.

Transposons: Jumping Genes

Transposons are genetic elements that promote their own proliferation and genetic movement. They insert into target DNA without a requirement for extensive nucleotide sequence homology. Although the details of the molecular mechanisms for the insertion of transposons into target DNA sites differs for the various transposons that have been studied to date, most of them share the property of being able to move into genomic DNA at many possible locations. Transposons that are present in plasmids can move from one bacterium to another as the plasmid is transmitted from one bacterium to another. In the case of self-transmissable conjugal plasmids that have a broad host range, any transposon present in the plasmid DNA can be widely disseminated in the bacterial world. When the transposon jumps from the plasmid to the bacterial genome, this constitutes a discontinuous event in the genome that could have evolutionary implications.

The distribution of some transposons in the genomes of representative enteric bacteria has been determined. The *E. coli* K12 genome contains many transposable elements. A "typical" strain harbors 31 known elements [seven copies of IS1, eight of IS2, five of IS3, one of IS4, and ten of IS5 (Timmons *et al.*, 1983)]. The elements are asymmetrically distributed along the *E. coli* K12 map, with a disproportionate number occuring in the neighborhood of the *lac* operon (Timmons *et al.*, 1984). Among various *E. coli* strains and other enterobacteria, there is variation in the numbers of copies of each element in the genomes. The number of IS1 sequences, for instance, ranged from zero to 30 in various *E. coli* strains (Hu and Deonier, 1981a; Nyman *et al.*, 1983), both in those that were standard laboratory stocks and in those that were more recently isolated from nature. In the most extreme example reported, of two *E. coli* strains that were isolated as coinhabitants of one animal gut, one isolate had no IS1 element, while the other carried about 30 copies of IS1 (Nyman, *et al.*, 1983). The transposons IS3 and IS5 also show variation in copy number and in their location with respect to other nucleotide sequences in the genomes of several *E. coli* strains, and variation was also seen in other enteric bacteria that did not correlate with their evolutionary relationships (Green *et al.*, 1984; Hu and Deonier, 1981a; Nyman *et al.*, 1981; Schoner and Schoner, 1981).

Not only eubacteria, but also archaebacteria harbor repeated sequences. *Halobacterium halobium* DNA contains more than 50 families of repeated sequences. These bacteria exhibited a high frequency of genomic variability as revealed by Southern hybridization when repeated sequences were used as probes (Sapienza *et al.*, 1982). It seems likely that the repeated sequences are mobile genetic elements that create genetic variation by their movement.

It seems clear that mobile genetic elements have the potential to contribute substantially to the generation of genetic diversity. Are there any functional consequences of the movement and change in location of copies of IS sequences? Under some circumstances, the answer is yes. For instance, insertion in front of a gene that lacks a functioning promoter can activate the gene either by providing a promoter that is located within the inserted sequence or by creating a promoter sequence at one of the join points [see examples in the review by Kleckner (1981)]. In another kind of event, an insertion sequence can reactivate a gene even when the insertion occurs too far from the gene to serve a promoter function. A cryptic gene in *E. coli* was reactivated by insertion of an IS1 sequence nearby but not in or immediately adjacent to the cryptic gene. The cryptic gene, *bgl*, is a gene that codes for a β-glucosidase enzyme, and is silent in the wild type but can be reactivated by a series of a few mutations. When a copy of IS1 was inserted near *bgl* in the wild type, the effect of the IS1 insertion was to reactivate the cryptic *bgl* gene (Reynolds *et al.*, 1981). In a similar case, function of a gene, *ebg*, carried by a plasmid vector was affected by insertions of the γδ insertion element near but not in the *ebg* gene (Stokes and Hall, 1984). Insertions near the *bop* gene of *Halobacterium halobium* inactivate the gene, and in one case a second insertion even further away from *bop* restores gene function (Pfeifer *et al.*, 1983). Clearly, if the insertion of movable genetic elements into certain genomic locations has an effect on expression of certain genes, then transposition activities of movable elements could be a significant source of discontinuous and sudden change.

One view of insertion sequences sees the mobile elements as "selfish" or, parasitic. An IS element is viewed as having no phenotype at the level of the host, but instead is replicating, and dispersing for no other purpose than to increase its own numbers (Doolittle and Sapienza, 1980; Orgel and Crick, 1980). However, phenotypes at the level of the organism have in fact been observed, both with respect to individual, nearby genes, as mentioned above, and also at the level of populations in respect to fitness to compete in a chemostat. In two of three strains tested, strains carrying the transposon Tn5 were fitter than transposon-less isogenic strains when subjected to changes in carbon source concentration in a chemostat setting. Function of the Tn5 transposase was required for the effect, but a specific molecular mechanism for the phenomenon is not yet known (Beil and Hartl, 1983). Similarly, transposition of the IS10 sequences of Tn10 conferred a fitness advantage to bacteria growing in a chemostat, suggesting that the mutator activity of insertion sequences plays an evolutionary role (Chao *et al.*, 1983).

Transposition activity can also move genes from one genome to another. In one study, the transposon IS1 was implicated as having the potential for causing transposition of genomic bacterial genes. The *arg*F gene of *E. coli* K12, the supernumerary duplicate of the *arg*I gene that was mentioned above, was found to reside in the genome within a 10 kb segment, flanked by two copies of the IS1 sequence (Hu and Deonier, 1981b; York and Stodolsky, 1981). It seems not unreasonable that the *arg*F gene and flanking DNA could have been transposed to their present position in the *E. coli* genome by action of the IS1 sequences.

In another study, the genes for catabolism of ribitol and arabitol, genes that are present in only some strains of *E. coli*, were found to reside within the vestige of what was once a transposon. Link and Reiner (1982) showed that the *rtl* and *dal* operons are flanked by 1.4-kb imperfect inverted repeat sequences, suggesting a transposon structure. In this case, the action of the transposon seems to have been to cause a genetic substitution. Alternate alleles exist at this genomic locus in *E. coli* (Link and Reiner, 1983; Woodward and Charles, 1983). Most strains are either ribitol–arabitol–positive or galactitol-positive, but not both. Mechanisms have been proposed that invoke the inverted repeat sequences as agents of implementing the substitution (Link and Reiner, 1983).

From consideration of these examples, it is not difficult to visualize an important role for transposons in the evolution of genomic bacterial genomic DNA. However, transposon-mediated events seem likely to have occurred relatively infrequently, since repeated and frequent transposon-facilitated rearrangement could change the entire topography of the genome, destroying genetic continuity. Still, even if they acted infrequently, transposons could play an important role by activating and inactivating expression of genes and by bringing to the bacterial genome from time to time segments of new DNA from some outside source, stoking the furnace with fresh coals, as it were, at infrequent but important intervals.

Plasmids: Incorporation into Genomic DNA

Plasmids also can create changes in the genome if the plasmid DNA is integrated into the genome. As long as plasmid DNA stays cytoplasmic, the genetic attributes carried by the plasmid must be thought of as transient, fated to be lost whenever the plasmid is lost from the cytoplasm [see review by Bennett and Richmond (1978)]. However, when plasmid genes are incorporated into the genome, a more permanent change has occurred. The gene has been acquired by the chromosome. We do not at this stage know why some genes are found in plasmids while others are not. There are many examples of incorporation of plasmid genes in bacteria. In the normal life cycle of temperate phages, the prophage DNA becomes a part of the genome. In the equilibrium between the F′ state and the Hfr state of *E. coli* donors, the F′ plasmid can integrate into the genome. This kind of lateral uptake and expulsion of segments of DNA in the bacterial genome seems to be a dynamic process that results in a high degree of variability among closely related genomes.

In the case of the temperate phage λ and its host *E. coli*, for instance, there are bits and pieces of DNA that are homologous to phage λ DNA sprinkled here and there in most *E. coli* genomes, whether the strains come from the laboratory or were isolated from nature (Anilionis and Riley, 1980; Harshman and Riley, 1980). Even enteric bacteria that are not sensitive to λ infection contain λ-like sequences in their genomes. Contemporary *S. typhimurium* is not sensitive to λ infection, yet its genome contains λ homologues (Riley and Anilionis, 1980). The fragments of λ-like sequences in the bacterial genomes presumably are the consequence of prior abortive infection with a one or more phages of the λ family, infections that entailed integration of all or part of the phage genome, perhaps terminated by imprecise excision of the prophage, leaving prophage fragments behind to reside in the genome as defective prophages. When λ phage DNA was used as a probe in Southern hybridization experiments in order to explore the existence of phage fragments in many different *E. coli* strains, highly variable patterns of genomic fragments of λ-like DNA sequences were visualized (Anilionis and Riley, 1980; Harshman and Riley, 1980). Either the variation could be a consequence of an accumulation of base sequence changes in the prophage fragments, changes that can be absorbed without constraint by the nonfunctional cryptic remnants of phage genes, or else the variation could come from heterogeneity in genetic location, or both.

Some of the phage fragments that reside in the *E. coli* genome encode intact expressed phage genes, and these genes contribute functions to the *E. coli* host or to other phages. A recombination function is contributed to *E. coli* by a set of λ-phage genes at one site (Kaiser and Murray, 1979), and a set of phage maturation functions are present at two other sites that are able to rescue defective superinfecting λ phages (Anilionis *et al.*, 1980; Kaiser, 1980; Espion *et al.*, 1983). One of these defective λ prophages, *qsr'*, and also the PA-2 prophage contain similar outer membrane protein genes that can replace the corresponding bacterial protein if the bacterial gene is defective (Highton *et al.*, 1985). Another defective prophage in the *E. coli* genome, *e*14, contributes a gene for a site specific recombinase that causes inversion between specific repeated sequences (Van de Putte *et al.*, 1984).

Thus, by picking up and incorporating fragments of phages and/or plasmids, sometimes with functional genes contained in the incorporated fragments, the bacterial genome undergoes acquisitive change. The extent to which the bacterial genome has been built up by a series of such integrations of fragments of external DNA is not known. It *is* known, however, that lysogens for many temperate phages have greater reproductive fitness than the corresponding non-lysogenic strains (Edlin *et al.*, 1977; Lin *et al.*, 1977). It is not possible at present to identify the sources of all of the genes in a bacterial genome, but perhaps the relationships of some smaller genomes, certain plasmid and transposon DNAs, provide models of how complex acquisition reactions take place sequentially. Analysis of homology relationships of segments of plasmid DNAs suggests that some plasmids have exchanged genetic segments during evolution and other plasmids have evolved by progressive, successive acquisitions of units of DNA, adding genes and growing in size with each addition. An example of apparent genetic exchange was analyzed in plasmids Rsc13 and pSM1, which are small derivatives formed from the large, closely related plasmids R1 and R100, respectively. Nucleotide sequences of a 2.9kb fragment from these two small plasmids were analyzed and were found to be 96% homologous except for a discrete segment of about 250bp that encoded a polypeptide of the *rep*A2 gene. This segment was only 44% homologous (Ryder *et al.*, 1982). Even though the similarity of base sequence was not high in this segment, the amino acid sequence and predicted conformation of the *rep*A polypeptide products from the two plasmids were very similar. The simplest interpretation of the results is that the *rep*A2 region of one of the two plasmids was replaced by recombination, substituting a functionally similar domain from another plasmid which was not as closely related as were the remaining nucleotides of Rsc13 and pSM1.

Other examples of apparent acquisition by recombination are shown in Fig. 9. Tn951 is a transposon that was found in a plasmid in the bacterium *Yersinia pseudotuberculosis*. Tn951 is a transposon that carries *lac* genes, and confers an ability to grow on lactose. Heteroduplex studies showed that Tn951 carries *lac* genes that are homologous with the *lac*I,′ *lac*Z and *lac*Y genes of the *E. coli lac* operon (Cornelis *et al.*, 1978). Further analysis of this transposon has shown that the ends of the transposon are highly similar to the ends of another transposon, Tn3 (Cornelis *et al.*, 1981). Between the Tn3-like termini there is a region of unknown origin next to the *E. coli*-like *lac* genes, followed by sequences identified as belonging to another transposon, Tn2501 (Michiels and Cornelis, 1984). Thus, Tn951 is a crazy quilt made up of fragments of more than three parental DNAs.

As another example of patchwork ancestry, analysis of the homology relationships of several mercury resistance transposons has shown that the relatively large multi-drug resistance and mercury resistance transposon, Tn2603, has homology relationships with other transposons such that it appears to have arisen by a series of sequential insertions, one of which was within another (Fig. 9) (Tanaka *et al.*, 1983). Perhaps some of the evolution of bacterial genomes proceeded in this discontinuous manner. Indirect evidence that processes of this kind were involved comes from comparative analysis of related bacterial genetic maps.

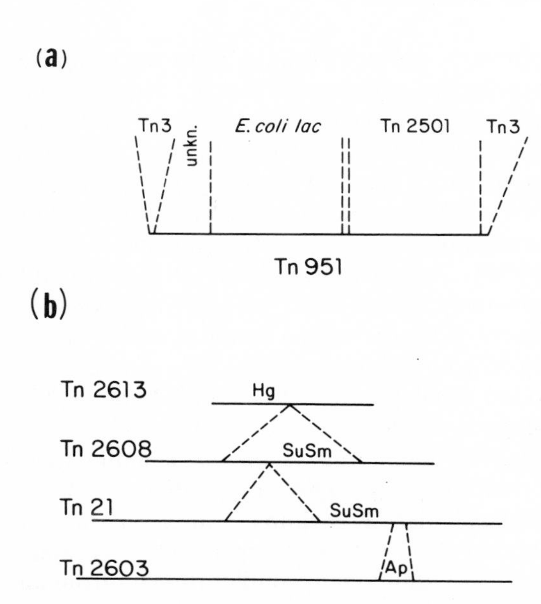

FIG. 9. Schematic representation of the genetic components in two transposons. (a) Tn951; (b) Tn2603. Dotted lines delineate sections of diverse ancestry.

Additions or Deletions within Genomic DNA

Although the genetic maps of related bacteria are very much alike, if one looks at them closely there are differences that may give clues to some of the kinds of genetic events that have occurred during divergence of the related bacteria.

The genetic map of *E. coli* C, although it contains many fewer genes, is very similar to that of *E. coli* K12 except that there is less genetic distance between markers that lie on either side of the region around map position 95 than there is in the same region of the genetic map of *E. coli* K12 (Wiman *et al.*, 1970; Bachmann, 1983). Perhaps *E. coli* K12 incorporated a segment of DNA in the course of divergence from strain C, or perhaps *E. coli* C lost a segment of DNA by deletion. The genetic map of *S. typhimurium* is more detailed than the map of *E. coli* C, so a comparison with the map of *E. coli* K12 provides more information. When the two maps were aligned with respect to those genes whose map positions are known in both bacteria, specific places were located in both maps where there was greater genetic distance in one map than in the other. The relationship can be represented either in terms of gaps in one genome relative to the other or in terms of loops of excess genetic distance

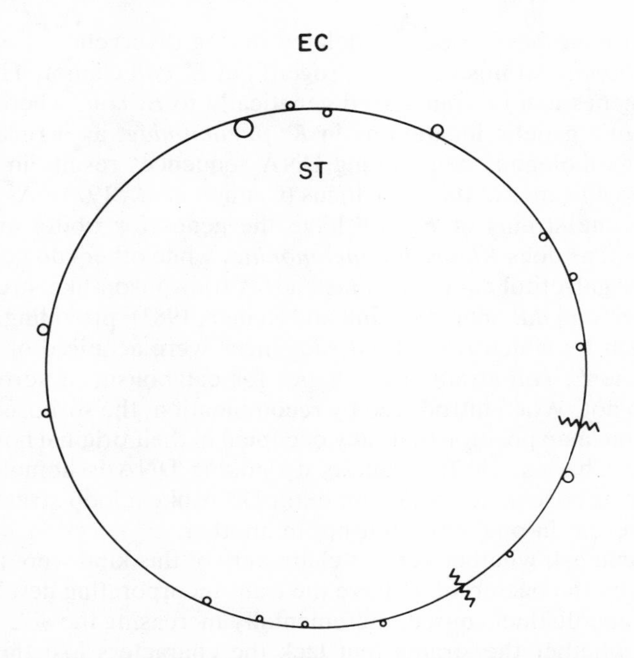

FIG. 10. Comparison of the genetic maps of *E. coli* and *S. typhimurium*. The uninterrupted parts of the circle represents linkage data that is congruent for the two genomes at a resolution of ±0.6 map units. Loops represent locations of greater genetic distance in one genome with respect to the other (for detail, see Riley and Anilionis, 1978).

in one map relative to the other. Figure 10 shows the configuration of the *E. coli* K12 map in relation to the *S. typhimurium* LT2 map in terms of loops of excess genetic distance in one genome or the other at different locations. At this rather insensitive level of analysis, there seem to be 15 loops that are greater than 25 kb in size, divided about evenly between the two genomes and accounting for about 10% of each genome. Do these loops correspond simply to localized abnormalities in recombination frequencies such that they give false information on the genetic distances between some genes, or do the loops correspond to actual differences in the amount of DNA in two genomes at these loci? In one case that has been tested, an *E. coli* loop contains DNA that appears to have no counterpart in the *S. typhimurium* genome as judged by the absence of hybridization even when relaxed conditions were employed (Lampel and Riley, 1982).

Some of the phenotypic differences between enteric bacteria are based on the presence or absence of certain genes. Multigene segments appear to have been added or deleted during divergence. For instance, *K. pneumoniae* strains can fix nitrogen, but *E. coli* cannot. The nitrogen fixation genes can be transferred genetically to *E. coli*, where they take up the same genetic location as in *K. pneumoniae*, as if recombination between homologous neighboring DNA sequences results in incorporation of the *nif* genes at the same locus (Cannon *et al.*, 1974). As mentioned earlier, some strains of *E. coli* have the genes for ribitol and arabitol catabolism, as does *Klebsiella pneumoniae*, while others do not, but have a gene for galactitol catabolism instead. A transposon-like structure contains the *rtl* and *dal* operons (Link and Reiner, 1983), providing a plausible mechanism by which *rtl* and *dal* sequences were acquired or lost. Similarly, some *E. coli* strains have genes for catabolism of sorbose, while others do not. When introduced by recombination, the sorbose genes take up the same map position that they occupied in their original home (Woodward and Charles, 1982), again as if flanking DNA is homologous, but the extra sorbose genes reside on extra DNA like a loop structure, DNA that is present in one strain but not in another.

We can ask whether genetic characters of this kind were at one time acquired by the bacteria that have the trait, incorporating new DNA from a genetically distinct source, incrementally increasing the size of the genome, or whether the strains that lack the characters lost this DNA by deletion after having once had it, decreasing the size of the genome in the process. Studies have shown that mutations of a few bases that create gaps within a bacterial gene are more often deletion mutations than they are insertions (deJong and Ryden, 1981). Whether or not this bias applies to events on a larger scale involving entire genes is not known.

There is reason to suppose that *S. typhimurium* suffered a deletion of its *lac* operon during its evolutionary history. Some enteric bacteria are *lac*-positive, some *lac*-negative, depending on whether they have an active genetic system for producing a β-galactosidase and a lactose transport system. The enterobacteria that are most distant from *E. coli*, the *Morganellae* and *Proteae*, are *lac*-negative, suggesting that the *lac* genes involved in enterobacteria (by addition to the genome?) after the time that *Morganellae* and *Proteae* split off. Most other enterobacteria are *lac* positive except for *Salmonella* species, *Edwardsiella*, and most *Shigella* species. *Shigellae* are viewed today as belonging to the *Escherichia coli* group. *Shigella dysenteriae* is phenotypically *lac* negative as a result of a few point mutations, *Shigella sonnei* is *lac*Z^+ and *lac*Y^-. Thus *Shigellae* are Lac^- because of one or a few point mutations. On the other hand, *Salmonella typhimurium* appears to be Lac^- because it lacks completely chromosomal DNA in the *lac* region. As mentioned above, Southern hybridization experiments with *E. coli lac* DNA and *S. typhimurium* DNA

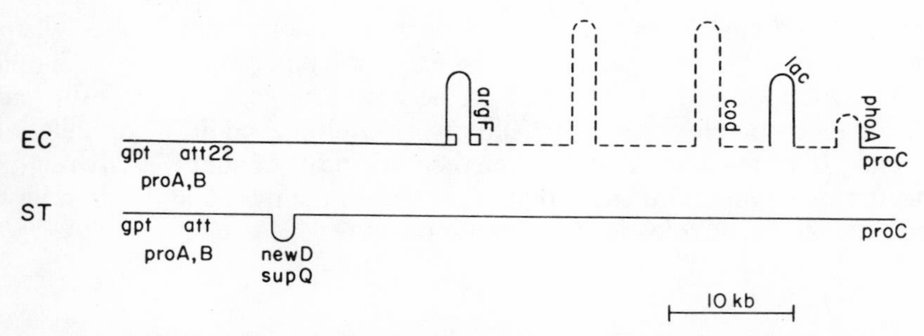

FIG. 11. Schematic representation of the relationships between *E. coli* and *S. typhimurium* DNAs between the *gpt* and *pro*C genes. Disparities in lengths of DNA are represented as loops of excess DNA in one of the genomes. Solid lines delineate regions for which there is some experimental information (see text); dotted lines show to scale the remainder of the excess length of *E. coli* DNA relative to *S. typhimurium* DNA, arranged in a hypothetical loop structure.

showed that *S. typhimurium* genomic DNA does not have sequences even partially homologous to *E. coli* DNA around the *lac* operon (Lampel and Riley, 1982; Buvinger *et al.*, 1984), thus the *lac* region could have been deleted from the *S. typhimurium* genome some time after diverging from the ancestor shared with *E. coli*.

The difference in the map distances between the *pro*B and *pro*C genes in the *S. typhimurium* and *E. coli* genomes amounts to about 75kb more DNA in *E. coli*. This difference is large compared to the size of the *lac* operon itself (6.2 kb). Experiments suggest that the 75 kb differential is comprised of more than one discontinuity. The *lac* operon lies near one end of the 75 kb segment in a section of nonhomologous DNA that is 12–13 kb long (Lampel and Riley, 1982). Near the other end of the 75 kb segment is the *arg*F region, containing the duplicate *arg*F gene with the structure of a transposon, flanked by two IS1 movable genetic elements (Hu and Deonier, 1981b; and York and Stodolsky, 1981). Nearby are the *sup*Q and *new*D genes of *S. typhimurium*. Genetic and hybridization experiments indicate that there are no counterpart genes in the *E. coli* genome (Fultz *et al.*, 1979; Riley *et al.*, 1984). The *sup*Q and *new*D genes of *S. typhimurium* may have been acquired from some outside source after divergence from *E. coli*. There are still other genetic discontinuities in the area. The *cod*A gene of *E. coli* maps in this region, but the corresponding *S. typhimurium* gene maps elsewhere. The features of the architecture of the *E. coli* and *S. typhimurium* genomes between the *gpt* and *pro*C genes are summarized in Fig. 11, showing the discontinuities schematically as loop structures. It seems likely that the kinds of relationships that have been found in the *lac* region of the genomes of *E. coli* and *S. typhimurium* will be found in other parts of their genomes where other mismatches in genetic distances exist. What may appear to be simple loops when viewed on the scale of the complete genome (Fig. 10) may be in fact complex structures involving multiple addition or deletion events. It seems likely that continued close study of evolutionarily related bacterial genomes will show that processes of addition and deletion have played important roles in their evolution.

SUMMARY

Variation has been introduced into bacterial genomes both by mutation and by the importation of genes from other sources. Both of these processes can involve discontinuous and potentially drastic changes. From the foregoing summary of information from comparative analyses of evolutionarily related bacterial DNAs, it is clear that, along with replacements of single base pairs, other kinds of changes have occurred in bacterial DNA in the course of evolution, changes involving rearrangement of nucleotide sequences both on a large scale, affecting many genes at once, and on a small scale, affecting only a few bases at a time. Internal rearrangements, nonhomologous recombination, and uptake of new DNA are the kinds of events that provide the opportunity for drastic changes to take place as a result of single genetic events.

Large-scale internal rearrangements affect some genes by cutting them at the join points; they change linkage relationships of many genes, and in some cases they may also affect the function of genes within the rearranged segments. The frequency and scope of internal rearrangements are constrained by stabilizing forces that have not been identified, conservative factors that tend to minimize the number of rearrangements and to maintain the similarity of related bacterial genetic maps.

Small-scale rearrangements also occur, such as addition or deletion of one or a few bases. Such changes are not confined to intergenic sequences, but occur in coding regions within the transcribed portion of evolving genes. When additions or deletions of bases occur in numbers that are not multiples of three, they can cause frameshifts in reading the nucleotide sequence code, providing the opportunity to make many changes in the amino acid sequence as a consequence of a single event at the DNA level. In such cases, the nucleotide sequences of a pair of evolutionarily related genes can be more closely related than are the amino acid sequences of the gene products.

Sometimes an addition to the genome involves segments of DNA that include intact foreign genes, such as the genes that are present in the remnants of cryptic prophage DNAs or genes that are transported as a part of movable genetic elements. Such genetic interactions create discontinuities in the DNA.

When carrying out comparative analyses of genes or genomes that are evolutionarily related, it is clear that sequence homologies, nucleotide replacements, and amino acid replacements only tell part of the story. Not all parts of the genome can be expected to have had an uncomplicated history of simple vertical descent. To the extent that discontinuities join together DNA sequences from different genetic backgrounds that have had different evolutionary histories, such discontinuities introduce complications into the analysis of evolutionarily related DNAs. Detection and identification of some kinds of discontinuities may be revealed by comparative analysis of nucleotide sequences. Location and identification of gaps pinpoints addition/deletion and frameshift events. In the case of an illegitimate recombination event that resulted in incorporation of a segment of DNA from another source, sharp boundaries might be found at junctions between segments of DNA that have different origins. Base composition in terms of overall percent GC, a change in base frequencies in the third position of codons, or a bias in codon usage patterns in terms of tRNA abundance could each differ in contiguous DNA segments if the origins of the segments of DNA differed and if the separate identities of the DNA segments have not yet been blurred by subsequent common histories.

As more is learned about the kinds of interactions and complexities that have taken place on the molecular level, as we catalog the locations, numbers, and types of discontinuities that exist in evolutionarily related genomes, the knowledge of these complexities will serve as important sources of information on the mechanisms of evolutionary change in molecular terms and on the ancestral relationships of the genomes. Construction of ancestral trees should be aided by the ability to infer from this kind of data the order in which rearrangements, additions, and deletions occurred in related genomes. Genes without discontinuities, genes that have experienced an essentially undisturbed history, undergoing mutation by single base changes only and experiencing nothing more exotic than straightforward intraspecific homologous recombination events, are ideal candidates to use for building and measuring against molecular clocks. Genes that have undergone discontinuous changes are less useful in that context but they are able to make unique contributions to our ever-growing appreciation of the complexities of the events that have acted on genomic DNA to bring about genetic diversity and evolutionary change.

REFERENCES

An, G., and Frisen, J. D., 1980, The nucleotide sequence of *tuf*B and four nearby tRNA structural genes of *Escherichia coli*, *Gene* **12:**33–39.

Anagnostopoulos, C., 1976, Genetic analysis of *Bacillus subtilis* strains carrying chromosomal rearrangements, in: *Modern Trends in Bacterial Transformation and Transfection* (A. Portoles, R. Lopez, and M. Espinosa, eds.), pp. 211–230, Elsevier/North-Holland Biomedical Press, Amsterdam.

Anderson, R. P., and Roth, J. R., 1977, Tandem genetic duplications in phage and bacteria, *Annu. Rev. Microbiol.* **31:**473–505.

Anilionis, A., and Riley, M., 1980, Conservation and variation of nucleotide sequences within related bacterial genomes: *Escherichia coli* strains, *J. Bacteriol.* **143:**355–365.

Anilionis, A., Ostapchuk, P., and Riley, M., 1980, Identification of a second cryptic lambdoid prophage locus in the *E. coli* K12 chromosome, *Mol. Gen. Genet.* **180:**479–481.

Bachmann, B. J., 1983, Linkage map of *Escherichia coli* K-12, edition 7, *Microbiol. Rev.* **47:**180–230.

Bachmann, B. J., Low, K. B., and Taylor, A. L., 1976, Recalibrated linkage map of *Escherichia coli* K-12, *Bacteriol. Rev.* **40:**116–167.

Beil, S. W., and Hartl, D. L., 1983, Evolution of transposons: Natural selection for Tn5 in *Escherichia coli* K12, *Genetics* **103:**581–592.

Bencini, P. A., Houghton, J. E., Hoover, T. A., Foltermann, K. F., Wild, J. R., O'Donovan, G. A., 1983, The DNA sequence of *arg*I from *Escherichia coli* K12, *Nucleic Acids Res.* **11:**8509–8518.

Bennett, P. M., and Richmond, M. H., 1978, Plasmids and their possible influence on bacterial evolution, in: *The Bacteria*, Vol. VI (L. N. Ornston and J. R. Sokatch, eds.), pp. 1–69, Academic Press, New York.

Buvinger, W., and Riley, M., 1985, The nucleotide sequence of *Klebsiella pneumoniae lac* genes, *J. Bacteriol*, **163.**

Calos, M. P., Galas, D., and Miller, J. H., 1978, Genetic studies of the *lac* repressor. VIII, DNA sequence change resulting from an intragenic duplication. *J. Mol. Biol.* **126:**865–869.

Cannon, F. C., Dixon, R. A., and Postgate, J. R., 1974, Chromosomal integration of *Klebsiella* nitrogen fixation genes in *Escherichia coli*, *J. Gen. Microbiol.* **80:**227–239.

Capage, M., and Hill, C. W., 1979, Preferential unequal recombination in the *gly*S region of the *Escherichia coli* chromosome, *J. Mol. Biol.* **127:**73–87.

Casadaban, M. J., 1976, Transposition and fusion of the *lac* genes to selected promoters in *Escherichia coli* using bacteriophage lambda and mu, *J. Mol. Biol.* **104:**541–555.

Casse, F., Pascal, M. -C., and Chippaux, M., 1973, Comparison between the chromosomal maps of *E. coli* and *S. typhimurium*. Length of the inverted segment in the *trp* region, *Mol. Gen. Genet.* **124:**253–257.

Chao, L., Vargas, C., Spear, B. B., and Cox, E. C., 1983, Transposable elements as mutator genes in evolution, *Nature* **303:**633–635.

Charlier, D., Crabeel, M., Cunin, R., and Glansdorff, N., 1979, Tandem and inverted repeats of arginine genes in *Escherichia coli*, *Mol. Gen. Genet.* **174:**75–88.

Charlier, D., Severne, Y., Zafarullah, M., and Glansdorff, N., 1983, Turn-on of inactive genes by promoter recruitment in *Escherichia coli*: Inverted repeats resulting in artificial divergent operon, *Genetics* **105:**469–488.

Chumley, F. G., and Roth, J. R., 1980, Rearrangement of the bacterial chromosome using Tn10 as a region of homology, *Genetics* **94:**1–14.

Coetzee, J. N., 1979, Genetic circularity of the *Proteus mirabilis* linkage map, *J. Gen Microbiol.* **110:**171–176.

Cornelis, G., Ghosal, D., and Saedler, H., 1978, Tn951: A new transposon carrying a lactose operon, *Mol. Gen. Genet.* **160:**215–224.

Cornelis, G., Sommer, H., and Saedler, H., 1981, Transposon Tn951 is defective and related to Tn3, *Mol. Gen. Genet.* **184:**241–248.

Dean, H. F., and Morgan, A. F., 1983, Integration of R91–5: Transposon 501 into the *Pseudomones putida* PPN chromosome and genetic circularity of the chromosomal map, *J. Bacteriol.* **153:**485–497.

DeJong, W. W., and Ryden, L., 1981, Cause of more frequent deletions than insertions in mutations and protein evolution, *Nature* **290:**157–159.

Doolittle, W. F., and Sapienza, C., 1980, Selfish Genes/The phenotype paradigm and genome evolution, *Nature* **284:**601–603.

Dykstra, C., Prasher, D., and Kushner, S. R., 1984, Physical and biochemical analysis of the cloned *rec*B and *rec*C genes of *Escherichia coli* K-12, *J. Bacteriol.* **157:**21–27.

Edlin, G., Lin, L., and Bitner, R., 1977, Reproductive fitness of P1, P2 and Mu lysogens, *J. Virol.* **21:**560–564.

Espion, D., Kaiser, K., and Dambly-Chandiere, C., 1983, A third defective lambdoid prophage of *Escherichia coli* K12 defined by the λ derivative λ qin 111, *J. Mol. Biol.* **170:**611–633.

Fultz, P. N., Kwoh, D. Y., and Kemper, J., 1979, *Salmonella typhimurium new*D and *Escherichia coli leu*C genes code for a functional isopropylmalate isomerase in *Salmonella typhimurium–Escherichia coli* hybrids, *J. Bacteriol.* **137:**1253–1262.

Goncharoff, P., and Nichols, B. P., 1984, Nucleotide sequence of *Escherichia coli pab*B indicates a common evolutionary origin of *p*-aminobenzoate synthetase and anthranilate synthetase, *J. Bacteriol.* **159:**57–62.

Grantham, R., Gautier, C., Gouy, M., and Mercier, R., 1981, Codon catalog usage is a genome strategy modulated for gene expressivity, *Nucleic Acid Res.* **9:**r43–r74.

Green, L., Miller, R. D., Dykhuizen, D. E., and Hartl, D. L., 1984, Distribution of DNA insertion element *Tn*5 in natural isolates of *Escherichia coli. Proc. Natl. Acad. Sci. USA* **81:**4500–4504.

Harshman, L., and Riley, M., 1980, Conservation and variation of nucleotide sequences in *Escherichia coli* strains isolated from nature, *J. Bacteriol.* **144:**560–568.

Hiestand-Nauer, R., and Iida, S., 1983, Sequence of the site-specific recombinase gene *cin* and of its substrates serving in the inversion of the C segment of bacteriophage P1, *EMBO J.* **2:**1733–1740.

Higgins, C. F., and Ames, G. F. L., 1981, Two periplasmic transport proteins which interact with a common membrane receptor show extensive homology: Complete nucleotide sequences, *Proc. Natl. Acad. Sci. USA* **78:**6038–6042.

Highton, P., Chang, Y., Marcotte, Jr., W., and Schnaitman, C., 1985, Evidence that the outer membrane protein gene *nmp*C of *Escherichia coli* K-12 lies within the defective *qsr'* prophage, *J. Bacteriol.* **161:**256–262.

Hill, C. W., and Harnish, B. W., 1981, Inversions between ribosomal RNA genes of *Escherichia coli*, *Proc. Natl. Acad. Sci. USA* **78:**7079–7072.

Hill, C. W., and Harnish, B. W., 1982, Transposition of a chromosomal segment bounded by redundant rRNA genes into other rRNA genes in *Escherichia coli*, *J. Bacteriol.* **149:**449–457.

Hill, C. W., Graftsrom, R. H., Harnish, B. W., and Hillman, B. S., 1977, Tandem duplications resulting from recombination between ribosomal RNA genes in *Escherichia coli*, *J. Mol. Biol.* **116:**407–428.

Holloway, B. W., Krishnapillai, V. and Morgan, A. F., 1979, Chromosomal genetics of *Pseudomonas*, *Microbiol. Rev.* **43:**73–102.

Hoover, T. A., Roof, W. D., Foltermann, K. F., O'Donovan, G. A., Bencini, D. A., and Wild, J. R., 1983, Nucleotide sequence of the structural gene (*pyr*B) that encodes the catalytic polypeptide of aspartate transcarbamoylase of *Escherichia coli*, *Proc. Natl. Acad. Sci. USA* **77**:2462–2466.

Hooykaas, P. J. J., Peerbolte, R., Regensburg-Tuink, A. J. G., deVries, P., and Schilperoort, R. A., 1982, A chromosomal linkage map of *Agrobacterium tumefaciens* and a comparison with the maps of *Rhizobium* spp, *Mol. Gen. Genet.* **188**:82–86.

Hori, H., 1976, Molecular evolution of 5S RNA, *Mol. Gen. Genet.* **145**:119–123.

Horowitz, N. H., 1965, The evolution of biochemical syntheses—Retrospect and prospect, in: *Evolving Genes and Proteins* (V. Bryson and H. J. Vogel, eds.), pp. 15–23, Academic Press, New York.

Horowitz, H., Christie, G. E., and Platt, T., 1982, Nucleotide sequence of the *trp*D gene encoding anthranilate synthetase component II of *E. coli*, *J. Mol. Biol.* **156**:245–256.

Houghton, J. E., Bencini, D. E., O'Donovan, G. A., and Wild, J. R., 1984, Protein differentiation: A comparison of aspartate transcarbamoylase and ornithine transcarbamoylase from *Escherichia coli* K-12. *Proc. Natl. Acad. Sci. USA* **81**:4864–4868.

Hu, M., and Deonier, R., 1981*a*, Comparison of IS1, IS2, and IS3 copy numbers in *Escherichia coli* strains K12, B and C, *Gene* **16**:161–170.

Hu, M., and Deonier, R., 1981*b*, Mapping of IS1 elements flanking the *arg*F gene region on the *Escherichia coli* K-12 chromosome, *Mol. Gen. Genet.* **186**:12–17.

Inokuchi, K., Mutoh, N., Matsuyama, S., and Mizushima, S., 1982, Primary structure of the *omp*F gene that codes for a major outer membrane protein of *Escherichia coli* K-12, *Nucleic Acids Res.* **10**:6957–6968.

Kaiser, K., 1980, The origin of Q-independent derivatives of phage λ. *Mol. Gen. Genet.* **179**:547–554.

Kaiser, K., and Murray, N. E., 1979, Physical characterization of the "Rac-prophage" in *E. coli* K12, *Mol. Gen. Genet.* **175**:159–174.

Kamp, D., and Kahmann, R., 1981, The relationship of two invertible segments in bactriophage Mu and *Salmonella typhimurium* DNA, *Mol. Gen. Genet.* **184**:564–566.

Kaplan, J. B., and Nichols, B. P., 1983, Nucleotide sequence of *Escherichia coli pab*A and its evolutionary relationship to *trp*(G)D, *J. Mol. Biol.* **168**:451–468.

Kleckner, N., 1981, Transposable elements in prokaryotes, *Annu. Rev. Genet.* **15**:341–404.

Konrad, E. B., 1977, Method for the isolation of *Escherichia coli* mutants with enhanced recombination between chromosomal duplications, *J. Bacteriol.* **137**:167–172.

Krikos, A., Mutoh, N., Boyd, A., and Simon, M. I., 1983, Sensory transducers of *E. coli* are composed of discrete structural and functional domains, *Cell* **33**:615–622.

Lampel, K. A., and Riley, M., 1982, Discontinuity of homology of *Escherichia coli* and *Salmonella typhimurium* DNA in the *lac* region, *Mol. Gen. Genet.* **186**:82–86.

Lin, L., Bitner, R., and Edlin, G., 1977, Increased reproductive fitness of *Escherichia coli* lambda lysogens, *J. Virol.* **21**:554–559.

Lin, R. J., Capage, M., and Hill, C. W., 1984, A repetitive DNA sequence, *rhs*, responsible for duplications within the *Escherichia coli* K12 chromosome, *J. Mol. Biol* **177**:1–18.

Link, C. D., and Reiner, A. M., 1982, Inverted repeats surround the ribitol–arabitol genes of *E. coli* C, *Nature* **298**:94–96.

Link, C. D., and Reiner, A. M., 1983, Genotypic exclusion: A novel relationship between the ribitol–arabitol and galactitol genes of *E. coli*, *Mol. Gen. Genet.* **189**:337–339.

Low, K. B., and Porter, D. D., 1978, Modes of gene transfer and recombination in bacteria, *Annu. Rev. Genet.* **12**:249–287.

Meyer, T. F., Mlawer, N., and So, M., 1982, Pilus expression in *N. gonorrhaeae* involves chromosome rearrangement, *Cell* **30**:45–52.

Meyer, T. F., Billyard, E., Haas, R., Storzbach, S., and So, M., 1984, Pilus genes of *Neisseria gonorrheae*: Chromosomal organization and DNA sequence, *Proc. Natl. Acad. Sci. USA* **81**:6110–6114.

Michiels, T., and Cornelis, G., 1984, Detection and characterization of Tn2501, a transposon included within the lactose transposon Tn951, *J. Bacteriol.* **158:**866–871.

Mizuno, T., Chou, M. Y., and Inouye, M., 1983, A comparative study on the genes for three porins of the *Escherichia coli* outer membrane: DNA sequence of the osmoregulated *omp*C gene, *J. Biol. Chem.* **258:**6932–6940.

Nichols, B. P., van Cleemput, M., and Yanofsky, C., 1981, Nucleotide sequence of *Escherichia coli trp*E anthranilate synthetase component I contains no tryptophan residues, *J. Mol. Biol.* **146:**45–54.

Nyman, K., Nakamura, K., Ohtsubo, H., and Ohtsubo, E., 1981, Distribution of the insertion element IS1 in Gram-negative bacteria, *Nature* **289:**602–12.

Nyman, K., Ohtsubo, H., Davison, D., and Ohtsubo, E., 1983, Distribution of insertion element IS1 in natural isolates of *Escherichia coli*, *Mol. Gen. Genet.* **189:**516–518.

Orgel, C. E., and Crick, F. H. C., 1980, Selfish DNA: The ultimate parasite, *Nature* **284:**604–607.

Overbeeke, N., Bergmans, H., van Mansfeld, F., and Lugtenberg, B., 1983, Complete nucleotide sequence of *pho*E, the structural gene for the phosphate limitation inducible outer membrane pore protein of *Escherichia coli* K12, *J. Mol. Biol.* **163:**513–532.

Parsot, C., Cossart, P., Saint-Girons, I., and Cohen, G. N., 1983, Nucleotide sequence of *thr*C and of the transcription termination region of the threonine operon in *Escherichia coli* K12, *Nucleic Acids Res.* **21:**7331–7345.

Pfeifer, F., Betlach, M., Martienssen, R., Friedman, J., and Boyer, H. W., 1983, Transposable elements of *Halobacterium halobium*, *Mol. Gen. Genet.* **191:**182–188.

Reeve, E. C. R., and Braithwaite, J. A., 1974, The lactose system in *Klebsiella aerogenes* V9A, 4. A comparison of the *lac* operons of *Klebsiella* and *Escherichia coli*, *Genet. Res.* **24:**323–331.

Reynolds, A. E., Felton, J., and Wright, A., 1981, Insertion of DNA activates the cryptic *bgl* operon in *E. coli* K12, *Nature* **293:**625–629.

Riley, M., and Anilionis, A., 1978, Evolution of the bacterial genome, *Annu. Rev. Microbiol.* **32:**519–560.

Riley, M., and Anilionis, A., 1980, Conservation and variation of nucleotide sequences within related bacterial genomes: Enterobacteria, *J. Bacteriol.* **143:**366–376.

Riley, M., O'Reilly, C., and McConnell, D., 1984, Physical map of *Salmonella typhimurium* LT2 DNA in the vicinity of the *pro*A gene, *J. Bacteriol.* **157:**655–657.

Rogolsky, M., 1970, The mapping of genes for spore formation on the chromosome of *Bacillus lichenoformis*, *Can. J. Microbiol.* **16:**595–600.

Ryder, T. B., Davison, D. B., Rosen, J. I., Ohtsubo, E., and Ohtsubo, E., 1982, Analysis of plasmid genome evolution based on nucleotide-sequence comparison of two related plasmids of *Escherichia coli*, *Gene* **17:**299–310.

Sammons, R. L., and Anagnostopoulos, C., 1982, Identification of a cloned DNA segment at a junction of chromosome regions involved in rearrangements in the *trp*E26 strains of *Bacillus subtilis*, FEMS Microbiol. Lett. **15:**265–268.

Sanderson, K. E., 1976, Genetic relatedness in the family Enterobacteriaceae, *Annu. Rev. Microbiol.* **30:**327–349.

Sapienza, C., Rose, M. R., and Doolittle, W. F., 1982, High-frequency genomic rearrangements involving archaebacterial repeat sequence elements, *Nature* **299:**182–185.

Schmid, M., and Roth, J. R., 1980, Circularization of transduced fragments: A mechanism for adding segments to the bacterial chromosome, *Genetics* **94:**15–29.

Schmid, M. B., and Roth, J. R., 1983*a*, Genetic methods for analysis and manipulation of inversion mutations in bacteria, *Genetics* **105:**517–537.

Schmid, M. B., and Roth, J. R., 1983*b*, Selection and end point distribution of bacterial inversion mutations, *Genetics* **105:**539–557.

Schneider, A. -M., and Anagnostopoulos, C., 1983, *Bacillus subtilis* strains carrying two non-tandem duplications of the *trp*E–*ilv*A and the *pur*B–*tre* regions of the chromosome, *J. Gen. Microbiol.* **129:**687–701.
Schoner, B., and Schoner, R. G., 1981, Distribution of the IS5 in bacteria, *Gene* **16:**347–352.
Schupp, T., Hutter, R., and Hopwood, D. A., 1975, Genetic recombination in *Nocardia mediterranei*, *J. Bacteriol.* **121:**128–136.
Simon, M., Zeig, J., Silverman, M., Mandel, G., and Doolittle, R., 1980, Phase variation: Evolution of a controlling element, *Science* **209:**1370–1374.
Stokes, H. W., and Hall, B. G., 1984, Topological repression of gene activity by a transposable element, *Proc. Natl. Acad. Sci. USA* **81:**6115–6119.
Straus, D. S., and Hoffman, G. R., 1975, Selection for a large genetic duplication in *Salmonella typhimurium*, *Genetics* **80:**227–237.
Straus, D. S., and Straus, L. D., 1976, Large overlapping tandem genetic duplications in *Salmonella typhimurium*, *J. Mol. Biol.* **103:**143–153.
Szekely, E., and Simon, M., 1983, DNA sequence adjacent to flagellar genes and evolution of flagellar-phase variation, *J. Bacteriol.* **155:**74–81.
Tanaka, M., Yamamoto, T., and Sawai, T., 1983, Evolution of complex resistance transposons from an ancestral mercury transposon, *J. Bacteriol.* **153:**1432–1438.
Timmons, M. S., Bogardus, A. M., and Deonier, R. C., 1983, Mapping of chromosomal IS5 elements that mediate type II F-prime plasmid excision in *Escherichia coli* K-12, *J. Bacteriol.* **153:**395–407.
Timmons, M. S., Spear, K., and Deonier, R. C., 1984, Insertion element IS*121* is near *pro*A in the chromosomes of *Escherichia coli* K-12 strains, *J. Bacteriol.* **160:**1175–1177.
Trowsdale, J., and Anagnostopoulos, C., 1976, Differences in the genetic structure of *Bacillus subtilis* strains carrying the *trp*E26 mutation and strain 168, *J. Bacteriol.* **126:**609–618.
Van de Putte, P., Plasterk, R., and Kuijpes, A., 1984, A Mu *gin* complementary function and an invertible DNA region in *E. coli* K12 are situated on the genetic element e14, *J. Bacteriol.* **158:**517–522.
Van Vliet, F., Jacobs, A., Piette, J., Gigot, D., Lauwreys, M., Pierard, A., and Glansdorff, N., 1984, Evolutionary divergence of genes for ornithine and aspartate carbamoyl-transferase—complete sequence and mode of regulation of the *Escherichia coli argF* gene; comparison of *argF* with *argI* and *pyrB*. *Nucleic Acids Res.* **12:**6277–6289.
Wiman, M., Bertani, G., Kelly, B., and Sasaki, I., 1970, Genetic map of *Escherichia coli* strain C, *Mol. Gen. Genet.* **107:**1–31.
Woodward, M. J., and Charles, H. P., 1982, Genes for L-sorbose utilization in *Escherichia coli*, *J. Gen. Microbiol.* **128:**1969–1980.
Woodward, M. J., and Charles, H. P., 1983, Polymorphism in *Escherichia coli*: *rtl atl* and *gat* regions behave as chromosomal alternatives, *J. Gen. Microbiol.* **129:**75–84.
Yanofsky, C., 1984, Comparisons of regulatory and structural regions of genes of tryptophan metabolism, *Mol. Biol. Evol.* **1:**143–161.
Yanofsky, C., Platt, T., Crawford, I. P., Nichols, B. P., Christie, G. E., Horowitz, H., van Cleemput, M., and Wu, A. M., 1981, The complete nucleotide sequence of the tryptophan operon of *Escherichia coli*, *Nucleic Acids Res.* **9:**6647–6668.
Yokota, T., Sugisaki, H., Takanami, M., and Kaziro, Y., 1980, The nucleotide sequence of the cloned *tuf*A gene of *Escherichia coli*, *Gene* **12:**25–31.
York, M. K., and Stodolsky, M., 1981, Characterization of P1 *arg*F derivatives from *Escherichia coli* K12 transuction, *Mol. Gen. Genet.* **181:**230–240.
Zengel, J. M., Archer, R. H., and Lindahl, L., 1984, The nucleotide sequence of the *Escherichia coli fus* gene, coding for elongation factor G, *Nucleic Acids Res.* **12:**2181–2192.

2

The Evolutionary Ecology of *Drosophila*

LAURENCE D. MUELLER

INTRODUCTION

In this review I have attempted to summarize a number of topics with which I am familiar that have significance for the field of evolutionary ecology. Since a good deal of this work includes laboratory studies, the following section provides a detailed description of the ecology of *Drosophila* (usually *melanogaster*) in the laboratory environment. This is followed by a section on the population dynamics of *Drosophila*, with particular attention paid to sources of bias that are inherent in certain experimental techniques. The next section considers the mechanisms and evolution of intra- and interspecific competition. This is followed by a section examining tradeoffs and correlations in *Drosophila* life-history traits and the consequences of density-dependent natural selection. The last section summarizes a rapidly growing literature on the habitat and oviposition preference of *Drosophila*. Several recent reviews touch upon areas of the ecology and evolution of *Drosophila* that I will not address (Parsons, 1980, 1981; Barker and Starmer, 1982). In this review I have attempted to develop a critical review of certain major topics and, I hope, unify certain disparate observations rather than be comprehensive in my citations. To all those whose work I have failed to cite I offer my apologies.

Often in the halls of academia or occasionally in print (King and Dawson, 1983) the relevance of laboratory research with *Drosophila* to the understanding of the natural ecology or evolution in this genus is

---

LAURENCE D. MUELLER • Department of Zoology and Program in Genetics, Washington State University, Pullman, Washington 99164.

questioned. Although it is perhaps ovbious to many, it may be worth emphasizing the role of controlled experiments on laboratory organisms. Laboratory studies of evolution tell us what the possible responses of a species may be; they do not necessarily tell us what has happened in nature or what will happen—only what is possible. This is not to say that trying to understand why *D. melanogaster* prefers to live in one garbage can over another is not a worthwhile endeavor; it should not be considered the *only* worthwhile endeavor. The obvious benefits of conducting laboratory experiments with *Drosophila* are that the evolutionary process can be examined in much finer detail than is possible "out of doors" and causation can be inferred since the experimenter is free to control all variables of interest.

It is true that the results obtained from any study of a "model" organism may have limited generality. This will probably be the case more often in studies of the evolution of *Drosophila* than with studies of its molecular genetics. As discussed in the section on life-history evolution, the evolution of density-dependent rates of population growth differ substantially between *D. melangaster* (Mueller and Ayala, 1981*d*) and *Escherichia coli* (Luckinbill, 1978). Despite these problems, the wealth of knowledge on the laboratory ecology of *Drosophila* and the facility of doing genetic experiments make species of this genus especially useful for experimental evolutionary ecology.

THE LABORATORY ECOLOGY OF *DROSOPHILA*

Since a large number of studies on the evolutionary biology of *Drosophila* are conducted in the laboratory environment, it is clearly of interest to understand the details of *Drosophila* ecology in these environments. Almost all detailed studies of *Drosophila* laboratory ecology have utilized *D. melanogaster*. Thus in this section references will be limited to *D. melanogaster* unless mentioned otherwise.

The laboratory environment may range from a small 1-oz. vial to large plexiglass cages of 6–8 ft^3, although the traditional culture consists of a $\frac{1}{2}$-pint milk bottle. In the vial and bottle cultures measured amounts of food are deposited on the bottom, while population cages often have several removable food cups, which can be changed at regular intervals. The food medium generally consists of sugars (sucrose, dextrose, molasses), carbohydrates (corn meal, flour), dead yeast, a mold inhibitor (propionic acid, tegosept), and agar to solidify the medium. In addition, the larvae and adults gain substantial nutrition from a growing yeast population. Because the yeast are growing during most experiments and the depth of the medium utilized is variable, it is hard to quantify the amount of food available for larvae and adults. Sang *et al.* (1949) have shown that the yeast population in *Drosophila* cultures where *Drosophila* are omitted increases exponentially for 5–8 days before reaching a maximum. There is a substantial decrease in the maximum yeast biomass when flies are added to these cultures and the decrease is roughly proportional to the number of files added. However, there is still substantial temporal variation in the biomass of live yeast in *Drosophila* cultures.

Simply increasing the volume of the food medium can increase larval survival (Sang, 1949*a*) and the number of adults produced (Sang, 1949*c*), although increases in the surface area are more effective than increases in the depth of the medium. In addition, cultures with live yeast are substantially more productive than cultures with dead yeast only (Sang, 1949*c*).

Mold can be a substantial problem in some *Drosophila* cultures. For instance, in the study by Mueller and Ayala (1981*a*) the density-dependent rates of population growth were estimated for 25 lines of *D. melanogaster* isogenic for different second chromosomes. One line (number 30) was so unproductive at the lowest density examined that many cultures were overgrown with mold before the end of the experiment. This resulted in an actual decline in the per-capita rate of population growth relative to higher densities. This ''Allee effect'' was not seen in other cultures unaffected by mold. Antimold agents can affect population dynamics. Sang (1949*c*) reports that the antimold agent nipagin added to *Drosophila* cultures results in a significant decline in productivity relative to cultures without nipagin.

Methods of maintaining populations will be discussed more fully in the next section. Briefly, most techniques involve moving adults to a fresh food source, since manipulation of eggs, larvae, or pupae is much more cumbersome. In some population cages adults are kept indefinitely in a single cage and food cups are changed at periodic intervals. The exact method of population maintenance is quite important if one is interested in modeling population dynamics. This will be discussed further in the following section.

The remainder of this section on *Drosophila* laboratory ecology will examine the egg, larval, pupal, and adult life stages, respectively.

Eggs

In studies of life-history evolution a distinction is usually made between the number of eggs an organism produces and the energy invested in these eggs. This distinction is especially important in species where the nutritional content of eggs may vary and this variation can translate into differential survival. In Bakker's (1961) study of larval competition he compared the average egg weight of a wild-type stock to a Bar-eyed stock. No significant differences were found. It should be noted that under competitive conditions the probability of survival and the final adult size of these two stocks can differ substantially. These differences are not due to an initial difference in the eggs. Prout (1985) has also shown that the viability of eggs is not dependent on their mothers' past history. Thus, even though factors such as larval density can affect female adult size and fecundity, they do not seem to affect egg viability. Although more data would be welcome, it seems reasonable to consider a *Drosophila* egg as being a standard unit of reproductive investment.

At 25°C eggs will hatch about 20 h after fertilization. However, the average time of egg hatch may depend on genotype. Bakker (1961) found a 45-min difference between his wild and Bar stocks.

Female *D. melanogaster* will lay unfertilized eggs, although at a much reduced rate (Alvarez and Fontdevila, 1981). Alvarez and Fontdevila (1981) also present data that indicate that for some female genotypes the proportion of unfertilized eggs they lay may increase with the age of the female.

Eggs may also fail to hatch due to the feeding activity of larvae. Larvae may bury eggs while feeding and the probability of this will depend on the density and age of the larvae. With day-old larvae less than 25% of all eggs are buried (Chiang and Hodson, 1950). However, with a dense culture of 5-day-old larvae nearly 100% of all eggs were buried (Chiang and Hodson, 1950). Chiang and Hodson further demonstrated that buried eggs rarely hatch. In both an agar medium and a yeast–water mixture fewer than 20% of buried eggs hatch, while controls showed 90% hatchability. This frequency increased to 50% if the egg was within a few hours of hatching when it was buried.

Larvae

The larval life stage is quite important in the laboratory ecology of *Drosophila* for a number of reasons. Although the effect is difficult to quantify, the larvae probably experience the most severe viability effects of crowded cultures (Sang, 1949*b*). In addition, the final size of the adult flies will be determined by the amount of food consumed by the larvae prior to pupation. Since female size is correlated with fecundity, nutrition of larvae has an important impact on this component of fitness. In crowded cultures *Drosophila* must cope with limited pupation sites, waste accumulation, and, perhaps most importantly, limited food. Several studies have looked specifically at the effect of limited food on larval survival (Chiang and Hodson, 1950; Bakker, 1961, 1969; Nunney, 1983). The difficulty in regulating the amount of food available in standard cultures has been overcome by using nonnutritive agar cultures onto which measured amounts of a yeast paste have been added. Since the yeast do not grow in such cultures, the food available for larvae can be determined accurately.

The process of larval competition for limited food will be described first with a model proposed by Nunney (1983) and then by experimental data that support the basic tenets of the model. Suppose a population consists of n competitive types with frequencies p_i $i = 1, \ldots, n$. These types can be different sexes and genotypes. These types can differ with respect to the minimum amount of food needed for pupation m_i, the relative feeding rate α_i, or the variance in feeding rate σ_i^2. Thus it is assumed that when food is limiting, the superior competitor will be the one that consumes food at the fastest rate (largest α_i) or is most efficient in turning food into biomass (smallest m_i). Interference between types is assumed to be unimportant in larval competition for food. In a population with limited food this model assumes that the amount of food that has been consumed by individuals of a given type by the time no food is left has a normal distribution with mean $\alpha_i Y/\overline{\alpha}$ and variance $(\sigma_i \alpha_i Y/\overline{\alpha})^2$, where $\overline{\alpha} = \sum p_i \alpha_i$ and Y is the initial amount of food per larva in the environment. Figure 1 shows the distribution of food consumed by individuals of competitive type i. Only those individuals that have consumed more than m_i grams of food will survive. Thus the probability of survival for type i competitors W_i is given by the shaded area in Fig. 1. If the population consists entirely of type i individuals, then $\alpha_i = \overline{\alpha}$ and W_i is independent of α. If a large number of inferior competitors, type j, are added to a pure population of type i individuals and Y is kept constant, then $\alpha_i/\overline{\alpha} > 1$ and W_i will increase. In a pure population each type i individual consumes Y grams of food on average. When types i and j compete for this same limited amount of food type, the faster rate of consumption of type i will allow them to consume more than Y grams of food on average by the time the food runs out. Similarly, type j individuals will have received less than Y grams of food on average. The opposite effect is also possible. If the amount of food per larva is kept the same but superior competitors are added to a population of type i individuals, W_i will decrease.

It is also obvious from Fig. 1 that if m_i is decreased and all other parameters are kept the same, then W_i will increase. The effect of σ_i on viability is more complicated. If m_i is less than the mean amount of food

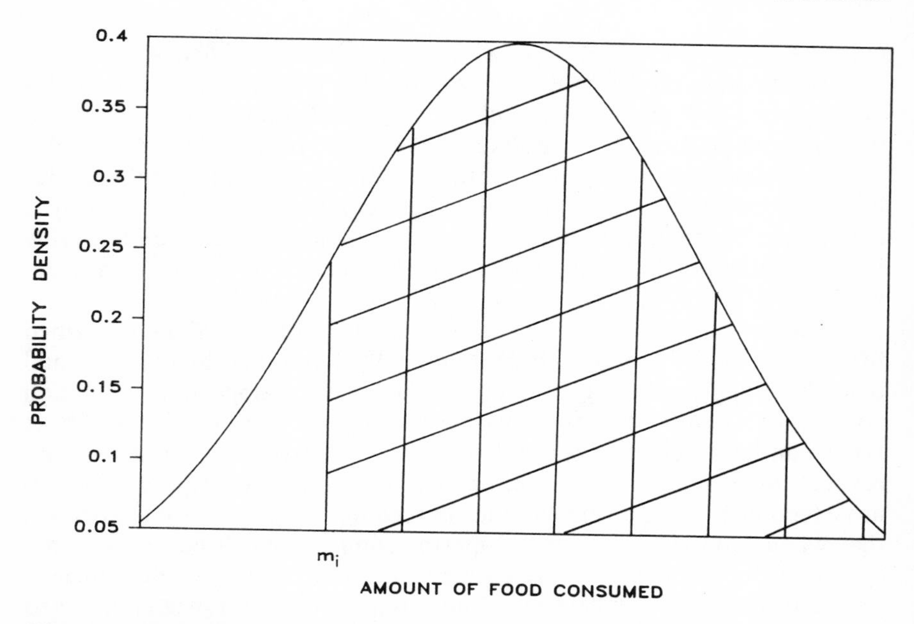

FIG. 1. A normal distribution describing the probability that a larva of competitive type i has consumed a certain amount of food in a population with n competitive types. The mean of the distribution is $\alpha_i Y/\bar{\alpha}$ and m_i is the minimum amount of food necessary for successful pupation by type i larve. The cross-hatched region represents the probability of a type i larva successfully pupating.

consumed, as shown in Fig. 1, then increasing σ_i will increase the area to the left of m_i and hence decrease W_i. However, when $m_i > \alpha_i Y/\bar{\alpha}$, σ_i will increase the area to the right of m_i and thus increase W_i. These conditions can be qualitatively interpreted as follows: when the average amount of food per larva is greater than the minimum needed for successful pupation, reducing the variance in food consumed by different individuals will ensure that all individuals consume more than the minimum amount. However, when the average amount of food per individual is less than the minimum amount, the only way to ensure some pupation is to have a few individuals consume much more than the average available and others consume much less, resulting in an increase in individual variation.

Figure 2 shows Bakker's (1961) data on the percent emergence of wild-type and Bar-eye flies as a function of food availability. These experiments all started with 200 first-instar larvae and the amount of food

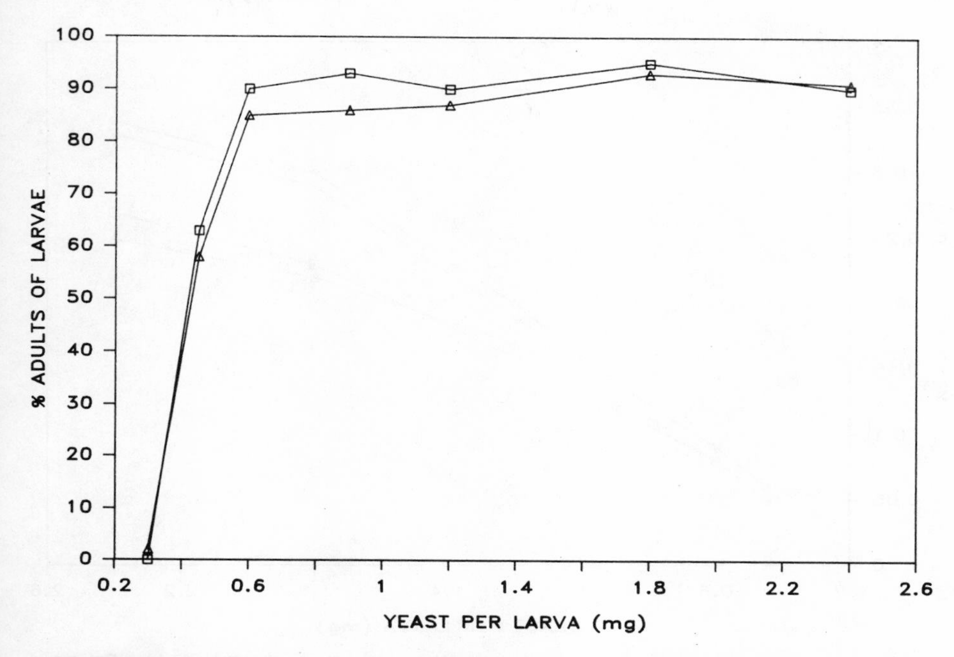

FIG. 2. The percent of all larvae that successfully pupate as a function of different food levels. The data are from Bakker (1961): (□) wild-type larvae, (△) Bar larvae.

used varied. Figure 3 gives the adult weights of males and females from each stock. Several results are immediately apparent: first, there is a minimum weight that males and females reach at about the same level of food availability; second, even after the maximum emergence rate has been recorded, the adult size continues to increase. These results alone do not eliminate the possibility that the competitive effects are due to interference rather than a simple scramble for the limited amounts of food. To test this possibility, Chiang and Hodson (1950) and Bakker (1961) looked at the percent emergence and adult weight by varying both the amount of food per larva and the density of larvae. In the range of densities examined in these experiments it was seen that the amount of food per larva was the most important determinant of percent emergence and adult weight. There was no increase noted in competitive effects by simply increasing the density while keeping the amount of food per larva constant.

If the above basic assumptions of larval competition are correct, we should be able to detect differential rates of growth for larvae with dif-

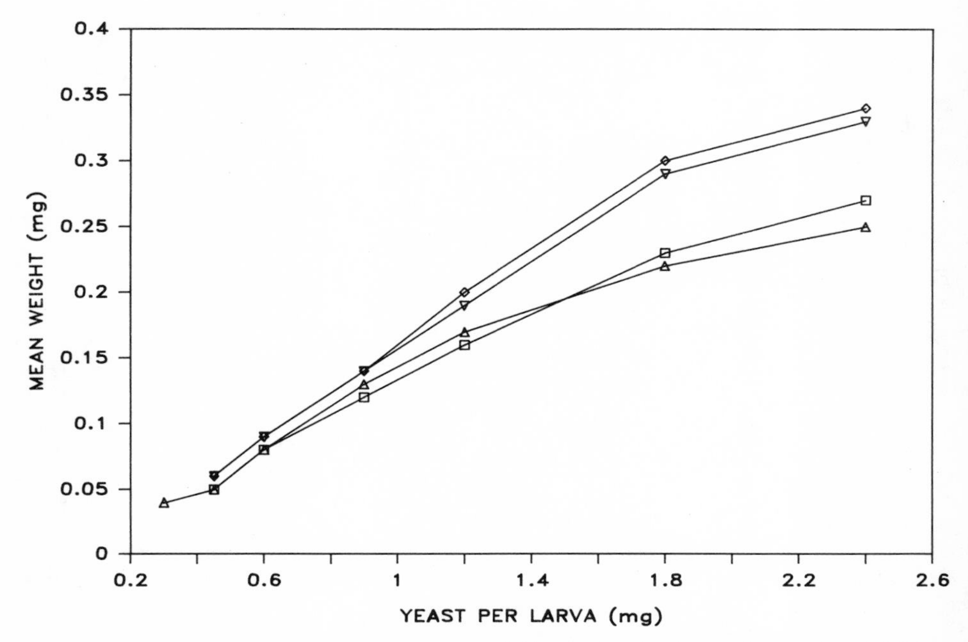

FIG. 3. Mean dry weight of adults as a function of different food levels. The data are from Bakker (1961): (◇) wild-type females, (▽) Bar females, (□) wild-type males, (△) Bar males.

ferent α_i. Bakker (1961) let larvae feed with unlimited food for periods of 39–60 h. At various times the larvae were transferred to nonnutritive medium; a portion of these were weighed and another portion were allowed to pupate. Limiting the length of feeding time produces similar effects as limiting the total amount of food. In Fig. 4 the difference between the wild and Bar populations in percent pupation is shown. Figure 5 shows the difference in larval weight for the same two populations as a function of feeding time. If the data in these two figures are reanalyzed, it turns out that the percent pupation is approximately the same for each population, given that larvae have achieved the same size. Since the results in Fig. 3 indicate that wild and Bar flies have approximately the same efficiency of converting food into *Drosophila* biomass, we can interpret the size differences in Fig. 5 as indicating a difference in the rate at which food is changed into biomass. From Fig. 5 it appears that Bar flies require approximately 3 h longer to reach the same larval size as a wild fly.

According to the model of larval competition, we would expect the

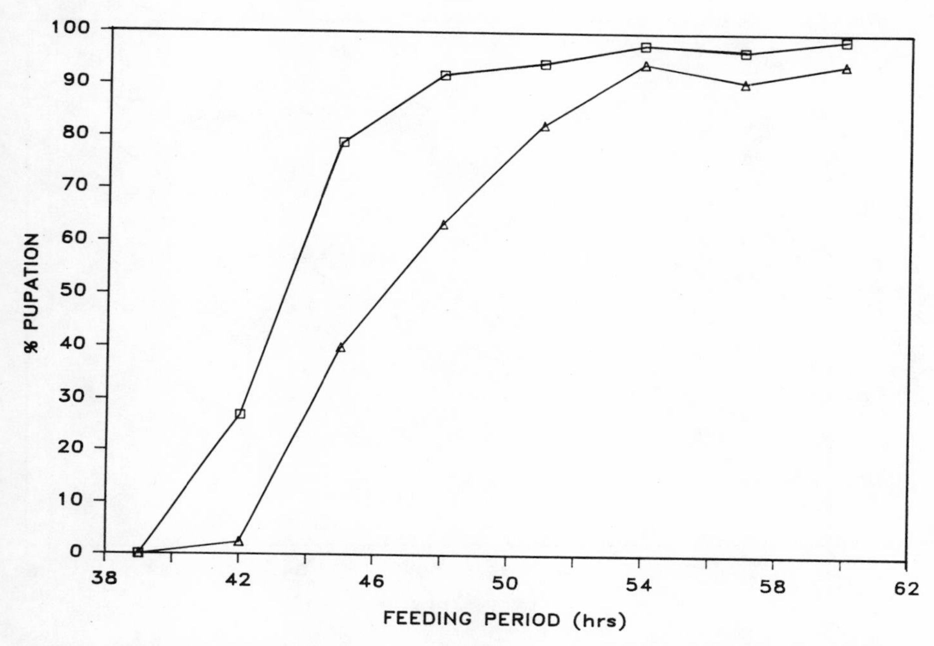

FIG. 4. The percent of all larvae that successfully pupate as a function of different feeding times. The data are from Bakker (1961); the symbols are the same as in Fig. 2.

wild flies to have a substantial competitive advantage relative to Bar flies due to their higher feeding rates. Bakker (1961) conducted larval viability experiments similar to those described previously with 100 wild larvae and 100 Bar larvae placed in vials with 0.42 mg yeast per larva. Only 40% of the larvae reached the adult stage, and of these, 85% were wild and 15% were Bar. It would seem reasonable that if differential feeding rates were the major reason for the differential competitive abilities, then this competitive difference could be eliminated by providing Bar larvae a slight headstart in feeding. As shown in Fig. 5, the Bar larvae generally lag 3 h behind the wild larvae in reaching a given weight. If Bar larvae are given a 3-h headstart, then by the time that the food is exhausted both larvae should be approximately the same size and hence have the same probability of successfully pupating. Bakker (1961) performed this experiment and found that among the emerging adults 59% were Bar.

Nunney (1983) has examined differential competitive abilities of males and females of the same population. Figure 6 shows Nunney's (1983) viability results as a function of food availability in addition to the

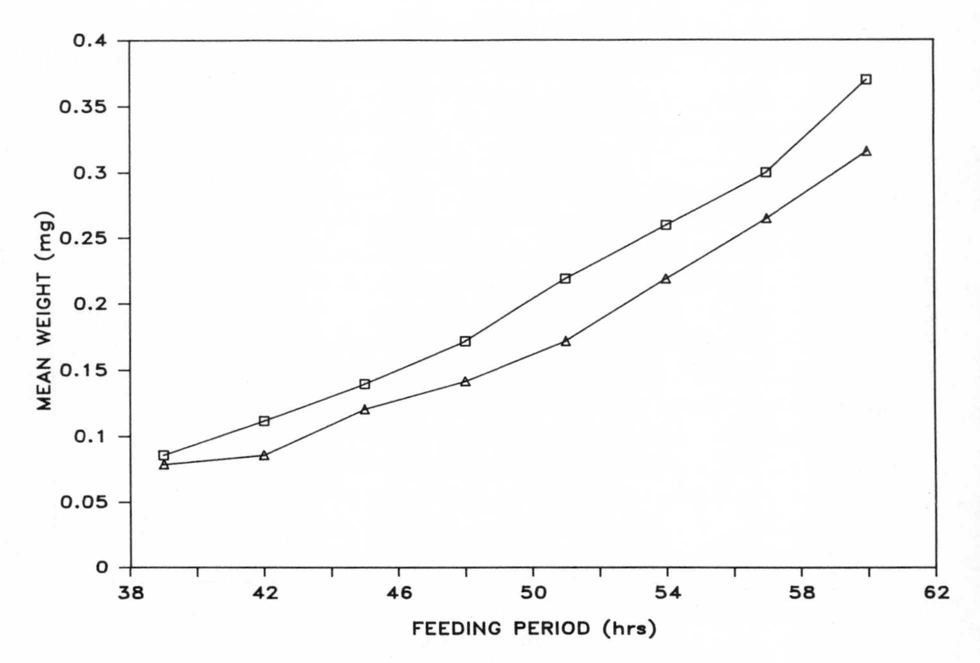

FIG. 5. The weight of larvae as a function of different feeding times. The data are from Bakker (1961); the symbols are the same as in Fig. 2.

least squares fit to the competition model outlined previously. Clearly, the fit is quite good. De Jong (1976) has examined a model similar to Nunney's (1983) in which she assumes that the amount of food consumed has a binomial distribution. Nunney (1983) demonstrates that his empirical results do not fit this model with the same degree of accuracy as the normal distribution model.

Pupae

Just as larval crowding and limited food will affect the final size of adults, they will also affect the size of pupae. Chiang and Hodson (1950) present data that show that pupal length declines with increasing larval density. There is an approximately linear relationship between pupal length and adult size as measured by wing length. Although not discussed by Chiang and Hodson, their data on adult wing length and pupal length show a clear difference between the male and female populations. The

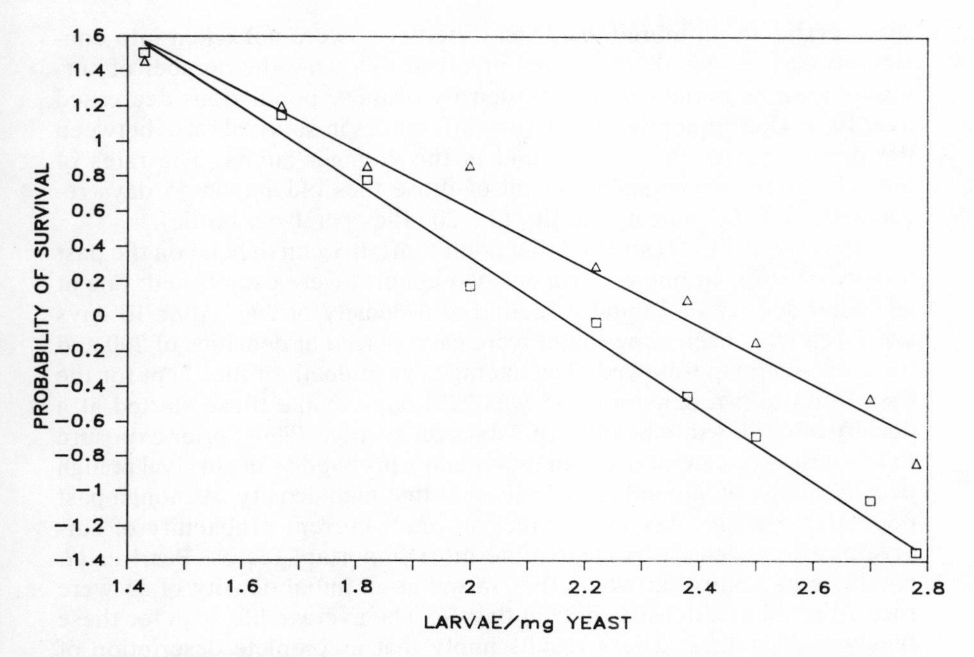

FIG. 6. The probability of survival, measured in standard normal deviates, versus food level. The lines are least squares estimate from the model described in the text. The data are from Nunney (1983): (△) Barton males, (□) Barton females.

relationship between wing length and pupal length is linear for males and females; however, there appear to be different parallel lines for each sex. Thus, females with the identical pupal length as males will develop into adults with longer wings.

Chiang and Hodson (1950) note two main sources of pupal mortality: first, drowning in the medium, and second, inadequate larval size leading to incomplete development. Pupation will often occur on the walls of the culture. In populations yielding large numbers of larvae, early-pupating individuals can become dislodged by larvae crawling up the sides of the culture to pupate. These dislodged pupae will often drown in the medium. In very crowded or old cultures food may be quite limiting and many pupae may fail to develop due to inadequate nutrition.

Adults

The ecological properties of most interest in adults are survival rates and female fecundity. Pearl *et al.* (1927) studied the density-dependent survivorship of an inbred line. Sex differences were not taken into consideration. Because these studies involved following the number of survivors from an initial cohort, the density of these populations decreased over time. Consequently, the largest differences in survival rates between the density treatments were found in the younger adults. The rates of mortality were very similar for all of those flies older than 55 days regardless of initial starting density (35–200 flies per 1-oz. bottle).

Pearl *et al.* (1927) showed that adult mortality can depend on the past history of a fly. In one experiment two cohorts were established: one at an initial density of 35 and a second at a density of 200. After 16 days survivors from each experiment were each placed at densities of 200 and their survivorship followed. The average age at death of flies kept for the first 16 days at a density of 35 was 22.8 days, while those started at a density of 200 lived to be only 19.7 days on average. Thus, prior exposure to low densities can increase an individual's probability of survival at high density relative to an individual raised at that high density. Although past population density has some effect on one's current probability of survival, current density is clearly the most important factor. Pearl *et al.* conducted a control in which flies raised at an initial density of 35 were reconstituted to a density of 35 at day 16. The average life span for these flies was 34.8 days. These results imply that a complete description of density-dependent growth in an age-structured population would require knowing not only the current age class sizes, but those for some time in the past. Although it is possible to develop such models in principle (Charlesworth, 1980), they are exceedingly difficult to analyze.

Chiang and Hodson (1950) have shown that female fecundity declines as the age and density of larvae in the culture increase. However, it is not clear from their experimental technique whether the females are actually laying fewer eggs or whether the larvae are burying eggs, which then remain uncounted. Even if females do not change the number of eggs they lay in the presence of larvae, they do change the location of eggs laid. Females lay a larger portion of their eggs on the walls of the culture as the density and age of larvae increase (Chiang and Hodson, 1950). Given the potentially high mortality of eggs buried by larvae, this behavioral flexibility in females is a useful adaptation. Chiang and Hodson have also examined female preference for laying eggs in medium differing in moisture and texture. In general females seem to prefer moist, rough surfaces as opposed to smooth, dry ones.

A number of studies have investigated the influence of crowding on female fecundity (Pearl, 1932; Bodenheimer, 1938; Robertson and Sang, 1944; Chiang and Hodson, 1950). All studies reveal that at very low densities (0.5–4 flies/cm^2) female fecundity is very sensitive to changes in

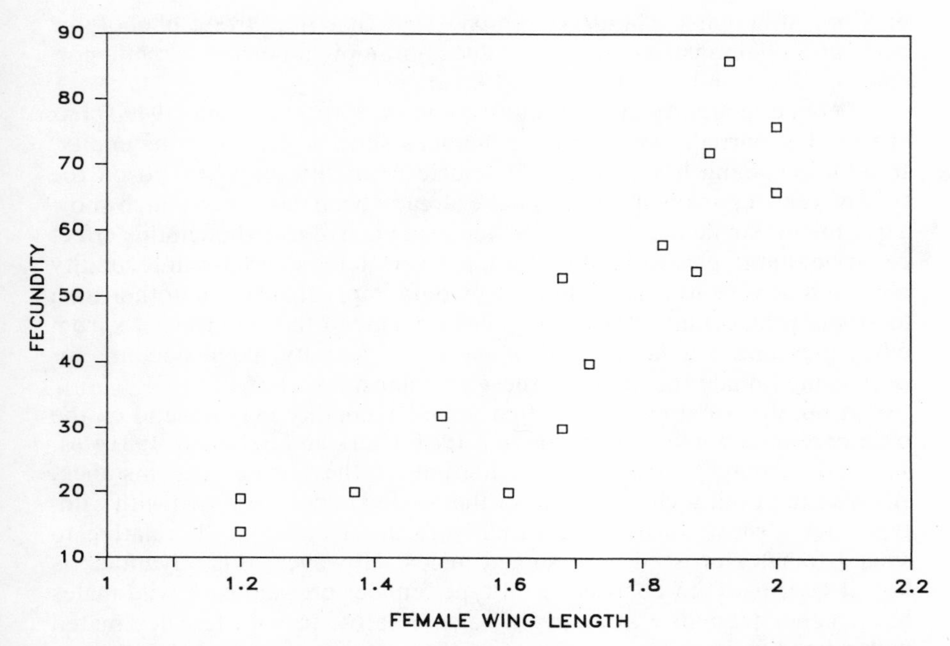

FIG. 7. Fecundity versus female adult size, as measured by wing length. [From Chiang and Hodson (1950).]

adult densities, while at densities greater than 10 flies/cm^2 female fecundity is nearly constant. Thus, studies of female fecundity carried out in 8-dram vials would yield quite different results as density changed from one mating pair to five, although all these densities are typically considered low.

It is apparent from the work considered earlier that in the absence of sufficient food the final size of adults can vary significantly. This size variation in females will inevitably lead to variation in fecundity. The results of Chiang and Hodson are reproduced in Fig. 7. Similar results have been reported by Robertson (1957) and Prout (1985). It should be noted that the females used in these studies are raised on suboptimal amounts of food or in crowded cultures.

In a quantitative genetic study of body size and egg production Robertson (1957) has found little correlation in the additive genetic components of these traits. However, these studies were conducted in uncrowded environments. The same studies conducted with limited food may show high additive genetic correlation between size and egg production, although it should be emphasized that the strong phenotypic correlation demonstrated in Fig. 7 does not imply a positive genetic correlation (Rose and Charlesworth, 1981*a*).

Female fecundity also depends on adult nutrition. Sang (1949*d*) has shown that partially starved adult females show a decline in fecundity. In addition, Sang has shown that female fecundity may depend on the type of yeast consumed. Studies have already been described which show a decline in female fecundity with increasing density. Although this effect could be due in part to limited food, the very large decrease in fecundity observed at very low adult densities would imply that factors other than food were important. Pearl (1932) has concluded that interference from other flies and a general level of increased activity might account for decreasing female fecundity at these low densities.

A number of studies show that female fecundity may depend on the male parent (Alvarez and Fontdevila, 1981; Clark and Feldman, 1981*a,b*). This effect may persist over the lifetime of the female. For instance, Alvarez and Fontdevila (1981) show that singed females crossed with wild-type males show an increase in early fecundity (days 1–3) relative to singed females crossed with singed males. However, this advantage is lost at later ages. In contrast, wild-type females crossed with wild males have higher fecundity at nearly all ages relative to wild females mated with singed males.

POPULATION DYNAMICS

Some of the earliest ecological work with *Drosophila* examined the dynamics of laboratory populations. Such studies have continued intermittently to the present. The development of research in this field provides a useful example of the interaction between theory and experiment. It also illustrates the problems that are encountered when inappropriate experimental regimes are used to test theory.

This section will begin with a discussion of the different methods of maintaining *Drosophila* populations. The discussion will be followed by a more detailed description of the serial transfer system (STS) and results from various studies. Finally, the ecological and evolutionary significance of population stability will be considered.

Population Maintenance

Some of the earliest and most cited work on the population dynamics of *Drosophila melanogaster* was carried out by Raymond Pearl and his co-workers (Pearl, 1927). Pearl's work is often used as the textbook example of logistic population growth. Pearl's method of maintaining cultures seems straightforward, but has a number of inherent problems. Into standard *Drosophila* cultures Pearl placed a small number of adults. At regular intervals the adult population were censused and returned to the original culture. Although such populations will initially show a rapid increase in adult numbers, unless the food supply is supplemented, the population will eventually go extinct. Rather than move the adult population to a fresh food source at regular intervals, Pearl added fresh food to the old culture as needed. Pearl (1927) notes that the "experiment is one in which an attempt is made to add food as the supply is used up. The technical difficulties of doing this satisfactorily with a *Drosophila* population are considerable, but with sufficient care they can be overcome in large degree." Clearly such a subjective protocol would weaken any theoretical results based on these observations. Sang (1949*d*) has noted these problems and concluded, "On the face of it, it is therefore difficult to understand how the results of such an experiment . . . fit a logistic for any reason other than the choice of the right time at which to add food and the right amount of it."

Due to these problems subsequent workers have attempted to replace food at regular intervals. This can be achieved in two ways: (1) the adult population is moved to a fresh food source or (2) the adult population is kept in the same culture with fresh food introduced at regular intervals. L'Heritier and Teissier (1933) describe an early population cage which exemplifies the second alternative. In this study new food cups were added every day. When a food cup was 20 days old it was removed. Thus at any time the population cage would have 20 food cups aged 1–20 days. The total amount of food available to the population and the rate of replenishment could thus be carefully controlled. Most population cages used today operate on the same principle (Ayala, 1968*b*).

The first method of renewing resources is exemplified by the serial transfer system (STS). Pearl (1927) actually describes a form of the STS, while refinements are described by Buzzati-Traverso (1955) and Ayala (1965*b*). In the STS there is a single adult population, which is allowed to lay eggs on fresh medium for a specified number of days, after which all surviving adults are transferred to a fresh culture and another round of egg laying commences. As adults start to emerge from the older cultures, they are collected and added to the egg-laying population of adults. Thus at any time interval the adult population consists of survivors from the previous time interval plus newly emerged adults from the old cultures. In Fig. 8 the STS is shown in which cultures are discarded when they are 4 time units old. The actual length of the time interval varies

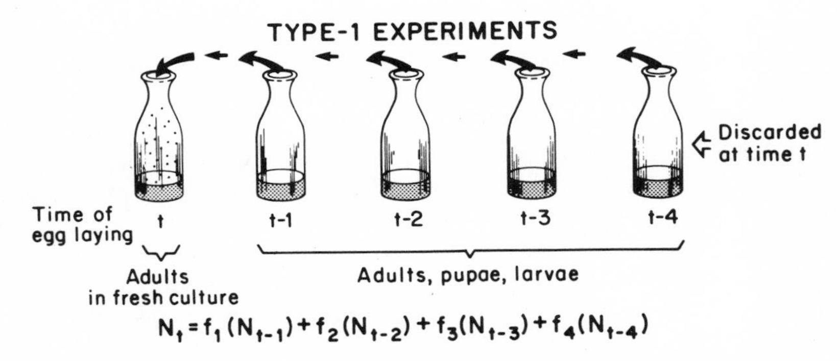

FIG. 8. The serial transfer system of maintaining *Drosophila* populations. [From Mueller and Ayala (1981*b*).] See text for additional details.

from 3 days (Buzzati-Traverso, 1955) to 1 week (Ayala *et al.*, 1973; Gilpin and Ayala, 1973; Thomas *et al.*, 1980; Pomerantz *et al.*, 1980; Hastings *et al.*, 1981; Mueller and Ayala, 1981*b*). The STS has the advantage of providing a regular supply of food and oviposition sites in addition to maintaining an adult population with overlapping generations. The disadvantage of the STS is that it is rather complicated to model (to be discussed further).

Since ecologists are most interested in examining models of population dynamics that are first-order difference equations, the following laboratory regime would seem to be the most appropriate. Adult flies are allowed 1–2 days for egg laying, after which time they are discarded. Twelve to 15 days later the newly emerged adults are collected and put in fresh cultures to start the next generation. This method of population maintenance is used commonly by population geneticists, since most population genetic models assume a fully discrete life cycle. Despite its widespread use by population geneticists, it is seldomly used by *Drosophila* population ecologists [for an exception see Prout (1985)]. The reason for this is not clear, although there may be an unconscious desire by ecologists to avoid the artificial termination of the adult life cycle. Undoubtedly, methods of population maintenance that preserve an adult population with overlapping generations would more closely resemble the "natural" ecology of *Drosophila*.

Modeling Population Dynamics of *Drosophila*

Ayala *et al.* (1973) were the first to describe a simple experimental technique which would allow density-dependent rates of population

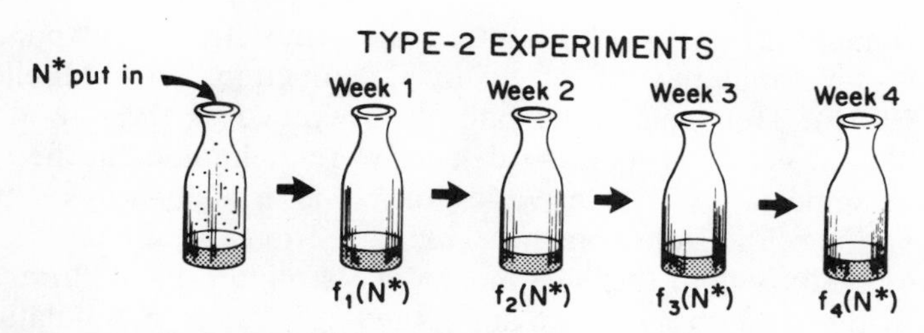

FIG. 9. The experimental protocol for estimating density-dependent rates of population growth in the STS [From Mueller and Ayala (1981*b*).]

growth to be determined in the serial transfer system. The procedure involves placing a specified number of adults N^* in a single culture. The survivors in this culture are counted 1 week (or any other time interval) later. The newly emerged progeny from this same culture are removed and counted at weekly intervals for the next 3 weeks. This experimental procedure is outlined in Fig. 9. It seems reasonable to expect rates of population growth in the STS to depend on the number of egg-laying adults that survive to the next generation, the total number of progeny produced from each week's egg laying, and the time at which these progeny emerge. This last point is especially important. Since progeny from 1 week of egg laying are being collected over a 3-week period and placed with the egg-laying adult population as soon as they emerge (see Fig. 8), it is reasonable to assume that a population in which most of the progeny emerge very early will grow at a faster rate than a population that has slowly developing progeny.

Referring to Fig. 8, one can see that the total number of egg-laying adults at time t, N_t, is simply the sum of the number surviving from the previous week, $f_1(N_{t-1})$, plus the number emerging from the 2-week old culture, $f_2(N_{t-2})$, plus the same numbers from the 3- and 4-week old cultures, $f_3(N_{t-3})$ and $f_4(N_{t-4})$. Each $f_i(N_{t-i})$ is some unspecified function describing the number of survivors or emerging progeny that come out of an i-weeks-old culture that initially had N_{t-i} adults laying eggs [see Mueller and Ayala (1981*b*) for more details]. Thus it is clear that in an STS, as in Fig. 8, N_t depends on the population size in four previous time intervals.

In the type 2 experiments shown in Fig. 9 it is obvious that each experiment provides an estimate of $f_1(N^*)$, $f_2(N^*)$, $f_3(N^*)$, and $f_4(N^*)$. These experiments can of course be replicated at each density N^* and carried out at a variety of densities to get a complete description of the population dynamics in the serial transfer system. Once the $f_i(N^*)$ have been estimated it is possible to obtain an asymptotic rate of population growth λ_{N^*} without having to specify the functions $f_i(\cdot)$ (Mueller and Ayala, 1981*b*). This asymptotic rate of growth λ_{N^*} is the rate at which the population would grow if the density were maintained in the vicinity of N^* for some time. The derivation of λ_{N^*} is in many ways similar to the derivation of rates of exponential increase from a Leslie matrix (Charlesworth, 1980), where the STS has an age structure for cultures rather than individuals. With the λ_{N^*} in hand, the total change in population size at a density N^* can be estimated as

$$N_t - N_{t-1} = \lambda_{N^*}N^* - N^* = N^*(\lambda_{N^*} - 1)$$

A basic assumption of logistic population growth is that per-capita rates of population growth decline linearly with increasing densities. It is precisely this assumption that can be tested with the λ_{N^*} values derived from the type 2 experiments just described (Mueller and Ayala, 1981b, 1982). It should be emphasized that when investigating first-order difference equations that relate λ_{N^*} and N^* one cannot imply that these same equations can be used as a general description of population dynamics in the STS. This is due to the fact that, in general, rates of population growth in the STS will depend on four previous time intervals rather than one. However, for purposes of studying some general relationships between density and rates of population growth it is more convenient to use the asymptotic rates of population growth.

This methodology is not without precedent. In Smith's (1963) study of the population dynamics of *Daphnia magna* he obtained density-dependent rates of population growth only after his experimental populations had reached a stable age distribution. Thus Smith was deriving asymptotic rates of population growth empirically by allowing his populations to remove the effects of the initial arbitrary age distribution. Mueller and Ayala (1981*b*) have achieved a similar result mathematically.

Several other studies (Ayala *et al.*, 1973; Pomerantz *et al.*, 1980; Thomas *et al.*, 1980; Hastings *et al.*, 1981) have used "net productivity" estimates from type 2 experiments to test models of population growth. The net productivity is simply the sum of the survivors and all emerging progeny minus the initial density N^* from a type 2 experiment. The main difference between the net productivity value and λ_{N^*} is that net productivity does not account for the time at which survivors or progeny are added to the egg-laying population of the STS. Thus, as an estimate of the rate of population growth in the STS net productivity is biased. At densities near carrying capacity this bias is small. In particular, at carrying capacity, since each individual only replaces itself, the time at which this occurs is immaterial and hence the bias should be zero. However, the bias will be greater at densities furthermost from carrying capacity. This bias is evident in the graphs of $N_{t+1} - N_t$ versus N_t seen in Pomerantz *et al.* (1980) and Hastings *et al.* (1981), which show a rapid rise in $N_{t+1} - N_t$ and an almost linear decline after reaching a peak. These curves should be compared to those in Mueller and Ayala (1981*b*). Hastings *et al.* (1981) argue that net productivity may be considered as useful a measurement of population dynamics as λ_{N^*}. Unfortunately, there is no direct theoretical connection between net productivity and rates of population growth in the STS as there is for λ_{N^*}. Consequently, net productivity should be regarded as a rough approximation to rates of population growth which will be biased upward at very low densities.

In Fig. 10 density-dependent rates of population growth for several populations of *D. melanogaster* are shown (Mueller and Ayala, 1981*a*). The discrete form of the theta model is

$$N_{t+1} - N_t = rN_t[1 - (K/N_t)^\theta]$$

When $\theta = 1$, the model is identical to the logistic model, and when $\theta < 1$, per-capita rates of population growth decline more rapidly than linear. From Fig. 10 it is clear that the per-capita rates of population growth for *D. melanogaster* are not linear functions of density as predicted by the logistic equation. This result appears to be quite general for *Drosophila*. It is of some interest to be able to understand the mechanism of intraspecific competition that produces these deviations from the logistic (Mueller and Ayala, 1981*b*; Haddon, 1982), although such an understanding is not essential to the development of a descriptive theory (Mueller and Ayala, 1982).

A preliminary attack on this problem would begin with an examination of density-dependent birth and death rates. In Fig. 11 are shown the probabilities of survival of males and females aged 7–14 days held at various densities for 1 week (Mueller, 1979). These rates change very slowly at low densities and would not seem of sufficient magnitude to explain the large decline in per-capita rates of population growth. Pearl *et al.* (1927) also present data on the average duration of adult life as a function of density. These data actually show increasing survivorship with increasing density over a range of low densities.

One can investigate the effects of limited food on female fecundity by using Bakker's data on food versus adult size and Robertson's data on female size versus fecundity. These results are shown in Fig. 12. Clearly, over the range of food levels tested the rate of decline in female fecundity is linear or slower than linear with decreasing food levels. It is

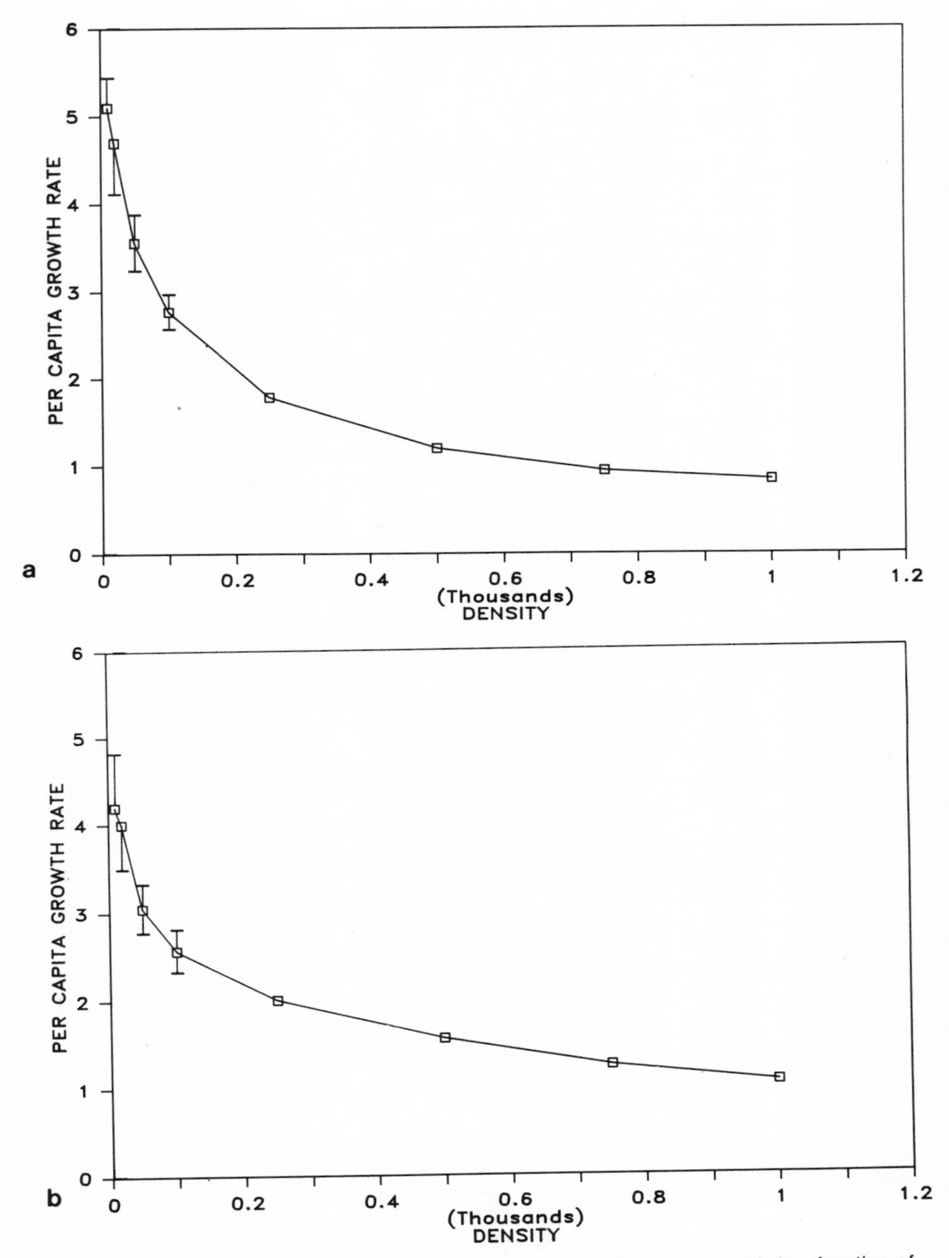

FIG. 10. Per-capita rates of population growth (with 95% confidence intervals) as a function of density for three populations of *D. melanogaster* isogenic for different second chromosomes. [From Mueller and Ayala (1981*a*).] No confidence intervals have been drawn when these were smaller than the width of the symbol. (a) Line 3, (b) line 18, (c) line 36.

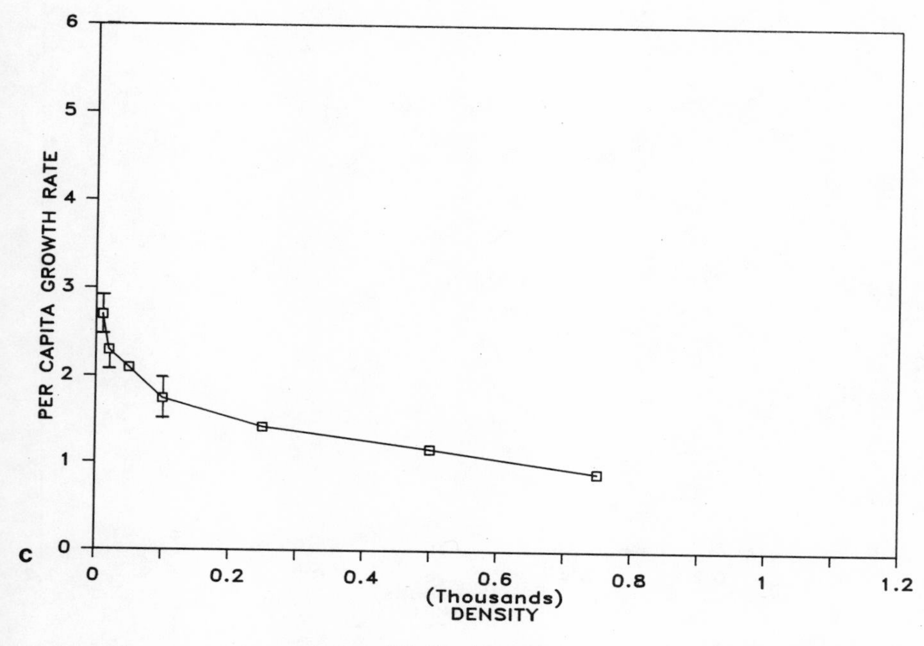

FIG. 10. (*Continued*)

unlikely that this phenomenon alone could account for the rapid decline in per-capita rates of population growth at low densities. Chiang and Hodson (1950) described dramatic declines in fecundity at densities of 10–78 flies per $\frac{1}{2}$-pint culture. These densities are at the lowest range examined in the STS. This evidence implicates rapidly declining female fecundity due to increasing adult density as the major cause of the faster than linear decline in the rates of population growth shown in Fig. 10.

Introduction of the parameter θ to the Lotka–Volterra competition model appears to result in a model that can more accurately describe the dynamics of competing populations of *D. pseudoobscura* and *D. willistoni* (Ayala *et al.*, 1973). Although the net productivity statistic was used in this paper, most density combinations were close to the two-species equilibrium. Consequently, the bias introduced by using net productivity as estimates of rates of population growth may not have been severe. For competing populations of *Drosophila*, θ is usually less than one. It would seem reasonable to suppose that interspecific competition would have the same effect on fecundity as intraspecific competition does. The size of emerging females will decline with increasing numbers of competitors,

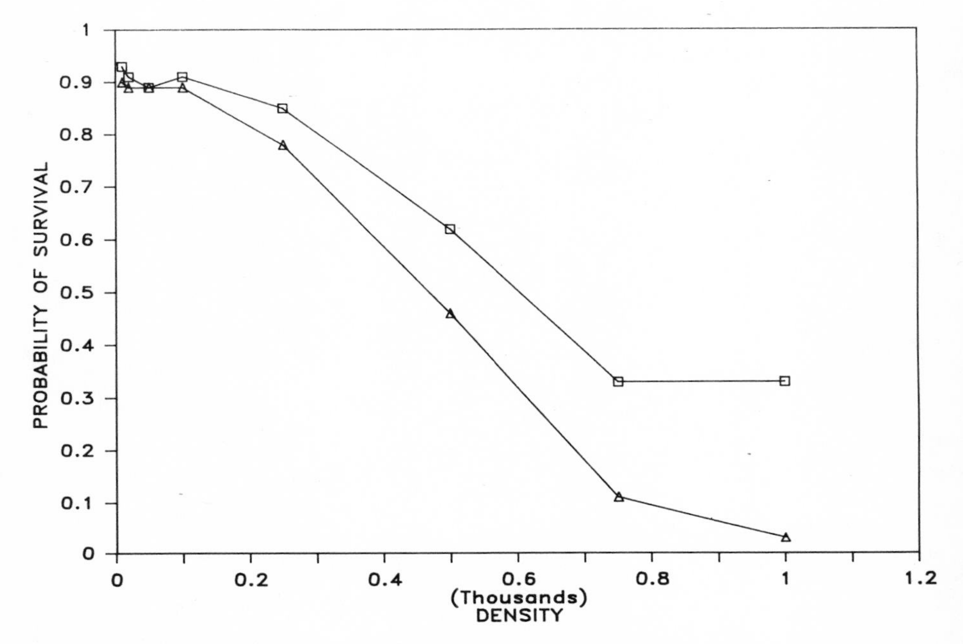

FIG. 11. The probability of surviving one week for (□) male and (△) female *D. melanogaster* as a function of the initial adult density (equal sex ratio). [From Mueller (1979).]

but the magnitude of this decline will depend not only on food availability, but also on the frequency of competing larvae and the relative competitive ability of these larvae (Bakker, 1961; Nunney, 1983).

Population Stability

For those populations in which reproduction takes place at discrete intervals, and are therefore well described by nonlinear difference equations, quantitative properties of density-dependent regulation can cause peculiar phenomena (May, 1974; May and Oster, 1976; Guckenheimer *et al.*, 1977). In particular, depending on the value of certain parameters, many difference equations can exhibit two-point, four-point, or higher order cycles or apparently chaotic population dynamics in completely deterministic environments. Most natural and laboratory populations will show some degree of fluctuation about an apparent equilibrium. However, it is often assumed that these fluctuations generally reflect some under-

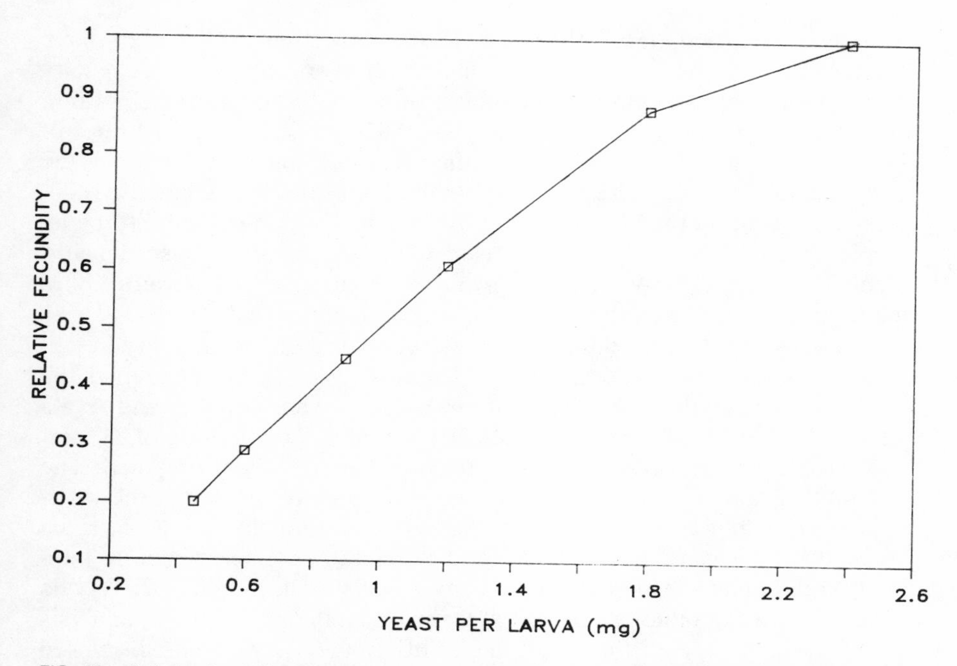

FIG. 12. Relative fecundity of wild-type *D. melanogaster* as a function of different larval food levels. The predictions are based on Bakker's (1961) data (see Fig. 3) relating adult size to food level and Robertson's (1957) data describing fecundity versus adult female size.

lying changes in the environment (e.g., temperature, moisture, or food availability). It is clearly of some importance to be able to separate variation due to environmental variability from that due to the intrinsic nature of density-dependent regulation.

Classification of a population as either possessing an asymptotically stable carrying capacity or a potentially cyclic or chaotic equilibrium requires knowledge of the linear dynamics in the vicinity of the carrying capacity [see Roughgarden (1979) for a review of the stability properties of difference equations]. In practice this has been accomplished with two different methods. The first method (Hassell *et al.*, 1976; Thomas *et al.*, 1980) has two levels of inference. At the first level a particular functional form is assumed to be an adequate description of the population dynamics. Thus, Hassell *et al.* (1976) use the first-order difference equation $N_{t+1} = \lambda N_t/(1 + aN_t)^{\beta}$ for a variety of species, while Thomas *et al.* use the theta model to analyze data from 27 species of *Drosophila*. At the second level of inference, parameters of the models must be estimated from direct observations of population growth rates (Thomas *et al.*, 1980) or by other less direct methods (Hassell *et al.*, 1976). Once these parameters have been estimated, the linear dynamics in the vicinity of the carrying capacity are obtained by the appropriate linearization of the nonlinear equation. The second method (Mueller and Ayala, 1981*c*) utilizes observed rates of population growth in the vicinity of the carrying capacity to estimate the local linear dynamics directly. This method avoids using a specific functional description of the population's dynamics, which is bound to be an approximation, but can only be used with populations whose density-dependent rates of growth can be measured directly. For laboratory populations of *Drosophila* this last qualification is not a restriction.

Two studies have addressed the problem of dynamic stability in laboratory populations of *Drosophila*. Thomas *et al.* (1980) examined the dynamics of 27 different species of *Drosophila*, while Mueller and Ayala (1981*c*) obtained data on 25 genetically different populations of *D. melanogaster*. In both studies the dynamics in the STS were obtained. The qualitative conclusions from these two studies were the same: the overwhelming majority of populations show asymptotically stable carrying capacities. There is little indication of potentially cyclic or chaotic population dynamics. Although Thomas *et al.* (1980) and Mueller and Ayala (1981*c*) used the same experimental techniques, their methods of analysis were quite different. Some of these differences have been discussed above. In addition, Thomas *et al.* used a linearization of a first-order difference equation to obtain the local linear dynamics of the STS. As discussed previously, a complete model of the STS will be a fourth-order (or higher) difference equation. Thus, the stability analysis of Thomas *et al.* is inappropriate for the STS. Given these very different methods of analysis, it seems surprising that they demonstrate a high degree of correspondence. It may be that the net productivity measurements near the carrying capacity of Thomas *et al.* are not biased estimates of rates of population growth and thus convey the correct information regarding population stability. A reanalysis of the data is clearly the only way in which to resolve this problem.

If the results on population stability obtained for laboratory *Drosophila* are at all relevant to natural populations, it is of interest to try to uncover a mechanism of natural selection that might promote population stability. Mechanisms involving both group (Thomas *et al.*, 1980) and individual selection (Heckel and Roughgarden, 1980; Turelli and Petry, 1980; Mueller and Ayala, 1981c) have been proposed. Thomas *et al.* have argued that as a population's stability eigenvalue increases in magnitude, the smallest population size reached at equilibrium decreases. As a consequence, the probability of chance extinction should also increase with increasing instability.

If one assumes that there is genetic variation within a population that affects parameters of growth equations (e.g., *r*, *k*, and θ of the theta model) and that fitness is density dependent, then individual selection may determine population stability. The existence of genetic variation for parameters of the logistic and theta models has been demonstrated for *D. melanogaster* (Mueller and Ayala, 1981*b*). The models of Heckel and Roughgarden (1980) and Turelli and Petry (1980) examined density-dependent selection in a variable environment with populations initially possessing a stable carrying capacity. Mueller and Ayala (1981*c*) numerically studied evolution in deterministic environments with populations that initially exhibited two-point cycles. Both Turelli and Petry and Mueller and Ayala saw that the evolution of population stability could be model dependent in particular settings. However, Turelli and Petry found that under the most general conditions which they considered, natural selection always resulted in stable population dynamics and this result was robust with regard to the particular density-dependent regulating function. Mueller and Ayala argued that there may be a strong negative correlation between parameters that determine population stability and that this correlation structure may constrain a population's evolution such that it usually has a stable carrying capacity.

The experimental paradigm for determining *Drosophila* population dynamics has utilized adults as the natural census stage. The reasons for this are that adults are the most conspicuous life stage of *Drosophila* and are the most easily manipulated. However, it must be remembered that even when the adult life stage is artificially shortened to a few days, all laboratory populations will have several immature stages prior to the adult census. In an important paper, Prout (1985) has shown that estimating population dynamics from observations of changes in the adult population size over a single generation may be inherently biased. At the root of this problem is the influence of larval density on adult fecundity. If female fecundity depends on larval density to any degree, then it is impossible to represent the number of adults at time t as a function of the adult number at time $t - 1$, even in populations with fully discrete generations (see section on life-history evolution). Under these conditions eggs are the natural life stage to census. Prout demonstrates that estimates of population stability at carrying capacity can also be biased when estimated from adult-to-adult transitions. If female fecundity does not change as a function of egg (or larval) density in the vicinity of carrying capacity, then the bias produced from adult data will be zero. However, even when the dependence of female fecundity on egg density is slight, the potential bias in estimates of population stability can be large.

Prout's results naturally call into question the results of Thomas *et al.* (1980) and Mueller and Ayala (1981*c*), since both used adult-to-adult transitions to estimate population stability. The degree to which the results of Thomas *et al.* and Mueller and Ayala may be in error will depend on the sensitivity of female fecundity to egg density in the serial transfer system. Such data have not been produced, although there is no reason to believe there will not be some dependence. The importance of Prout's theoretical results are bolstered by his own experimental data, which indicate that the eigenvalue governing stability in his experimental populations may be less than -1. This result was obtained from a first-order difference equation describing the change in the number of eggs at time t, e_t:

$$e_{t+1} = F(e_t)S(e_t)e_t$$

where $F(e_t)$ describes the effect of egg density on adult fecundity and $S(e_t)$ describes the relationship between egg-to-adult survival and egg density. Prout conducted separate experiments to estimate the functions $F(e_t)$ and $S(e_t)$; his analysis explicitly assumed that adult density has no effect on fecundity. Prout's data actually indicate a slight decline in fecundity with increasing adult density. It can be shown that even this weak dependence of adult fecundity on adult density will result in eigenvalues larger than those obtained by Prout. Consequently, the eigenvalues obtained by Prout are biased in a direction opposite to the bias present in Thomas *et al.* and Mueller and Ayala. Unfortunately, the only way to avoid the bias introduced by ignoring the effects of larval and adult density on fecundity is to observe directly the change in number of eggs from one generation to the next.

Several studies claim to have uncovered evidence of cycling in *Drosophila* populations. Shorrocks (1970) studied 14 populations of *D. melanogaster*. These populations reproduced at discrete intervals. Three populations showed strong positive autocorrelations, although the lag differed for each population. These lags occurred at two, three, and four generations. Many populations did not show any signs of a strong autocorrelation. Thus, although a few populations may have exhibited some cycling, it is certainly not a consistent phenomenon even with replicated experimental populations.

Nogueś (1977) maintained *D. subobscura* in an STS-like system. This population exhibited pronounced oscillatory behavior. These cycles were probably environmentally induced. During peaks of population growth, culture bottles tended to become overgrown with mold. Such contamination may have precipitated the abrupt population decline, resulting in the cyclical behavior.

COMPETITION

Interspecific competition has assumed a major role in the theory of ecology (Roughgarden, 1979). Competition may place limits on species diversity (Abrams, 1976) and be responsible for resource partitioning (Roughgarden, 1976). When competition is especially intense, natural selection may favor genotypes that will avoid competition, resulting in a coevolution-structured community (Rummel and Roughgarden, 1983). Despite this body of work, the importance of competition in structuring natural populations has been questioned recently (Connor and Simberloff, 1979; Strong *et al.*, 1979; Connell, 1980). Although not all communities show evidence of interspecific competition, a recent review (Schoener, 1983) concludes that competition was found in 76% of the species examined in over 150 field studies.

Competition in Natural Populations

Evidence for competition in natural populations of *Drosophila* has been found by Atkinson (1979) and Grimaldi and Jaenike (1984). Atkinson collected fruits and vegetables from local markets and measured the wing length of flies emerging from these natural substrates. He also recorded the total number of adults emerging from a particular fruit and used this number as an indicator of larval densities. Atkinson found that within a particular type of fruit or vegetable there was a significant decline in the wing length of *D. melanogaster* with increasing larval densities. There was also a negative correlation between the wing length of *D. simulans, immigrans*, and *hydei* and the number of emerging *melanogaster*. These results provide evidence of both intra- and interspecific competition.

Grimaldi and Jaenike (1984) collected mushrooms infested with dipteran larvae and cut these in half. One half was kept in an enclosed jar and the second half was supplemented with previously frozen mushroom of the same species. *Drosophila falleni, putrida*, and *testacea* showed significant increases in the number of emerging adults when supplement was provided. These three species and *D. recens* also showed an increase in size (as measured by thorax length) in the supplemented jars. It is clear that for the mycophagous *Drosophila*, resources are limiting, although it is not possible to distinguish between intra- and interspecific competition.

Competition in the Laboratory

Laboratory studies of competition with *Drosophila* have examined both intergenotypic and interspecific effects. Environmental variables such as temperature have been shown to affect interspecific (Ayala, 1966, 1971; Bierbaum and Ayala, 1985) and intraspecific (Mather and Cooke, 1962) competitive abilities. Equal numbers of *D. willistoni* and *D. pseudoobscura* were used to initiate populations kept at 18, 19, 20, 21, 22, and 25°C by Bierbaum and Ayala. At 25°C the tropical species, *D. willistoni*, competitively excluded the temperate species, *D. pseudoobscura*, while the opposite was true at 18°C. At intermediate temperatures (19–22°C) the two species coexisted for a period of 50 weeks.

Even within the homogenized environment that makes up the laboratory *Drosophila* culture it has been possible to document differences between species in habitat utilization. Merrell (1951) has shown that in competition *D. melanogaster* does better in fresh food relative to *D. funebris*, which seems to have an advantage in older food. In population cages the two species are able to coexist because the food cups naturally pass through stages that are preferred by each species.

The sibling species *D. melanogaster* and *D. simulans* also show differential utilization of experimental culture medium: *D. simulans* seems to prefer to oviposit on the center of the medium and on older surfaces relative to *D. melanogaster*, which prefers the edges of fresh medium (Moore, 1952*b*). Barker (1971) notes that *D. simulans* larvae are more often found in the lower half of the medium. The differences in habitat utilization *could* play a substantial role in competitive interactions. This point was experimentally demonstrated by Moore (1952*b*). Populations of both species were allowed to lay eggs on old and fresh food surfaces. In one treatment the next generation was started with eggs laid on the edges of the fresh food, while the second treatment utilized eggs from the center of the old food. The populations started with fresh edges showed a rapid rise in the frequency of *D. melanogaster*, with the ultimate exclusion of *D. simulans*. The populations started with old centers resulted in the fixation of *D. simulans* in one case and its near fixation in a second replicate after 13 generations.

Most laboratory studies of intra- and interspecific competition have used similar experimental techniques. Usually the effects of competition on the number of adults produced in one generation will be determined over a range of mixtures of two competing types with the total initial density of adults or larvae held constant. This procedure can be expanded by also varying the total density. Statistical methods for analyzing these experimental data are given in Mather and Caligari (1981). In some experiments the effects of competition have been measured by determining the productivity of each competing species in the serial transfer system (Gilpin, 1974; Goodman, 1979). It is interesting that many seemingly confusing features of these experiments can be understood in light of the simplified model of larval competition presented in the preceeding section.

One consistent result from both intra- and interspecific competition studies is that the performance of a line (or species) in pure culture is not a good predictor of that line's competitive ability (Lewontin, 1955; Lewontin and Matsuo, 1963; Gale, 1964). For instance, Gale (1964) studied the competitive abilities of three lines of *D. melanogaster*. In pure culture the vestigial line had the lowest viability, but in competition it had the highest viability. A second consistent observation is that the effects of competition can be ameliorated by increasing the amount of food or decreasing the density of the population (Lewontin and Matsuo, 1963; Gale, 1964; Miller, 1964; Barker and Podger, 1970).

Consider two competing populations A and B which differ only in the parameters α and m of the larval competition model considered in the preceding section, yet have a common variance. Further assume that $\alpha_A > \alpha_B$ and $m_A > m_B$. Thus, population A is competitively superior to B, but requires more food to pupate successfully. When viabilities are determined for each population separately it is clear that population B will have a higher viability (in Fig. 1, m_B would be to the left of m_A). When these two populations compete for limited food the average individual in populations A and B will not receive the same amount of food. Instead, individuals from population A will receive more food (since they are consuming it at a faster rate) and the distribution of food consumed by A individuals will be shifted to the right of the distribution for B individuals. If this shift in the relative positions of the two distributions is sufficiently large (that is, α_A is sufficiently greater than α_B), the viability of individuals in population A will be greater than those in population B. This form of competition clearly requires no active interference or facilitation among competing types as hypothesized by Lewontin (1955). This discussion does not preclude other types of competitive interactions: it merely demonstrates that many observations of competition in complicated environments can be accounted for by a simple model of competition for limited food. Sulzbach (1980) reports no improvement in the competitive abilities of two strains of *D. melanogaster* that have competed as larvae for 23 generations relative to stocks kept in isolation. These results are not surprising if one assumes that the major component of competitive interactions in *D. melanogaster* is the ability to compete for limited food. Both the controls and experimental populations used by Sulzbach were kept at the same density and thus experienced the same levels of food availability. Thus all populations had approximately the same intensity of selection for improved ability to survive on limited food. Consequently, the evolutionary improvement of a population to intraspecific competition would also result in improved interspecific competition and *vice versa*.

The major features of more detailed studies of competition can also be understood in terms of the larval feeding model. Caligari (1980) examined the competitive abilities of two inbred lines of *D. melanogaster*: Wellington (Well) and a line selected for high chaeta number (6CL). In monocultures, 96, 72, 48, and 24 eggs of each line were placed in vials with measured amounts of yeast. The mixed cultures contained these same numbers of one line and enough eggs from the second line to make the total density 96. The 6CL monocultures had 18–29% survivorship at all densities tested. The Well monocultures had 63, 60, 42, and 23% survival at densities of 24, 48, 72, and 96 eggs, respectively. The most reasonable interpretation of these results is that the 6CL line has a genetically determined low viability which is independent of food availability. There is no difference between the number of emerging Well adults in the competition tests and the numbers emerging in the monocultures. However, the weight of the adult Well flies does decrease in the competition tests. These results can be interpreted in light of the larval competition model and by assuming that the genetic death [hard selection of Wallace (1970)] of the 6CL line occurs early in larval life. In the competition tests with 24 and 48 6CL eggs and 72 and 48 Well eggs, so few 6CL larvae survive genetic death that the survival of Well is not appreciably affected. In the tests with 72 6CL eggs and 24 Well eggs, only 16–24 6CL will survive genetic death. This leaves a total density of 40–48 larvae. In the density range of 24–48 eggs the survivorship of the Well larvae is relatively constant. This would correspond to the flat region in Fig. 2. However, as is evident from Fig. 3, the weight of adults can still be changing when adult survivorship is near its maximum.

A large number of studies have examined the evolution of competitive ability in *Drosophila* (Moore, 1952*a*; Seaton and Antonovics, 1967; Ayala, 1969; Futuyma, 1970; Barker, 1973; Sulzbach and Emlen, 1979; Pruzan-Hotchkiss, Perele *et al.*, 1980; Sulzbach, 1980; Bierbaum and Ayala, 1985). Moore (1952*a*) reports finding one line of *D. simulans* out of 20 that was not competitively excluded by *D. melanogaster* while both species were maintained in population cages. This line was superior to stocks of *simulans* in its competitive ability with *melanogaster*. Ayala (1969) maintained *D. nebulosa* and *D. serrata* together by the serial transfer system. After 22 weeks the numbers of *D. nebulosa* increased and this was due to an increase in the competitive ability of this species (Ayala, 1969). After only three generations of selection Seaton and Antonovics (1967) recorded increases in the lines of *D. melanogaster* relative to their unselected stocks. Despite these positive findings, numerous studies have recorded no changes in competitive ability (Futuyma, 1970; Pruzan-Hotchkiss *et al.*, 1980; Sulzbach, 1980). It is worth noting that the studies reporting negative findings often contain more replicates.

Bierbaum and Ayala (1986) conducted experiments with *D. willistoni* and *pseudoobscura* to test hypotheses in the coevolution of interspecific competitors. The major evolutionary hypothesis they tested was proposed by Pimentel *et al.* (1965), which states that when two species compete, natural selection will most strongly favor an increase in the competitive ability of the less frequent species. Such a process may be responsible for the reversals of competitive dominance seen by Ayala (1966). To test the Pimentel hypothesis, Bierbaum and Ayala (1986) established three different selection regimes: in each both species were maintained in a serial transfer system and the total density of the egg-laying adult population was kept at 300 flies. The three treatments differed in their frequency of *D. willistoni*, which was either 20, 50, or 80%. The *D. willistoni* adults for each generation came from the competition cultures, while the *pseudoobscura* came from single-species stocks. Thus, only the *D. willistoni* population could adapt to the competitive environment. Eight independent populations were maintained in this fashion and allowed to respond to the competitive environment for about 30 generations.

The expectations of Pimentel *et al.* (1965) were not met. The lines kept at 50 and 20% *willistoni* had evolved a lower competitive ability than the 80% lines. Bierbaum and Ayala conclude that the primary adaptations of these populations may be to density-dependent intraspecific interactions, and that interspecific interactions play a minor role.

The conclusions of Bierbaum and Ayala are supported by quite different evidence from Wijsman (1984). In this very interesting study the effects of new mutations on the intra- and interspecific competitive abilities of two lines of *D. melanogaster* and one line of *D. simulans* were determined. Lines were treated with ethylmethanesulfonate and the inter- and intraspecific competitive abilities of all pairwise comparisons of treated and untreated lines were determined. Competitive ability was estimated by following the numbers of each competing type in population cages over many generations. Such a procedure was possible for the two lines of *melanogaster*, since one line had a compound third chromosome which causes it to be reproductively isolated from populations with normal third chromosomes. Although inter- and intraspecific competitive ability were reduced in the newly mutated populations, these mutations had a significantly larger effect on intraspecific competitive ability than they did on interspecific competitive ability. These results may have significant implications for the evolution of intra- versus interspecific competitive ability.

Clark (1979) reports experimental evidence that interspecific competition may alter allele frequency dynamics. Clark studied a fourth-chromosome polymorphism in *D. melanogaster*. He determined the equilibrium allele frequencies in pure populations of *melanogaster* and in populations where *D. simulans* was maintained at a frequency of one-third or two-thirds. The allele frequency equilibrium reached in the two populations with *D. simulans* were similar to each other but different from that reached in the pure *melanogaster* population.

Island Biogeography

MacArthur and Wilson (1967) first suggested that the number of species on an island may represent a balance between immigration and extinction. Although the vast majority of empirical work in this field has been conducted with natural populations, laboratory populations of *Drosophila* offer a tremendous opportunity to study this process.

Wallace (1975) has conducted the only detailed study of island biogeography with *Drosophila* to my knowledge. Wallace's experimental technique was very precise and not surprisingly he was able to obtain much more detailed information than is possible in field studies. The regularity of migration episodes and the maintenance of island populations actually allows, and probably requires, a much different description of the immigration and extinction process than that given by MacArthur and Wilson (1967). I will try to outline how one might model the immigration and extinction processes on these islands, after describing the experimental techniques of Wallace.

Migration took place inside a large cage called "the island machine." On the bottom of the cage were numerous holes to which were attached shell vials with food. At regular intervals 18 species of *Drosophila* were placed in the island machine. Migration was said to occur when a fly entered a particular vial. Four preidentified vials represented the migrants for one of 30 different islands. Each group of four vials is called a unit. Rates of immigration were controlled by the duration of time that vials were attached to the machine. "Distant" islands were left on the machine for only 4 h, while ''near'' islands were attached to the machine overnight. The units of each island can be thought of as different niches. At the same time that migration occurred, a specified number of females from each unit were transferred to fresh vials. On some islands the vials of each unit were maintained independently; on others there was some exchange of flies between vials of the same unit. Units were maintained until there were no flies to begin the next generation (extinction) or until 24 time units had passed (1 time unit = 3–4 weeks).

Using only species-specific extinction and immigration rates, one can describe the number of units occupied by species i on a particular island, $X^{(i)}(t)$, at any time. Such a derivation will assume that there is mixing of individuals between the vials of a single unit and the immigration and extinction probabilities apply to whole units. At any time t only immigration events that have occurred in the last 24 time units will be of interest. Let the total number of such events be $N^{(i)}(t)$. The time to extinction for each successful immigration event is Y_j, $j = 1, \ldots, N^{(i)}(t)$, and $Y_j \leq 24$. Let the time at which the jth immigration event took place be Y_j, $t - 24 \leq \mathrm{Y}_j < t$. Then $X^{(i)}(t)$ is given by

$$X^{(i)}(t) = \sum_{j=1}^{N^{(i)}(t)} w(t - \mathrm{Y}_j, Y_j)$$

where $w(t - \mathrm{Y}_j, Y_j) = 1$ if $0 < t - \mathrm{Y}_j \leq Y_j$ or 0 otherwise. The statistical problem is then to find the distribution of $X^{(i)}(t)$ assuming $N^{(i)}(t)$ has a binomial distribution and Y_j has a truncated geometric distribution. Once this distribution is known, the probability that $X^{(i}(t) = 0$ can be calculated and thus the total expected number of species on the island can be determined. This model provides even more detailed information. It would give the expected number of units that should be occupied by each species and the variance of this expected value. Such expectations could be compared to the observed values from Wallace's data set to determine if immigration and extinction probabilities alone can adequately predict the species composition of islands.

Wallace's (1975) data indicate that the immigration and extinction probabilities can depend substantially on the current species composition of an island. Such findings invalidate the considerations in the previous paragraphs. However, Wallace's data indicate that it is largely the presence or absence of *D. melanogaster* that affects the success of the other species. In the absence of this dominant species the extinction and immigration rates of other species may be less sensitive to the *Drosophila* community composition.

LIFE-HISTORY EVOLUTION

Life histories involve the timing of reproduction, maturation rate, and the duration of life. The theory of life-history evolution attempts to explain how natural selection might affect these various traits within certain constraints. These constraints or tradeoffs assume that limited energy will prevent the limitless improvement of life-history traits. Natural selection may fine tune a particular life-history character, such as increasing early fecundity, but only at the expense of some other life-history component (i.e., decreased late fecundity or survival). These ideas were articulated early by Pearl (1928) and have found their way into many recent formulations of life-history evolution (Gadgil and Bossert, 1970; Roughgarden, 1971; Taylor *et al.*, 1974; Stearns, 1976; Charlesworth, 1980; Iwasa and Teramoto, 1980; Mueller and Ayala, 1981*c*; Schaffer, 1983).

Empirical verification of the tradeoff hypothesis has usually involved the estimation of correlations between life-history traits from a collection of genetically or phenotypically different populations. The design of these studies is such that only estimates of phenotypic correlations are possible. The arguments presented above emphasize the changes in life-history traits during evolution. Predictions about this process require knowledge of additive genetic correlations. Unfortunately, phenotypic correlations not only may differ quantitatively, but also may be of a different sign than the additive genetic correlations (Rose and Charlesworth, 1981*a*). Phenotypic correlations of life-history traits by themselves may give totally misleading predictions concerning evolutionary dynamics. Phenotypic correlations of life-history traits also seem to depend on the degree of inbreeding in the experimental populations.

Covariation in Life Histories

The following studies were all performed with *D. melanogaster*. Using vestigial and wild-type laboratory stock, Alpatov (1932) found a consistent negative correlation between mean daily egg production and the life span of a female. However, his method of calculating these statistics may have introduced some spurious correlation (Gowen and Johnson, 1946). Gowen and Johnson (1946) examined a large number of laboratory and inbred strains. Although the average per-capita fecundity was positively correlated with life span over all lines, when egg production over a constant period was compared to longevity there was a negative correlation.

In a series of experiments Giesel and co-workers examined correlations for inbred (Giesel, 1979; Giesel and Zettler, 1980; Giesel *et al.*, 1982*a,b*) and some outbred populations of *D. melanogaster* (Giesel *et al.*, 1982*a,b*) and *D. simulans* (Murphy *et al.*, 1983). In general these studies have shown positive correlations for a host of life-history traits. For instance, the correlation among lines for reproduction at age 2 days and age of death was 0.719 and 0.578 for inbred and outbred lines, respectively (Giesel *et al.*, 1982*b*). The most complete quantitative genetic analysis of life-history traits was performed by Rose and Charlesworth (1981*a*) on flies sampled from a large outbred population. This study yielded a number of interesting observations: (1) there is substantial additive genetic variance for female fecundity at all ages and (2) there is sometimes little relationship between phenotypic and additive genetic correlations. The first observation has also been made for *D. simulans* (Murphy *et al.*, 1984). The phenotypic correlation between egg laying at days 11–15 and longevity is 0.210, while the additive genetic correlation is −0.712 (Rose and Charlesworth, 1981*a*). In contrast to many of the results from Giesel and co-workers, Rose and Charlesworth obtained negative additive genetic correlations for many traits. For instance, egg laying from days 1 to 5 is negatively correlated with longevity, last day of egg laying, egg laying days 6–10, and egg laying days 11–15 (Rose and Charlesworth, 1981*a*). The biological significance of these negative correlations was tested by conducting artificial and natural selection for early and late fitness components (Rose and Charlesworth, 1981*b*). In some cases the selection regimes produced no detectable direct or indirect results. However, natural selection for late fitness components resulted in increased late female fecundity, longevity, and duration of female reproduction, with a concomitant decrease in early female fecundity and mean egg-laying rate (Rose and Charlesworth, 1981*b*). The soundness of these results has been bolstered by their recent successful duplication (Rose, 1984*a*). The reduction in early fecundity exhibited by these populations appears to be due to a significant reduction in ovary weight in young flies (Rose *et al.*, 1984).

Work in this field has served to highlight problems that occur with the use of inbred lines as experimental material. It appears that highly inbred lines tend to exhibit correlations in fitness components that are uncharacteristic of outbred populations. Simmons *et al.* (1980) have noted a qualitative difference between newly arisen and old mutations in heterozygous condition. Viability of newly arisen mutants appeared to be uncorrelated with other fitness components in *D. melanogaster*, whereas mutations from equilibrium populations that decrease viability are compensated for by increases in other components of fitness. Perhaps of more interest is the observation of Hiraizumi (1961). He obtained rates of development and estimates of fertility for various second-and third-chromosome homozygotes and heterozygotes. These components of fitness were negatively correlated for normal or high-fitness genotypes, but showed a positive correlation for low-fitness genotypes. A similar phenomenon was also seen by Mueller and Ayala (1981*d*). They looked at the correlation between density-dependent rates of population growth at high and low densities for a number of *D. melanogaster* populations isogenic for whole second chromosomes. These correlations were consistently positive. However, when genetically heterogeneous populations were allowed to evolve at high and low densities the resulting populations exhibited a tradeoff in density-dependent rates of population growth.

The usefulness of inbred lines for extracting evolutionary information on life-history phenomena was rigorously tested by Rose (1984*b*). Using the same populations that he had utilized in his earlier studies, Rose established 30 inbred lines by full-sib mating. Longevity and age-specific fecundities were determined for five females from each of the 30 lines. The correlations among the lines between early and late fecundity and fecundity at any age and longevity were consistently positive. As noted previously, many additive genetic correlations measured in the outbred population used to initiate this study are negative. This work shows concisely that there need be no concordance between patterns of life-history covariation derived from inbred and outbred samples of the same population.

The finding of large amounts of additive genetic variance for life-history traits by Charlesworth and Rose (1981*a*) provides useful information for predicting evolution in controlled laboratory environments. However, the usefulness of this information for gaining insights into natural populations is reduced by the existence of genotype–environment correlations. Such conditions have been observed for life-history traits of *D. simulans* at several temperatures (Murphy *et al.*, 1984) and for bristle number in *D. melanogaster* measured at several temperatures and densities (Gupta and Lewontin, 1982).

Density-Dependent Natural Selection

The theory of density-dependent selection has its origins in the works of MacArthur (1962) and MacArthur and Wilson (1967). In these works the notion that density-dependent rates of population growth could be viewed as measurements of fitness was introduced. It was also asserted that extreme environmental conditions might lead to the evolution of different population characteristics. Thus, populations kept at low densities by density-dependent mortality (and hence having abundant resources) should evolve a high intrinsic rate of growth r, but be incapable of superior preformance at high densities. In contrast, populations usually living at high density and thus experiencing strong competition for limiting resources) should evolve high intraspecific competitive ability and enhance their carrying capacity K.

The basic ideas put forth by MacArthur and Wilson were made more explicit with the formulation of several mathematical theories (Charlesworth, 1971; King and Anderson, 1971; Roughgarden, 1971; Clarke, 1972). For instance, Roughgarden (1971) assumes that the fitness of the ijth genotype W_{ij} is a linear function of total population size N,

$$W_{ij} = 1 + r_{ij} - (r_{ij}N/K_{ij})$$

in which the values of r and K vary among genotypes. If it is assumed that an initial population is polymorphic for genotypes that show a tradeoff (i.e., genotypes with high r's have low K's and *vice versa*), the outcome of evolution in Roughgarden's model is dependent on the environment. In stable environments the genotype with the largest value of K ultimately becomes established and all other eliminated. When the population is often below its carrying capacity due to frequent episodes of density-independent mortality, the genotype with the highest r value is favored. Thus, according to this model, evolution favors the genotype that makes the highest per-capita contribution to population growth at either high or low densities, depending on environmental conditions.

Most empirical studies have dealt with the predictions of the verbal theory of r- and K-selection (Pianka, 1970, 1972; Gadgil and Solbrig, 1972). These theories have argued that phenotypes that should be correlated with high r's or increased competitive ability will also respond to density-dependent selection. Thus, r-selected phenotypes will be smaller in size, with shorter generation times, and will be more semelparous than K-selected species, which will tend to be large, long-lived, and iteroparous. It should be noted that the verbal theory represents a substantial jump from the mathematical theory. The verbal theory has used the qualitative interpretation of parameters, such as r and K of the logistic equation, to translate theoretical results of discrete-generation models to populations with overlapping generations. More detailed examination of evolution in populations with age structure and density-dependent regulation yields results at odds with the verbal predictions. For instance, natural selection will favor increased survivorship and fecundity at earlier ages in density-regulated populations (Charlesworth, 1980). If fecundity is an increasing function of size, then selection for delayed reproduction can depend on the life stage affected by density-dependent regulation (Charlesworth, 1980). Iwasa and Teramoto (1980) have shown that the evolution of high fecundity versus high juvenile survival can depend on whether fecundity or preadult survival is subject to density-dependent regulation. Thus, not only is the presence or absence of density regulation important to life-history evolution, but the details of the manner in which it operates can be important.

Templeton and Johnston (1982) present an example of K-selection in a natural population which is at odds with predictions of the verbal theory. Templeton and Johnston studied the abnormal abdomen polymorphism in natural populations of *D. mercatorum*. Abnormal abdomen shows a number of pleiotropic effects, including increased early egg production and decreased longevity. During 1981 there was a severe drought, which resulted in marked declines in the size of many *mercatorum* populations and a decline in the longevity of adult flies. Since these flies breed in rotting pads of the cactus *Opuntia megacantha*, the number of available pads and the length of time they were suitable for *Drosophila* was greatly decreased as a result of the drought. During this same period there was also a marked increase in the frequency of abnormal abdomen in many populations. It appears that under these new ecological conditions the decrement in fitness caused by the reduced longevity of abnormal abdomen was more than balanced by the increase in fitness due to high early fecundity resulting in the net increase of abnormal abdomen. Thus, in an environment in which the carrying capacity was reduced, and hence overpopulated with *Drosophila*, evolution favored traits typically associated with r-selection. Templeton and Johnston correctly point out that pleiotropic genetic effects can result in life-history evolution that deviates substantially from what might be considered optimal.

The predictions of density-dependent natural selection clearly hinge on fitness being equivalent to the per-capita contribution to rates of population growth. Obviously, to test this prediction directly one must have a collection of genotypes whose density-dependent rates of population growth can be measured and compared to an independent measurement of fitness. Such an experiment is possible with populations of *D. melanogaster* isogenic for whole second chromosomes. The fitness of such genotypes relative to wild-type flies has been determined several times by direct observations of changes in genotype frequencies (Tracey and Ayala, 1974; Seager and Ayala, 1982). These experiments have been conducted in population cages at high densities. If one eliminates the data for sterile populations from Tracey and Ayala's study, the average relative fitness of second-chromosome homozygotes is 0.23. The relative fitness of these genotypes as estimated from density-dependent rates of popu-

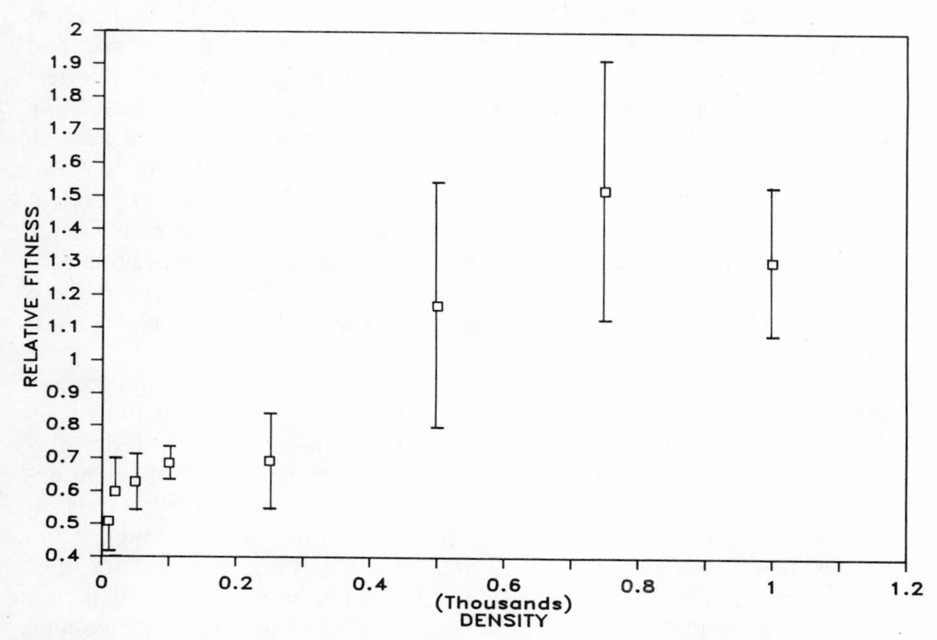

FIG. 13. Fitness (±95% confidence interval) as a function of density of 24 lines of *D. melanogaster* isogenic for different second chromosomes relative to a wild-type population. The calculations were based on density-dependent rates of population growth as explained in Mueller and Ayala (1981*a*).

lation growth are given in Mueller and Ayala (1981*a*) and reproduced in Fig. 13. These data show that at low densities there is a substantial difference between the isogenic lines and the wild-type population in rates of population growth that gradually diminish at higher densities.

Because the average isogenic line and wild-type population have similar rates of population growth at high densities, they also have similar carrying capacities. Mourão *et al.* (1972) obtained a similar result with *D. willistoni*. For 15 lines isogenic for whole second chromosomes they determined fitness relative to wild-type flies and carrying capacity in the serial transfer system. The correlation between fitness and the carrying capacity was −0.16 and not significantly different from zero.

These observations can be explained if one assumes that the isogenic lines differ most in density-independent fitness components of fecundity or larval viability and that the density-dependent functions that determine larval survival or female fecundity are quite similar for all populations (Prout, 1980; Mueller and Ayala, 1981*a*). With these assumptions it is expected that at low densities the large differences in density-independent components will produce large differences in number of offspring and adults produced by the various lines. Such differences will be manifest in the rates of population growth. At high densities one arrives at quite different predictions. As long as most lines produce more than enough eggs and larvae than can survive to become adults, due to limited food, pupation sites, etc., then the assumption of similar density-dependent regulating functions for each line will result in similar numbers of adults appearing at these high densities. A phenomenon similar to the one outlined here may also explain the lack of correlation between competitive and noncompetitive measures of fitness observed by Haymer and Hartl (1983).

This work shows that in general some caution must be made when equating density-dependent rates of population growth with fitness. A related question is: in genetically variable populations, how will density-dependent rates of population growth respond to natural selection at extreme densities? As argued previously, inbred lines may not provide information useful for answering this question. Consequently, Mueller and Ayala (1981*d*) have carried out natural selection at high and low densities on a genetically variable population of *D. melanogaster*. After eight generations of selection, rates of population growth at one low and two high densities were measured for three low-density populations (*r*-selected) and three high-density populations (*K*-selected). At the two high densities the *K*-selected populations had significantly higher rates of population growth and net productivity, while at the low density the *r*-selected populations had the higher values, although only the difference for net productivity was statistically significant (Mueller and Ayala, 1981*d*). These results are in marked contrast to similar experiments with *E. coli* (Luckinbill, 1978). In Luckinbill's experiment the *K*-selected populations showed higher rates of population growth at all densities relative to the *r*-selected populations.

Additional work has recently been completed on the selected populations of Mueller and Ayala (Bierbaum *et al.*, 1986). Adult survival and female fecundity were examined for *r*- and *K*-selected flies raised under uncrowded conditions. There were no substantial differences between the different selected populations for these traits. Density-dependent larval survival was also examined and the weights of the emerging adults recorded. At the lowest density tested the *r*-selected larvae had a higher survival rate (0.927 versus 0.910), but the difference was not statistically significant. At two higher densities the *K*-selected larvae had significantly higher survival rates. In addition, adult male and female *K*-selected flies emerging in these high-density vials were heavier than their *r*-selected counterparts. As discussed in the section on the laboratory ecology of *Drosophila*, there is a strong correlation between female size and fecundity for partially starved flies (Alpatov, 1932; Robertson, 1957). Using Robertson's data, it can be estimated that the *r*-selected females raised at the highest larval density should suffer an 11% reduction in lifetime fecundity relative to *K*-selected females raised under the same conditions. Thus, the higher rates of population growth exhibited by the *K*-selected populations (Mueller and Ayala, 1981*d*) may be due to both increased larval survival and female fecundity.

One aspect of the Mueller and Ayala (1981*d*) and Bierbaum *et al.* (1986) results is that the differences between the selected populations were quite small and difficult to detect at low densities, but at high densities the differences were easily detected. It may be that the selective pressure on the *r*-selected lines is not as strong as on the *K*-selected lines and thus the magnitude of response of the *r*-selected lines is diminished. It is also possible that there are inherent asymmetries in the response that populations can demonstrate. For instance, larval survival at low densities is greater than 0.90 in both populations and clearly cannot be greater than 1.0. Thus, to increase survivorship 1 or 2% may require substantial genetic and physiological changes. Such asymmetries have been seen previously with other characters (Falconer, 1981). Lastly, it is possible that as a species *Drosophila melanogaster* has experienced conditions similar to those experienced by the *r*-selected populations. Consequently, there could be less additive genetic variation for the traits under selection in these environments.

Several other studies have documented an increase in the carrying capacity of laboratory populations of *D. melanogaster* (Buzzati-Traverso, 1955; Ayala, 1965*a*, 1968*a*). Thus, it is likely that the process of adaptation observed by Mueller and Ayala (1981*d*) and Bierbaum *et al.* (1986) is a repeatable and consistent phenomenon in *D. melanogaster*.

Three other studies with *Drosophila* have attempted to observe directly the consequences of density-dependent natural selection (Taylor and Condra, 1980; Barclay and Gregory, 1981, 1982). Taylor and Condra conducted *r*- and *K*-selection with *D. pseudoobscura*. Consistent with theory, a decline in the development time of the *r*-selected population and increased adult survival of *K*-selected flies were observed. They saw no differences in intrinsic rate of increase, carrying capacity, body size, fecundity, and timing of reproduction. The *r*-selected populations were maintained by using the first 100 adults to start the next generation; thus, development time was under strong selection. Both adult survival and fecundity were measured under uncrowded conditions. The fecundity results are consistent with Bierbaum *et al.* (1986). Taylor and Condra did not examine density-dependent rates of population growth directly, although they attempted to estimate the intrinsic rate of increase from Euler's equation and the observed birth and death reates. Thus, it is possible that if they had measured fecundity, larval survival, or rates of population growth at high densities, different results might have been obtained.

Taylor and Condra (1980) conclude that their selected populations have similar carrying capacities, since there are similar numbers of progeny produced from 2 weeks of egg laying by specific numbers of females (2–15). Their test is not a reliable indicator of the carrying capacity for two reasons. First, the equilibrium population size will depend not only on the number of progeny produced, but also on the mortality of adults. Taylor and Condra's own data indicate potential differences in adult mortality, yet their method of estimating carrying capacity does not take adult mortality into account. Second, the density conditions under which the experimental females were raised were not varied. As the work of Bierbaum *et al.* (1986) indicates, there may be substantial differences in fecundity of females raised at high and low densities.

Two other studies have examined life-history evolution in *Drosophila melanogaster* (Barclay and Gregory, 1981, 1982). Barclay and Gregory (1981) examined life-history evolution in populations that were subject to different combinations of larval and adult mortality. Although their description of the experimental techniques is not entirely clear, it seems that populations were given fresh food at 5-week intervals. The suitability of *Drosophila* medium for oviposition and the number of larvae it can sustain change drastically in 5 weeks. Barclay and Gregory would, at regular intervals of 1 or 2 weeks, remove adults and/or larvae from these cultures. Thus, in addition to the different regimes of adult and juvenile mortality, flies had to contend with a rapidly deteriorating environment. It is not at all obvious whether the most important evolutionary force in these experiments was the mortality imposed by the experimenters or the ability of larvae and adults to tolerate and survive the putrid conditions of the 5-week-old cultures. The second set of experiments by Barclay and Gregory (1982) also contain methodological problems which leave one somewhat skeptical of their results. In this study they again maintained cultures for a very long time. Due to the aging of the food there was probably little difference between their fast (changed every 20 days) and slow (changed every 40 days) food rotations. They also attempted to impose different levels of adult mortality. Unfortunately, this was not done in a controlled fashion; instead, a frog (*Hyla regilla*), with no apparent ecological relationship to *D. melanogaster*, was put in the population cage.

Most models of density-dependent natural selection have relied on very general descriptions of the effects of density on rate of population growth. This generality is often achieved with a model that is less realistic for many specific populations (Levins, 1966). Such problems led Stearns (1977) to point out that the parameter K of the logistic equation is unrelated in any simple or obvious way to life-history traits that can be measured in many populations. Current efforts to understand density-dependent natural selection in *Drosophila* are limited largely by the absence of models that incorporate important ecological aspects of *Drosophila* biology. Although such models would be of limited applicability, they could be extremely useful for interpreting the results of studies such as that of Bierbaum *et al.* (1986). Such a model would make possible the study of the interaction between density and larval competitive ability (Bakker, 1961, 1969; Nunney, 1983). Although competitive ability should change according to the verbal theory of r- and K-selection, it is seldom explicitly included in theories of density-dependent natural selection [for exceptions see Christiansen and Loeschcke (1980), Anderson and Arnold (1983), and Asmussen (1983)]. Results of Bierbaum *et al.* have also suggested that adult size, and therefore female fecundity, may change as a result of density-dependent natural selection. A complete model of density-dependent natural selection would also incorporate this aspect of *Drosophila* biology. Thus the most desirable theory would incorporate these relevant aspects of *Drosophila* ecology as functions of parameters that could be empirically estimated.

Progress toward this goal has been made recently by L. D. Mueller (in preparation). In this model the number of eggs at time $t + 1$, e_{t+1}, is related to the number in the previous generation by

$$e_{t+1} = F(ve_t)W(ve_t)ve_t$$

The model assumes that density-independent mortality v leaves ve_t larvae in the population. Due to limited resources (e.g., food), there is additional mortality $W(ve_t)$, which results in $W(ve_t)ve_t$ adults to produce the next generation of eggs. The density-dependent mortality function $W(ve_t)$ is precisely the normal truncation function described by Nunney (1983) and outlined here in the section on the laboratory ecology of *Drosophila*. The value of this function depends on the competitive abilities α_i of the various types in the population and the minimum amount of food required for successful pupation m_i. The recursion is completed by specifying the adult per-capita fecundity $F(ve_t)$, which is also density-dependent. This function assumes that adult size, and therefore female fecundity, increases with the amount of food consumed in excess of the minimum m_i. Such an assumption is warranted given the data in Fig. 7 and Robertson's (1957) results. This model can be used to investigate the consequences of natural selection on the evolution of intraspecific competitive ability α_i and the efficiency of food utilization m_i. This model predicts that natural selection favors increasing α and decreasing m. The intensity of selection for these parameters increases with increasing population size. Thus, they can be thought of as components of K-selection. However, changing α does not affect the equilibrium population size, while decreasing m increases the carrying capacity. Lastly, decreasing m not only will result in increased larval survival at a given density, but also will increase the fecundity (and size) of females emerging at a given density. Such results are in accord with the empirical results of Bierbaum *et al.* (1986). Finally, this model has as an additional asset parameters that can all be empirically estimated.

OVIPOSITION AND HABITAT CHOICE

Importance of Oviposition and Habitat Choice

Maintenance of Genetic Polymorphisms

Concomitant with the revelation of the large amounts of genetic variation in natural populations during the mid and late 1960s was an interest in developing theoretical explanations for the maintenance of this genetic variation. Population genetic models that incorporated different selection regimes in a variety of niches or habitats with some specified pattern of dispersal could maintain genetic variation even in the face of directional selection within each habitat. Karlin (1981) has recently provided a comprehensive review of the theory in this field. Three important characteristics of these models are: (1) the selection regime in each deme, (2) the specification of random mating either within each deme, in a common area, or in some mixture of both, and (3) a description of migration from one deme to every other. For instance, the island model (Wright, 1943) assumes that each deme receives the same fraction of migrants, while the Levene (1953) model permits a different, but constant, fraction of the adult population to disperse to each deme. Mating can occur entirely within a common area (Levine, 1953) or a fraction may stay in their deme to mate (Deakin, 1966). It is interesting that despite the variety of migration and mating structures considered in the review by Karlin (1981), observations obtained from *Drosophila* research suggest a new model. Jaenike's (1982) data on larval and adult conditioning effects on female oviposition behavior suggest the following life cycle. Viability selection acts on egg to adult survival. Although the larval environment does not appear to affect oviposition behavior, adult conditioning does. Thus, newly emerged adults may linger in their larval habitat and become conditioned to either prefer, avoid, or be indifferent to the larval substrate. Mating then occurs in a common area and migration back to the available demes depends on the larval habitat. Although it was not motivated by specific observations, the assumption was made by Jaenike that the number of adults emerging from a niche was a function of the fitnesses of various genotypes, that is, that hard selection was operating. Jaenike's model demonstrated how environmentally induced habituation to a normally repellant substrate (niche) or an induced preference can weaken the conditions necessary for new alleles to increase when rare. Since individuals with new alleles demonstrate increased viability in the repellent substrate, the population can be thought of as adapting to a secondary host.

As discussed by Jaenike, environmentally induced dispersal and hard selection will result in the change over time of the proportion of eggs laid in (or individuals dispersing to) each environment, even in the absence of genetic variation. The dynamics of this change is described by a linear fractional equation which depends on the viability of individuals in niches a and b (W_a, W_b) and the probabilities of females dispersing to niche b, given that they were raised in niche a (c) or b (c'). It is possible that if enough time lapses, the proportion of eggs laid in each niche will approach an equilibrium. However, if

$$[c'W_b - W_a(1 + c)]^2 + 4cW_a(W_b - W_a) < 0$$

or

$$[c'W_b - W_a(1 + c)]^2 = 0$$

the number of eggs laid in each niche will not reach an equilibrium, but rather will oscillate in a regular fashion. Clearly, Jaenike's analysis of evolution in this system is predicated on the resident population having reached a stable equilibrium with respect to the number of eggs laid in each niche. The genetic model of Jaenike has recently been extended to include soft selection and an arbitrary number of niches (M. Moody and L. D. Mueller, in preparation).

The efficacy of a heterogeneous environment and habitat selection in maintaining genetic variation has been demonstrated by Jones and Probert (1980). They studied the dynamics of the white-eye allele w in *D. simulans* in constant and heterogeneous environments. White-eye flies (*w*/*w*) avoid bright white light and move toward dim red light. Wild-type flies seem to mate less well in dim red light. Control population cages with either bright white light or dim red light were established. In each case the *w* allele was eliminated after 30 weeks, although the rate of elimination was slower in the dim red cage. An experimental cage was established that gave flies access to areas of bright white light and dim red light. The frequency of the *w* allele was quite higher in the dim red area after just a few generations. After 20 weeks the frequency of the *w* allele was still 0.32. It is quite possible that the *w* allele would have been eliminated had this experiment been continued; however, it is clear that the rate of decline of the *w* allele was much slower in the experimental cage than in either control.

Promotion of Sympatric Speciation

The idea that insects may become locally adapted to specific hosts and preferentially seek out this host is a long-standing one. Indeed, Hopkin's (1917) host selection principle stated that "an insect species which breeds in two or more hosts will prefer to continue to breed in the host to which is has become adapted." Just as geographic isolation has been invoked as a first step in the genetic differentiation of species (Mayr, 1963), it has been suggested that ecological isolation due to host preferences may serve a similar function (Thorpe, 1945). Maynard Smith (1966) showed that even in the face of random mating in each generation, the presence of niches that favored alternative genotypes and the appropriate migration schemes could maintain a genetic polymorphism. The model of Maynard Smith is actually a special case of the model developed by Moody and Mueller. Maynard Smith went on to argue that once such a polymorphism was established, the evolution of isolating mechanisms, by a number of different means, should be favored.

There are a variety of examples of potential host races that have led or are leading to speciation (White, 1978). Some of the most convincing evidence comes from Bush's work (1974) on *Rhagloetis pomonella*. However, Jaenike (1981) and Futuyma and Mayer (1980) point out that it is important to distinguish between host-associated sibling species and host races. In the later case gene flow is restricted solely or primarily because of differential host preference. Jaenike (1981) provides a list of criteria which he believes are necessary for demonstrating the existence of host races.

Oviposition Site Preference

Although oviposition site preference is a very labile behavior in *Drosophila*, progress has been made in studying environmental factors that affect this behavior. Chiang and Hodson (1950) showed that *D. melanogaster* females prefer to lay eggs on a scored surface rather than a smooth surface. Once egg laying has begun on a fresh surface, the deposition of eggs may not be random. Females of *D. pseudoobscura* and *D. melanogaster* tend to lay eggs in close proximity to other eggs (Del Solar and Palomino, 1966; Del Solar, 1968). Del Solar (1968) has also shown that there is genetic variation for this gregarious egg-laying behavior and has been successful in selecting lines of *D. pseudoobscura* that show very strong and weak propensities to aggregate their eggs.

The behavior of the neotropical species *D. flavopilosa* is opposite to the gregarious egg-laying behavior of *D. pseudoobscura* and *D. melanogaster* (Brncic, 1966). The larvae of this species develop in the flowers of the solanaceous plant *Cestrum pargui* and females lay only one egg per flower.

In addition to a preference for laying eggs near previously laid eggs, females of *D. melanogaster* seem to prefer to lay eggs in medium already colonized by larvae even if these are of a different species (Del Solar and Palomino, 1966). The effects of larval preconditioning on viability have also been studied, but the results have been equivocal.

An important component of many natural habitats of *Drosophila* is the concentration of ethanol. Indeed, the tolerance of both adults and larvae to ethanol can vary greatly between species (McKenzie and Parsons, 1972). Ethanol can also affect oviposition preference. McKenzie and Parsons (1972) report that female *D. simulans* rejected alcohol-impregnated sites, while *D. melanogaster* showed a slight preference for these same sites. Richmond and Gerking (1979) have carried out a detailed analysis of oviposition site preference in 14 species of *Drosophila* as a function of ethanol concentration. Richmond and Gerking offered females two different substrates for oviposition: agar–cornmeal–molasses medium with 9% ethanol by volume and the same medium with distilled water as the additive. For each species the percent of all eggs laid on the ethanol medium was tabulated. There was significant heterogeneity among species. The preference for the ethanol medium ranged from 7.9% for *D. paulistorum* to 97.4% for one strain of *D. melanogaster*. Unlike the study of McKenzie and Parsons, Richmond and Gerking found no significant differences between *D. simulans* and four strains of *D. melanogaster*.

The results of Richmond and Gerking depart substantially from those reported by Jaenike (1982). Jaenike utilized females of *D. melanogaster* raised on standard cornmeal–agar–molasses in oviposition preference tests. When these females were provided with a choice of standard medium and standard medium plus 7% ethanol, only 24% of all eggs were laid on the ethanol medium. It is quite difficult to explain these discrepancies in results other than to assume that some subtle difference in experimental technique is responsible. It is possible that the temperature of the medium may have differed at the time the ethanol was added. The final concentration of ethanol in the medium could thus be much less than reported. There may have also been genetic differences between the lines used in each study. Regardless of the cause, these results demonstrate that caution should be exercised when results from different labs are being compared.

Given the evolutionary significance of habitat or oviposition site preference, it is not surprising that a number of studies with *Drosophila* have been initiated to uncover environmental and genetic variation for these traits. Takamura (1980) studied the oviposition preferences of *D. melanogaster* when offered a soft (standard medium) versus a hard surface (paper). Forty isofemale lines from two natural populations showed significant variation for preference of hard versus soft medium. Takamura conducted a similar experiment on two populations of flies collected in wineries. One population was derived from eggs collected from grape seeds; the other was derived from eggs laid on grape sacs. In preference tests those flies derived from seeds laid a significantly larger proportion of their eggs on the paper surface. Thus, the hard grape seeds and soft sacs may represent different ecological niches for *D. melanogaster* and genetically based oviposition site preferences may be resulting in genetic differentiation of these subpopulations.

A substantial contribution to the empirical foundation of oviposition site preference in *Drosophila* has been made by Jaenike. In his first major work (Jaenike, 1982) the oviposition site preference of *Drosophila melanogaster, pseudoobscura, immigrans*, and *recens* was tested. Females used in this study were characterized by having developed in either standard or experimental medium as either larvae, adults, or both. The experimental medium had one of the following substances added to standard medium: NaCl, ethanol, ethyl acetate, lactic acid, piperidine, or peppermint oil. Females were placed in vials with a single slide containing the standard medium and one with the experimental medium. The percent of all eggs laid on the experimental medium was recorded. The effect of each experimental substance on a given life stage can be characterized by comparing the proportion of eggs laid on the experimental medium when the particular life stage was raised on standard medium (c) versus this same proportion when the life stage was raised on experimental medium (c'). One very consistent result was that exposure of larvae to the experimental substances had no effect on the oviposition preference of adults; however, exposure of adults did have an effect in a number of cases. Females that show an active preference ($c > 0.5$) or aversion ($c < 0.5$) to a particular substance are said to be habituated if $c > c' > 0.5$ or $c < c' < 0.5$. Females may show an induced preference or an induced aversion if $c' > 0.5$ and $c' > c$ or $c' < 0.5$ and $c' < c$, respectively. *Drosophila melanogaster, pseudoobscura*, and *immigrans* raised on standard medium all show an aversion to peppermint oil, but when these flies have experienced peppermint oil as adults, *D. pseudoobscura* and *immigrans* show a significant habituation effect and *melanogaster* shows an induced preference. It is also clear from Jaenike's results that oviposition site choice is not always adaptive. For instance, *D. recens* is indifferent to 0.025% peppermint oil, whether or not there has been prior conditioning, even though a 0.05% concentration is lethal to adults.

The larval environment consistently has no effect on the oviposition preference of females. However, the larval environment may have a subsequent effect on adult oviposition choice if these adults linger in the larval environment for some time after emergence. Under these conditions models that assume that adult migration is dependent on the larval environment may still be justified.

Jaenike (1983) has performed similar experiments on *D. melanogaster* using several "natural" larval habitats: apples, tomatoes, bananas, and squash. The choice by females followed the same patterns previously seen: the larval environment had no appreciable effect on oviposition preference, while adult exposure to either apples or tomatoes enhanced the preference for these fruits as oviposition sites. Jaenike (1983) conducted a second set of experiments in which larval *D. melanogaster* were raised on apples, oranges, grapes, tomatoes, or onions. Females were then given a choice of all five foods for oviposition. One interesting result from this study was that adult exposure to one food (e.g., apple) resulted in an increase in the preference for another food (e.g., grape). Jaenike has called this phenomenon cross-induction.

It is logical to study the degree to which genetic differences might affect oviposition preference. Jaenike and Grimaldi (1983) studied the oviposition preference of 13 strains of *D. tripunctata* and four strains of *D. putrida* raised under identical conditions. Females were given a choice of lettuce, tomato, and three species of mushroom. The oviposition preference of *D. tripunctata* shows a significant food by strain interaction, while *D. putrida* shows no such interaction. These results indicate that

D. tripunctata has the appropriate genetic variation for host plant differentiation.

Although it appears that adult conditioning can affect female oviposition choice in an artificial situation, it would be of interest to know if such conditioning can be detected in the field. Jaenike (1986) has conducted a series of experiments with *D. melanogaster* to collect such information. His study only addresses the lability of food preference, so some caution must be exercised in extrapolating these results to oviposition preference. Adult flies were conditioned on tomatoes, grapes, and standard medium. These flies were then marked and released at field sites and the frequency of capture on either tomato or grape baits was recorded. Twenty-four hours after conditioning and release, attraction to grape or tomato in the field still depended on the prior conditioning. Those flies that were raised on grapes showed an increased preference for this bait. These results are in qualitative accord with Jaenike (1983), in which the frequency of eggs laid on grape nearly tripled in those females raised on grape rather than other foods. When the flies were again tested 4 days after conditioning there was no detectable difference in their preference for grape or tomato baits in the field. Thus, environmental conditioning of food preference is a labile behavior, with the effect vanishing between 1 and 4 days. It is interesting that laboratory studies of conditioned behavior show that *D. melanogaster* will exhibit these effects for up to 24h (Tempiel *et al.*, 1983). One should not conclude that any correlation between the larval environment and adult food preference will be completely lost in a single day, since those adults returning to their larval environment due to conditioning will have this preference reinforced and thus continue to prefer (or avoid) their larval environment for a few more days. These results do imply that one would expect a gradual decay in the correlation between food preference and larval habitat.

Habitat Choice in the Field

Certainly the relevance of genetically or environmentally induced habitat choice is greatest if these effects can be demonstrated in the field. Consequently, a large number of studies have been aimed at determining the degree of habitat choice in natural populations. The various studies sometimes differ substantially, which makes comparisons difficult. Some factors which seem to be important in habitat choice are sex, genotype, previous experience, food used in baits, and species of yeast. This review of habitat choice will proceed by considering each of these factors.

Previous Experience

Several studies have already been considered which convincingly demonstrate that adult conditioning can have a substantial effect on subsequent oviposition and habitat choice. In addition, this conditioning effect appears to be lost in 1–4 days if not reinforced, at least for *D. melanogaster*. Several studies have indicated the lack of detectable habitat preferences in the field (Jaenike, 1978; Atkinson and Miller, 1980; Turelli, Coyne and Prout, 1984). Jaenike (1986) argues that this may largely be a consequence of conditioning that results in less than perfect fidelity. Suppose that there are two habitats *A* and *B* in which *Drosophila* breed and feed. Assume also that the choice exhibited for either *A* or *B* is determined in part by prior conditioning. The standard field experiment consists in collecting flies in habitat *A* or *B*, and marking and releasing them, and then recapturing flies in the same two areas. Clearly, if flies that were originally caught in area *A* are almost always recaught in area *A*, then some degree of habitat preference can be inferred. However, in the sample of flies originally caught in area *A* there will be some that actually prefer area *B* due to less than perfect conditioning or genetic effects. This same error of misclassification occurs among the recaptured flies. If one knew the genotype and prior conditioning of released flies, then the sort of errors just described would only occur among the recaptured population. In the Jaenike (1986) field study prior conditioning and genotypes were known for released flies and thus the ability to detect differences in habitat choice is much greater than in previous studies. Jaenike (1986) shows how this prior information can allow one to detect statistically significant differences with sample sizes an order of magnitude less than those required by standard methods.

Food Used in Baits

From a number of studies considered previously it is clear that the attractiveness of certain foods varies with species, sex, and genotype and also depends on the prior experience of an individual. These facts imply that it will be difficult to make inferences concerning the absolute or relative abundance of species of *Drosophila* by just enumerating the species caught in a particular trap.

The choice of food used in traps also has important consequences for habitat choice experiments. Most studies can be divided into two categories: those that use food unlikely to be encountered by *Drosophila* out of doors (Taylor and Powell, 1978; Atkinson and Miller, 1980) and those that use foods that represent natural oviposition and larval habitats

(Turelli *et al.*, 1984). When "unnatural" foods are used, the habitats under consideration are really different geographic localities chosen by the experimenter for reasons that may or may not be important in habitat selection by *Drosophila*. Those studies may fail to detect "habitat" choice for the geographic areas chosen even when there may be highly specific preferences for the unknown and unobserved oviposition sites. Similarly, these experiments may uncover a preference for the habitat in which a fly is initially captured even though utilization of oviposition sites and foods may be identical across habitats.

Such problems as those just discussed are clearly reduced when natural oviposition sites can be identified and utilized in baits. Turelli *et al.* (1984) studied habitat choice of *D. melanogaster/simulans* and *pseudoobscura* in fruit orchards. Two experiments studied attraction to known breeding sites: oranges and grapefruits in one, figs and peaches in the other. Turelli *et al.* quantify their results with an index of choice $\Delta = P_{x|x} - P_{x|y}$, where $P_{x|y}$ is the probability of recapture in habitat x given that the fly was initially caught in y, and $P_{x|x}$ is similarly defined. Thus, $-1 \leq \Delta \leq 1$, with $\Delta = 1$ implying absolute habitat fidelity and $\Delta = -1$ implying absolute avoidance. Their results indicate a small but significant fidelity for the pooled *D. melanogaster/simulans* sample ($\Delta = 0.087$) on figs and peaches, but *pseudoobscura* shows no preference or negative Δ. This last result is in sharp contrast to the habitat preference ($\Delta = 0.185$–0.306) Taylor and Powell (1978) found for *D. pseudoobscura*. Given the differences in techniques utilized by Taylor and Powell and by Turelli *et al.*, one can only conjecture about the reasons for the different results. As mentioned previously, it might happen that despite the preference for geographic localities observed by Taylor and Powell (1978), there may be no difference in underlying preference for oviposition sites. If one assumes that the results of Taylor and Powell imply differences in oviposition preference, then one might conjecture that their study sites represent different but relatively stable oviposition sites. This situation can be contrasted with the orchards studied by Turelli *et al.*, which have an abundance of different oviposition sites over short distances. If adult habitat preference is largely governed by prior experience, then a difference in oviposition site availability could explain the different results from these field studies.

Sex

Differences between sexes in habitat choice have been seen by Richardson and Johnston (1975), Shorrocks and Nigro (1981), and Jaenike (1985, 1986). For instance, Jaenike (1986) saw that prior adult conditioning could affect subsequent attraction to baits in the field. *Drosophila melanogaster* males and females were both affected, although there were quantitative differences in the response. In contrast, *D. tripunctata* males were unaffected by prior experience, while females showed a marked effect: that is, those females raised on tomatoes showed a preference for tomato relative to controls.

Genotype

Several lines of evidence indicate that there are genetic differences between flies found in different habitats (Stalker, 1976; Taylor and Powell, 1977) and that genotypes differ in their habitat preference (Kekić *et al.*, 1980; Jaenike and Grimaldi, 1983; Jaenike, 1986). Kekić *et al.* (1980) report habitat preference for individuals of *D. subobscura* caught in light and dark areas. Flies from each area were brought into the laboratory and the phototactic behavior of the F_1's was observed in a light maze. Offspring from flies caught in light areas were more positively phototactic than those caught in dark areas. As discussed previously, Jaenike and Grimaldi (1983) reported that oviposition preference varied significantly among strains of *D. tripunctata*, but not in strains of *D. putrida*. Similarly, for *D. tripunctata* that had been preconditioned and released in the field, strain genotype had a significant effect on habitat preference.

Yeasts

Differential attraction to yeasts has been demonstrated in *D. buzzatii* and *aldrichi* (Barker *et al.*, 1981*a,b*), *pseudoobscura* and *persimilis* (Dobzhansky *et al.*, 1956; Klaczko *et al.*, 1985), *azteca, occidentalis, pinicola, miranda, melanogaster, californica, immigrans*, and *victoria* (Carson *et al.*, 1956), and a variety of tropical species (DaCunha *et al.*, 1957). Although these results indicate that preference tests should standardize the type of yeast used, it is not clear how important this phenomenon may be in nature. In many hosts a large number of yeast species have been isolated (Carson *et al.*, 1956; de Cunha *et al.*, 1957; Heed *et al.*, 1976; Starmer *et al.*, 1976, 1981). This lack of correspondence between yeasts and habitats means that preferences for different yeasts need not translate into habitat preference. Heed *et al.* (1976) conclude that yeasts alone do not limit certain desert *Drosophila* to particular cacti. *Drosophila pachea*, for instance, is found almost exclusively in senita cactus rots (Fellows and Heed, 1972). This is apparently due to *D. pachea*'s nutritional requirement for a sterol found in the senita cactus. However, the profile of yeast species found in the gut of adult *D. pachea* is similar to the species' composition of the senita cactus (Starmer *et al.*, 1976, 1981). It should be noted that several species of yeast were misclassified or lumped together in Starmer *et al.* (1976); however, a more precise classification of these species was carried out by Starmer *et al.* (1981). Thus, although *D. pachea* is a specialist on senita cactus, it seems to be a generalist with regard to available yeast species.

SUMMARY

In addition to the many empirical findings of interest on their own, this review has attempted to outline the importance of research on a model organism. The depth of knowledge of the ecology and genetics of *Drosophila* permit a more detailed analysis of evolutionary phenomena. A few examples that illustrate this point follow.

A large number of seemingly unrelated observations from intra- and interspecific competition experiments can be understood after considering the details of *Drosophila* larval competition. Repeated observations show faster than linear declines in per-capita rates of population growth in laboratory populations of *Drosophila*. This can be explained, at least partly, by the great sensitivity of female fecundity to changes in adult density. Even determining population dynamics in simplified laboratory environments can require a detailed knowledge of *Drosophila* ecology. As shown by Prout (1985), the life stages affected by changes in density can have significant effects on estimates of rates of population growth. Density-dependent natural selection may result in increased larval viability and female fecundity in *Drosophila*. These predictions follow only after considering the details of *Drosophila* ecology.

Many of the recent results in life-history evolution, density-dependent natural selection, the evolution of competitive ability, and oviposition preference suggest and will be followed by more detailed studies. It is hoped that future studies will attempt to integrate findings from past research and gather more information on the genetic basis of these traits. *Drosophila* should continue to be a unique research tool for some of the more interesting questions in evolutionary ecology.

ACKNOWLEDGMENTS

I wish to express my appreciation to J. S. F. Barker, T. J. Bierbaum, M. Clegg, J. Coyne, J. T. Giesel, J. Jaenike, R. C. Lewontin, P. Parsons,

J. R. Powell, T. Prout, M. R. Rose, B. Shorrocks, C. Taylor, A. Templeton, and M. Turelli for providing me with reprints and unpublished manuscripts. For their many helpful comments and encouragement during the course of this project I thank F. J. Ayala, G. Hammond, C. J. Krieks, and B. Wallace.

This work was supported by NIH grant GM 28016 to M. W. Feldman, NIH National Service Award GM 07310 from the National Institute of General Medical Sciences, and funds provided by Washington State University.

REFERENCES

Abrams, P., 1976, Limiting similarity and the form of the competition coefficient, *Theor. Popul. Biol.* **8**:355–375.

Alpatov, W. W., 1932, Egg production in *Drosophila melanogaster* and some factors which influence it, *J. Exp. Zool.* **63**:85–111.

Alvarez, G., and Fontdevila, A., 1981, Effect of the singed locus on the egg production curve of *Drosophila melanogaster, Can. J. Genet. Cytol.* **23**:327–336.

Anderson, W. W., and Arnold, J., 1983, Density-regulated selection with genotypic interactions, *Am. Nat.* **121**:649–655.

Asmussen, M. A., 1983, Density-dependent selection incorporating intraspecific competition. II. A diploid model, *Genetics* **103**:335–350.

Atkinson, W. D., 1979, A field investigation of larval competition in domestic *Drosophila, J. Anim. Ecol.* **48**:91–102.

Atkinson, W. D., and Miller, J. A., 1980, Lack of habitat choice in a natural population of *Drosophila subobscura, Heredity* **44**:193–199.

Ayala, F. J., 1965*a*, Evolution of fitness in experimental populations of *Drosophila serrata, Science* **150**:903–905.

Ayala, F. J., 1965*b*, Relative fitness of populations of *Drosophila serrata* and *Drosophila birchii, Genetics* **51**:527–544.

Ayala, F. J., 1966, Reversals of dominance in competing species of *Drosophila, Am. Nat.* **100**:81–83.

Ayala, F. J., 1968*a*, Genotype, environment, and population numbers, *Science* **162**:1453–1459.

Ayala, F. J., 1968*b*, An inexpensive and versatile population cage, *Drosophila Information Service* **43**:185.

Ayala, F. J., 1969, Evolution of fitness. IV. Genetic evolution of interspecific competitive ability in *Drosophila, Genetics* **61**:737–747.

Ayala, F. J., 1971, Competition between strains of *Drosophila willistoni* and *D. pseudoobscura, Experientia* **27**:343.

Ayala, F. J., Gilpin, M. E., and Ehrenfeld, J. G., 1973, Competition between species: Theoretical models and experimental tests, *Theor. Popul. Biol.* **4**:331–356.

Bakker, K., 1961, An analysis of factors which determine success in competition for food among larvae of *Drosophila melanogaster, Arch. Neerl. Zool.* **14**:200–281.

Bakker, K., 1969, Selection for rate of growth and its influence on competitive ability of larvae of *Drosophila melanogaster, Neth. J. Zool.* **19**:541–595.

Barclay, H. J., and Gregory, P. T., 1981, An experimental test of models predicting life-history characteristics, *Am. Nat.* **117**:944–961.

Barclay, H. J., and Gregory, P. T., 1982, An experimental test of life history evolution using *Drosophila melanogaster* and *Hyla regilla, Am. Nat.* **120**:26–40.

Barker, J. S. F., 1971, Ecological differences and competitive interaction between *Drosophila melanogaster* and *Drosophila simulans* in small laboratory populations, *Oecologia* **8**:134–156.

Barker, J. S. F., and Podger, R. N., 1970, Interspecific competition between *Drosophila melanogaster* and *Drosophila simulans*: Effects of larval density on viability, development period and adult body weight, *Ecology* **51**:170–189.

Barker, J. S. F., and Starmer, W. T., eds., 1982, *Ecological Genetics and Evolution: The Cactus–Yeast–Drosophila Model System*. Academic Press, New York.

Barker, J. S. F., Parker, G. J., Toll, G. L., and Widders, P. R., 1981*a*, Attraction of *Drosophila buzzatii* and *D. aldrichi* to species of yeasts isolated from their natural environment. I. Laboratory experiments, *Aust. J. Biol. Sci.* **34**:593–612.

Barker, J. S. F., Toll, G. L., East, P. D., and Widders, P. R., 1981*b*, Attraction of *Drosophila buzzatii* and *D. aldrichi* to species of yeasts isolated from their natural environment. II. Field experiments, *Aust. J. Biol. Sci.* **34**:613–624.

Bierbaum, T. J., and Ayala, F. J., 1986, The coevolution of interspecific competitors: An experimental test in *Drosophila*, manuscript.

Bierbaum, T. J., Mueller, L. D., and Ayala, F. J., 1986, Density-dependent evolution of life history characteristics in *Drosophila melanogaster*, manuscript.

Bodenheimer, F. S., 1938, *Problems of Animal Ecology*, London.

Brncic, D., 1966, Ecological and cytogenetic studies of *Drosophila flavopilosa*, a neotropical species living in *Cestrum* flowers, *Evolution* **20**:16–29.

Bush, G. L., 1974, The mechanism of sympatric host race formation in the true fruit flies (*Tephritidae*), in: *Genetic Mechanisms of Speciation in Insects* (M. J. D. White, ed.), pp. 3–23, Australian and New Zealand Book Co., Sydney.

Buzzati-Traverso, A. A., 1955, Evolutionary changes in components of fitness and other polygenic traits in *Drosophila melanogaster* populations, *Heredity* **9**:153–186.

Caligari, P. D. S., 1980, Competitive interactions in *Drosophila melanogaster, Heredity* **45**:219–231.

Carson, H. L., Knapp, E. P., and Phaff, H. J., 1956, Studies on the ecology of *Drosophila* in the Yosemite region of California. III. The yeast flora of the natural breeding sites of some species of *Drosophila, Ecology* **37**:538–544.

Charlesworth, B., 1971, Selection in density-regulated populations, *Ecology* **52**:469–474.

Charlesworth, B., 1980, *Evolution in Age-Structured Populations*, Cambridge University Press, Cambridge.

Chiang, H. C., and Hodson, A. C., 1950, An analytical study of population growth in *Drosophila melanogaster, Ecol. Monogr.* **20**:173–206.

Christiansen, F. B., and Loeschcke, V., 1980, Evolution and intraspecific exploitative competition. I. One-locus theory for small additive gene effects, *Theor. Popul. Biol.* **18**:297–313.

Clark, A., 1979, The effects of interspecific competition on the dynamics of a polymorphism in an experimental population of *Drosophila melanogaster Genetics* **92**:1315–1328.

Clark, A. G., and Feldman, M. W., 1981*a*, Density-dependent fertility selection in experimental populations of *Drosophila melanogaster, Genetics* **98**:849–869.

Clark, A. G., and Feldman, M. W., 1981*b*, The estimation of epistasis in components of fitness in experimental populations of *Drosophila melanogaster* II. Assessment of meiotic drive, viability, fecundity and sexual selection, *Heredity* **46**:347–377.

Clarke, B., 1972, Density-dependent selection, *Am. Nat.* **106:**1–13.

Connell, J. H., 1980, Diversity and the coevolution of competitors, or the ghost of competition past, *Oikos* **35:**131–138.

Connor, E. F., and Simberloff, D., 1979, The assembly of species communities: Chance or competition, *Ecology* **60:**1132–1140.

Da Cunha, A. B., Shehata, A. M. E., and de Oliveira, W., 1957, A study of the diets and nutritional preferences of tropical species of *Drosophila, Ecology* **38:**98–106.

Deakin, M. A. B., 1966, Sufficient conditions for genetic polymorphism, *Am. Nat.* **100:**690–692.

De Jong, G., 1976, A model of competition for food. I. Frequency-dependent viabilities, *Am. Nat.* **121:**67–93.

Del Solar, E., 1968, Selection for and against gregariousness in the choice of oviposition sites by *Drosophila pseudoobscura, Genetics* **58:**275–282.

Del Solar, E., and Palomino, H., 1966, Choice of oviposition in *Drosophila melanogaster, Am. Nat.* **100:**127–133.

Dobzhansky, T., Cooper, D. M., Phaff, H. J., Knapp, E. P., and Carson, H. L., 1956, Studies on the ecology of *Drosophila* in the Yosemite region of California. IV. Differential attraction of species of *Drosophila* to different species of yeasts, *Ecology* **37:**544–550.

Falconer, D. S., 1981, *Introduction to Quantitative Genetics*, 2nd ed., Longman, London.

Fellows, D. P., and Heed, W. B., 1972, Factors affecting host plant selection in desert-adapted cactiphilic *Drosophila, Ecology* **53:**850–858.

Futuyma, D. J., 1970, Variation in genetic response in laboratory populations of *Drosophila, Am. Nat.* **104:**239–252.

Futuyma, D. J., and Mayer, G. C., 1980, Nonallopatric speciation in animals, *Syst. Zool.* **29:**254–271.

Gadgil, M., and Bossert, W., 1970, Life historical consequences of natural selection, *Am. Nat.* **104:**1–24.

Gadgil, M., and Solbrig, O. T., 1972, The concept of *r*- and *K*-selection: Evidence from wild flowers and some theoretical considerations, *Am. Nat.* **106:**14–31.

Gale, J. S., 1964, Competition between three lines of *Drosophila melanogaster, Heredity* **19:**681–699.

Giesel, J. T., 1979, Genetic co-variation of survivorship and other fitness indices in *Drosophila melanogaster, Exp. Gerontol.* **14:**323–328.

Giesel, J. T., and Zettler, E. E., 1980, Genetic correlations of life historical parameters and certain fitness indicies in *Drosophila melanogaster*: r_m, r_s, diet breadth, *Oecologia* **47:**299–302.

Giesel, J. T., Murphy, P. A., and Manlove, M. N., 1982*a*, The influence of temperature on genetic interrelationships of life history traits in a population of *Drosophila melanogaster*: What tangled data sets we weave, *Am. Nat.* **119:**464–479.

Giesel, J. T., Murphy, P. A., and Manlove, M. N., 1982*b*, An investigation of the effects of temperature on the genetic organization of life history indices in three populations of *Drosophila melanogaster*, in: *Evolution and Genetics of Life Histories* (H. Dingle and J. P. Hegmann, eds.), pp. 189–207, Springer, New York.

Gilpin, M. E., 1974, Intraspecific competition between *Drosophila* larvae in serial transfer systems, *Ecology* **55:**1154–1159.

Gilpin, M. E., and Ayala, F. J., 1973, Global models of growth and competition, *Proc. Natl. Acad. Sci. USA* **70:**3590–3593.

Goodman, D., 1979, Competitive hierarchies in laboratory *Drosophila, Evolution* **33:**207–219.

Gowen, J. W., and Johnsen, L. E., 1946, Section on genetics and evolution on the mechanism of heterosis. I. Metabolic capacity of different races of *Drosophila melanogaster* for egg production, *Am. Nat.* **80:**149–179.

Grimaldi, D., and Jaenike, J., 1984, Competition in natural populations of mycophagous *Drosophila, Ecology* **65:**1113–1120.

Guckenheimer, J., Oster, G., and Ipaktchi, 1977, The dynamics of density dependent population models, *J. Math. Biol.* **4:**101–147.

Gupta, A. P., and Lewontin, R. C., 1982, A study of reaction norms in natural populations of *Drosophila pseudoobscura, Evolution* **36:**934–948.

Haddon, M., 1982, Frequency dependent competitive success in an age-structured model, *Am. Nat.* **120:**405–410.

Hassell, M., Lawton, J., and May, R. M., 1976, Pattern of dynamical behavior in single-species populations, *J. Anim. Ecol.* **45:**471–486.

Hastings, A., Serradilla, J. M., and Ayala, F. J., 1981, Boundary-layer model for the population dynamics of single species, *Proc. Natl. Acad. Sci. USA* **78:**1972–1975.

Haymer, D. S., and Hartl, D. L., 1983, The experimental assessment of fitness in *Drosophila*. II. A comparison of competitive and noncompetitive measures, *Genetics* **104:**343–352.

Heckel, D., and Roughgarden, J., 1980, A species near its equilibrium size in a fluctuating environment can evolve a lower intrinsic rate of increase, *Proc. Natl. Acad. Sci. USA* **77:**7497–7500.

Heed, W. B., Starmer, W. T., Miranda, M., Miller, M. W., and Phaff, H. J., 1976, An analysis of the yeast flora associated with cactophilic *Drosophila* and their host plants in the Sonoran Desert and its relation to temperate and tropical associations, *Ecology* **57:**151–160.

Hiraizumi, Y., 1961, Negative correlation between rate of development and female fertility in *Drosophila melanogaster, Genetics* **46:**615–624.

Hopkins, A. D., 1917, A discussion of C. G. Hewitt's paper on "Insect Behavior," *J. Econ. Entomol.* **10:**92–93.

Iwasa, Y., and Teramoto, E., 1980, A criterion of life history evolution based on density dependent selection, *J. Theor. Biol.* **84:**545–566.

Jaenike, J., 1978, Host selection by mycophagous *Drosophila, Ecology* **59:**1286–1288.

Jaenike, J., 1981, Criteria for ascertaining the existence of host races, *Am. Nat.* **117:**830–834.

Jaenike, J., 1982, Environment modification of oviposition behavior in *Drosophila, Am. Nat.* **119:**784–802.

Jaenike, J., 1983, Induction of host preference in *Drosophila melanogaster, Oecologia* **58:**320–325.

Jaenike, J., 1985, Genetic and environmental determinants of food preference in *Drosophila tripunctata, Evolution* **39:**362–369.

Jaenike, J., 1986, Intraspecific variation in food preference of *Drosophila, Biol. J. Linn. Soc.*, in press.

Jaenike, J., and Grimaldi, D., 1983, Genetic variation for host preference within and among populations of *Drosophila tripunctata, Evolution* **37:**1023–1033.

Jones, J. S., and Probert, R. F., 1980, Habitat selection maintains a deleterious allele in a heterogeneous environment, *Nature* **287:**632–633.

Karlin, S., 1981, Classifications of selection–migration structures and conditions for a protected polymorphism, in: *Evolutionary Biology*, Vol. 14 (M. K. Hecht, B. Wallace, and G. T. Prance, eds.), pp. 61–204, Plenum Press, New York.

Kekić, V., Taylor, C. E., and Andjelković, M., 1980, Habitat choice and resource specialization by *Drosophila subobscura, Genetika* **12:**219–225.

King, C. E., and Anderson, W. W., 1971, Age-specific selection. II. The interaction between *r* and *K* during population growth, *Am. Nat.* **105:**137–156.

King, C. E., and Dawson, P. S., eds., 1983, *Population Biology: Retrospect* and *Prospect*, Columbia University Press, New York.

Klaczko, L. B., Powell, J. R., and Taylor, C. E., 1983, *Drosophila* baits: Yeasts and species attracted, *Oecologia* **59:**411–413.

Levene, H., 1953, Genetic equilibrium when more than one ecological niche is available, *Am. Nat.* **87:**331–333.

Levins, R., 1966, The strategy of model building in population biology, *Am. Sci.* **54:**421–431.

Lewontin, R. C., 1955, The effects of population density and composition on viability in *Drosophila melanogaster, Evolution* **9:**27–41.

Lewontin, R. C., and Matsuo, Y., 1963, Interaction of genotypes determining viability in *Drosophila busckii, Proc. Natl. Acad. Sci. USA* **49:**270–278.

L'Heritier, P., and Teissier, G., 1933, Etude d'une population de Drosophiles en equilibre, *C. R. Acad. Sci. Fr.* **197:**1765–1767.

Luckinbill, L. S., 1978, *r*- and *K*-selection in experimental populations of *Escherichia coli, Science* **202:**1201–1203.

MacArthur, R. H., 1962, Some generalized theorems of natural selection, *Proc. Natl. Acad. Sci. USA* **48:**1893–1897.

MacArthur, R. H., and Wilson, E. O., 1967, *The Theory of Island Biogeography*, Princeton University Press, Princeton.

Mather, K., and Caligari, P. D. S., 1981, Competitive interactions in *Drosophila melanogaster*. II. Measurement of competition, *Heredity* **46:**239–254.

Mather, K., and Cooke, P., 1962, Differences in competitive ability between genotypes of *Drosophila, Heredity* **17:**381–407.

May, R. M., 1974, Biological populations with non-overlapping generations: Stable points, stable cycles and chaos, *Science* **186:**645–647.

May, R. M., and Oster, G., 1976, Bifurcations and dynamic complexity in simple ecological models, *Am. Nat.* **110:**573–599.

Maynard-Smith, J., 1966, Sympatric speciation, *Am. Nat.* **100:**637–650.

Mayr, E., 1963, *Animal Species and Evolution*, Harvard University Press, Cambridge.

McKenzie, J. A., and Parsons, P. A., 1972, Alcohol tolerance: An ecological parameter in the relative success of *Drosophila melanogaster* and *Drosophila simulans, Oecologia* **10:**373–388.

Merrell, D. J., 1951, Interspecific competition between *Drosophila funebris* and *Drosophila melanogaster, Am. Nat.* **85:**159–169.

Miller, R. S., 1964, Larval competition in *Drosophila melanogaster* and *D. simulans, Ecology* **45:**132–148.

Moore, J. A., 1952*a*, Competition between *Drosophila melanogaster* and *Drosophila simulans*. II. The improvement of competitive ability through selection, *Proc. Natl. Acad. Sci. USA* **38:**813–817.

Moore, J. A., 1952*b*, Competition between *Drosophila melanogaster* and *Drosophila simulans*. I. Population cage experiments, *Evolution* **6:**407–420.

Mourão, C. A., Ayala, F. J., and Anderson, W. W., 1972, Darwinian fitness and adaptedness in experimental populations of *Drosophila willistoni, Genetica* **43:**552–574.

Mueller, L. D., 1979, Fitness and density dependence in *Drosophila melanogaster*, Ph. D. Thesis, University of California, Davis.

Mueller, L. D., and Ayala, F. J., 1981*a*, Fitness and density dependent population growth in *Drosophila melanogaster, Genetics* **97:**667–677.

Mueller, L. D., and Ayala, F. J., 1981*b*, Dynamics of single-species population growth: Experimental and statistical analysis, *Theor. Popul. Biol.* **20:**101–117.

Mueller, L. D., and Ayala, F. J., 1981*c*, Dynamics of single-species population growth: Stability or chaos? *Ecology* **62:**1148–1154.

Mueller, L. D., and Ayala, F. J., 1981*d*, Trade-off between *r*-selection and *K*-selection in *Drosophila* populations, *Proc. Natl. Acad. Sci. USA* **78:**1303–1305.

Mueller, L. D., and Ayala, F. J., 1982, Population dynamics in the serial transfer system: Comments on Haddon's model, *Am. Nat.* **120:**548–550.

Murphy, P. A., Giesel, J. T., and Manlove, M. N., 1983. Temperature effects on life history variation in *Drosophila simulans, Evolution* **37:**1181–1192.

Nogueś, R. M. 1977, Population size fluctuations in the evolution of experimental cultures of *Drosophila subobscura, Evolution* **31:**200–213.

Nunney, L., 1983, Sex differences in larval competition in *Drosophila melanogaster*: The testing of a competition model and its relevance to frequency dependent selection, *Am. Nat.* **121:**67–93.

Parsons, P. A., 1980, Isofemale strains and evolutionary strategies in natural populations, in: *Evolutionary Biology*, Vol. 13 (M. K. Hecht, W. C. Steere, and B. Wallace, eds.), pp. 175–217, Plenum Press, New York.

Parsons, P. A., 1981, Evolutionary ecology of Australian *Drosophila*, a species analysis in: *Evolutionary Biology*, Vol. 14 (M. K. Hecht and G. T. Prance, eds.), pp. 297–349, Plenum Press, New York.

Pearl, R., 1927, The growth of populations, *Q. Rev. Biol.* **2:**532–548.

Pearl, R., 1928, *The Rate of Living*, Knopf, New York.

Pearl, R., 1932, The influence of density of population upon egg production in *Drosophila melanogaster, J. Exp. Zool.* **63:**57–83.

Pearl, R., Miner, J. R., and Parker, S. L., 1927, Experimental studies on the duration of life. XI. Density of population and life duration in *Drosophila, Am. Nat.* **61:**289–317.

Pianka, E. R., 1970, On *r*- and *K*-selection, *Am. Nat.* **104:**592–596.

Pianka, E. R., 1972, *r* and *K* selection or *b* and *d* selection? *Am. Nat.* **106**:581–588.

Pimentel, D., Feinberg, E. H., Wood, P. W., and Hayes, J. T., 1965, Selection, spatial distribution and the co-existence of competing fly species, *Am. Nat.* **99:**97–109.

Pomerantz, M. J., Thomas, W. R., and Gilpin, M. E., 1980, Asymmetries in population growth regulated by intraspecific competition: Empirical studies and model tests, *Oecologia* **47:**311–322.

Prout, T., 1980, Some relationships between density-independent and density-dependent population growth, in: *Evolutionary Biology*, Vol. 13 (M. K. Hecht, W. C. Steere, and B. Wallace, eds.), pp. 1–68, Plenum Press, New York.

Prout, T., 1985, The delayed effect on adult fertility of immature crowding: Population dynamics, *Am. Nat.*, in press.

Pruzan-Hotchkiss, A., Perelle, I. B., Hotchkiss, F. H. C. and Ehrman, L., 1980, Altered competition between two reproductively isolated strains of *Drosophila melanogaster, Evolution* **34:**445–452.

Richardson, R. H., and Johnston, J. S., 1975, Ecological specialization of Hawaiian *Drosophila, Oecologia* **21:**193–204.

Richmond, R. C., and Gerking, J. L., 1979, Oviposition site preference in *Drosophila, Behav. Genet.* **9:**233–241.

Robertson, F. W., 1957, Studies in quantitative inheritance. XI. Genetic and environemtnal correlation between body size and egg production in *Drosophila melanogaster, J. Genet.* **55:**428–443.

Robertson, R. W., and Sang, F. H., 1944, The ecological determinants of population growth in a *Drosophila* culture, *Proc. Roy. Soc. Lond. B* **132:**258–291.

Rose, M. R., 1984*a*, Laboratory evolution of postponed senescence in *Drosophila melanogaster, Evolution* **38**:516–526.

Rose, M. R., 1984*b*, Genetic covariation in *Drosophila* life history: Untangling the data, *Am. Nat.* **123**:565–569.

Rose, M. R., and Charlesworth, B., 1981*a*, Genetics of life history in *Drosophila melanogaster*. I. Sib. analysis of adult females, *Genetics* **97**:173–186.

Rose, M. R., and Charlesworth, B., 1981*b*, Genetics of life history in *Drosophila melanogaster*. II. Exploratory selection experiments, *Genetics* **97**:187–196.

Rose, M. R., Dorey, M. L., Coyle, A. M., and Service, P. M., 1984, The morphology of postponed senescence in *Drosophila melanogaster, Can. J. Zool.* **62**:1576–1580.

Roughgarden, J., 1971, Density dependent natural selection, *Ecology* **52**:453–468.

Roughgarden, J., 1976, Resource partitioning among competing species—A coevolutionary approach, *Theor. Popul. Biol.* **9**:388–424.

Roughgarden, J., 1979, *Theory of Population Genetics and Evolutionary Ecology: An Introduction*, MacMillan, New York.

Rummel, J. D., and Roughgarden, J., 1983, Some differences between invasion-structured and coevolutionary-structured competitive communities: A preliminary theoretical analysis, *Oikos* **41**:477–486.

Sang, J. H., 1949*a*, The ecological determinants of population growth in a *Drosophila* culture. III. Larval and pupal survival, *Physiol. Zool.* **22**:183–202.

Sang, J. H., 1949*b*, The ecological determinants of population growth in a *Drosophila* culture. IV. The significance of successive batches of larvae, *Physiol. Zool.* **22**:202–210.

Sang, J. H., 1949*c*, The ecological determinants of population growth in a *Drosophila* culture. V. The adult population count, *Physiol. Zool.* **22**:210–223.

Sang, J. H., 1949*d*, Population growth in *Drosophila* cultures, *Biol. Rev.* **25**:188–219.

Sang, J. H., McDonald, J. M., and Gordon, C., 1949, The ecological determinants of population growth in a *Drosophila* culture. VI. The total population count, *Physiol. Zool.* **22**:223–235.

Schaffer, W. M., 1983, The application of optimal control theory to the general life history problem, *Am. Nat.* **121**:418–431.

Schoener, T. W., 1983, Field experiments on interspecific competition, *Am. Nat.* **122**:240–285.

Seager, R. D., and Ayala, F. J., 1982, Chromosome interactions in *Drosophila melanogaster*. I. Viability studies, *Genetics* **102**:467–483.

Seaton, A. P. C., and Antonovics, J., 1967, Population inter-relationships. I. Evolution in mixtures of *Drosophila* mutants, *Heredity* **22**:19–33.

Shorrocks, B., 1970, Population fluctuations in the fruit fly (*Drosophila melanogaster*) maintained in the laboratory, *J. Anim. Ecol.* **39**:229–253.

Shorrocks, B., and Nigro, L., 1981, Microdistribution and habitat selection in *Drosophila subobscura* Collin, *Biol. J. Linn. Soc.* **16**:293–301.

Simmons, M. J., Preston, C. R., and Engels, W. R., 1980, Pleiotropic effects on fitness of mutations affecting viability in *Drosophila melanogaster, Genetics* **94**:467–475.

Smith, F. E., 1963, Population dynamics in *Daphnia magna* and a new model for population growth, *Ecology* **44**:651–663.

Stalker, H. D., 1976, Chromosome studies in wild populations of *D. melanogaster, Genetics* **82**:323–347.

Starmer, W. T., Heed, W. B., Miranda, M., Miller, M. W., and Phaff, H. J., 1976, The ecology of yeast flora associated with cactophilic *Drosophila* and their host plants in the Sonoran Desert, *Microbiol. Ecol.* **3**:11–30.

Starmer, W. T., Phaff, H. J., Miranda, M., Miller, M. W., and Heed, W. B., 1981, The yeast flora associated with the decaying stems of columnar cacti and *Drosophila* in North America, in: *Evolutionary Biology*, Vol. 14 (M. K. Hecht, B. Wallace, G. T. Prance, eds.), pp. 269–295, Plenum Press, New York.

Stearns, S. C., 1976, Life history tactics: A review of the ideas, *Q. Rev. Biol.* **51**:3–47.

Stearns, S. C., 1977, The evolution of life history traits: A critique of the theory and a review of the data, *Annu. Rev. Ecol. Syst.* **8**:145–171.

Strong, D., Szyska, L., and Simberloff, D., 1979, Tests of community-wide character displacement against null hypothesis, *Evolution* **33**:897–913.

Sulzbach, D. S., 1980, Selection for competitive ability: Negative results in *Drosophila*, *Evolution* **34**:431–436.

Sulzbach, D. S., and Emlen, J. M., 1979, Evolution of competitive ability in mixtures of *Drosophila melanogaster*: Populations with an initial asymmetry, *Evolution* **33**:1138–1149.

Takamura, T., 1980, Behavior genetics of choice of oviposition site in *Drosophila melanogaster*. II. Analysis of natural population, *Jpn. J. Genet.* **55**:91–97.

Taylor, C. E., and Condra, C., 1980, *r*- and *K*-selection in *Drosophila pseudoobscura*, *Evolution* **34**:1183–1193.

Taylor, C. E., and Powell, J. R., 1977, Microgeographic differentiation of chromosomal and enzyme polymorphisms in *Drosophila persimilis*, *Genetics* **85**:681–695.

Taylor, C. E., and Powell, J. R., 1978, Habitat choice in natural populations of *Drosophila*, *Oecologia* **37**:69–75.

Taylor, H. M., Gourley, R. S., Lawrence, C. E., and Kaplan, R. S., 1974, Natural selection of life history attributes: An analytical approach, *Theor. Popul. Biol.* **5**:104–122.

Tempiel, B. L., Bonini, N., Dawson, D. R., and Quinn, W. G., 1983, Reward learning in normal and mutant *Drosophila*, *Proc. Natl. Acad. Sci. USA* **80**:1482–1486.

Templeton, A. R., and Johnston, J. S., 1982, Life history evolution under pleiotropy and *K*-selection in a natural population of *Drosophila mercatorum*, in: *Ecological Genetics and Evolution: The Cactus–Yeast–Drosophila Model System* (J. S. F. Barker and W. T. Starmer, eds.), pp. 225–239, Academic Press, New York.

Thomas, W. R., Pomerantz, M. J. and Gilpin, M. E., 1980, Chaos, asymmetric growth and group selection for dynamical stability, *Ecology* **61**:1312–1320.

Thorpe, W. H., 1945, The evolutionary significance of habitat selection, *J. Anim. Ecol.* **14**:67–70.

Tracey, M. L., and Ayala, F. J., 1974, Genetic load in natural populations: Is it compatible with the hypothesis that many polymorphisms are maintained by natural selection? *Genetics* **77**:569–589.

Turelli, M., and Petry, D., 1980, Density-dependent selection in a random environment: An evolutionary process that can maintain stable population dynamics, *Proc. Natl. Acad. Sci. USA* **77**:7501–7505.

Turelli, M., Coyne, J. A., and Prout, T., 1984, Habitat and food choice in orchard populations of *Drosophila*, *Biol. J. Linn. Soc.* **22**:95–106.

Wallace, B., 1970, *Genetic Load*, Prentice-Hall, Englewood Cliffs, New Jersey.

Wallace, B., 1975, The biogeography of laboratory islands, *Evolution* **29**:622–635.

White, M. J. D., 1978, *Modes of Speciation*, Freeman, San Francisco.

Wijsman, E. M., 1984, The effect of mutagenesis on competitive ability in *Drosophila*, *Evolution* **38**:571–581.

Wright, S., 1943, Isolation by distance, *Genetics* **28**:114–138.

3

Rank-Order Selection and the Analysis of Data Obtained by *ClB*-like Procedures

BRUCE WALLACE

and

CHARLES E. BLOHOWIAK

INTRODUCTION

Before the application of gel electrophoresis and histochemical staining techniques to the analysis of genetic variation in populations (Harris, 1966; Lewontin and Hubby, 1966), *ClB*-like techniques (Muller, 1928) provided the bulk of all data on genetic variation within populations of many *Drosophila* species. Lewontin (1974), in a chapter entitled, "The Struggle to Measure Variation," presents an excellent account of the contrasting types of information yielded by these two analytical procedures.

Beginning with studies by Sturtevant (1937) and Dobzhansky and Queal (1938), autosomal inversions and appropriate mutant genes have been used in special breeding programs to reveal concealed genetic variation in natural populations of *Drosophila*. Reviews of these techniques, the species within which they have been developed, and the special pur-

BRUCE WALLACE • Department of Biology, Virginia Polytechnic Institute and State University, Blacksburg, Virginia 24061. CHARLES E. BLOHOWIAK • Computing Center, Virginia Polytechnic Institute and State University, Blacksburg, Virginia 24061.

poses for which they have been used may be found in Wallace (1968*a*, 1981).

Probably the most extensive use of the modified *ClB* techniques involved several laboratory populations of *D. melanogaster* that were once maintained at Cold Spring Harbor, New York. Sizeable samples of second chromosomes were obtained from flies taken as eggs from these populations at frequent intervals for more than 7 years and analyzed by the *CyL/Pm* (*Curly Lobe/Plum*) technique that is available for that species (Wallace, 1950, 1951, 1952, 1956, 1959*a*, 1968*a*, pp. 225ff, 1981, pp. 326ff; Wallace and King, 1951, 1952; Dobzhansky and Wallace, 1953; Wallace and Madden, 1953). These data are to be reanalyzed in the present chapter. Two reasons can be cited in justifying the new analysis: first, as we shall see immediately below, our understanding of the role that variation in fitness (or in its components viability and competitive ability) plays in populations has evolved during the past three or four decades; second, whatever the sophistication of mathematical or molecular approaches to evolutionary problems, the final test of any suggestion will be that of survival of a given class of individuals. The *ClB*-like techniques provide that test.

EVOLVING FITNESS CONCEPTS

Although the term was first used by Muller only in 1950 and defined by Crow in 1958, the early studies of genetic variation were in fact studies of the genetic loads of populations; this fact is confirmed by Haldane's (1937) anticipation of much that Muller (1950) discovered independently some dozen years later.

The chief measure provided by the *ClB*-like procedure is that of the relative egg-to-adult viability of wild-type flies that are either homozygous for chromosomes sampled from populations or heterozygous for random combinations of these same chromosomes. With considerable extra effort, the fertility of these wild-type flies—homozygotes or heterozygotes—can also be measured. The flies can be examined as well for abnormal phenotypes; these latter observations are unavoidably subjective. Understandably, then, the bulk of all information provided by the *ClB* technique concerns the viability of wild-type flies relative to a standardized, genetically marked class of flies and, through that standard, to each other.

Because viability is an important component of fitness, high viability as revealed by the modified *ClB* technique has been equated to high fitness. Unlike all other phenotypic traits, optimal fitness is the highest fitness; there is no intermediate optimum in this case. Genetic load, as defined by Crow (1958), is the proportionate decrease in mean fitness of a population relative to that possessed by the genotype exhibiting highest fitness. In order not to have a genetic load, a population must exhibit no variation in fitness; if such variation exists, the average must of necessity fall below the maximum.

Although the measure of fitness within a population is, by convention, calculated relative to the maximum fitness (1.00) exhibited by the optimal genotype of that population, the average fitness of one population is often compared with that of another in order to create interpopulation rankings. Only one justification exists for the use of *intra*population averages of relative viabilities in creating *inter*population rankings: all populations of a given species must be genetically nearly identical; each misses perfection for essentially trivial reasons. The discovery of tremendous allozyme variation severely undercut this justification.

Genetic load theory has little relevance to the ecological problems faced by a population. Both numbers and kinds of individuals must be considered in understanding the ecological genetics of populations. Wallace (1968*b*) showed that a population can grow in size while its genetic load grows as well. Any improvement in the adaptive value (= fitness) of an already superior heterozygote (such as might occur during the evolution of dominance or improving a mimic pattern) must, as a consequence, increase the segregational load (one of many identifiable genetic loads) of a population.

The correlation between the genetic load borne by a population and its (numerical) size—one of the measures of a population's well-being—is poor, at best; an even stronger statement can be, and has been, made (Wallace, 1982*a*): To the extent that the ability of a population to thin its own numbers allows an overcrowded population to meet the limited carrying capacity of its environment (and hence to continue existing), a phenotypic load with respect to competitive ability (increases) population fitness. Here, population fitness is assumed to include persistence, a suggestion made originally by Thoday (1953). If a phenotypic load with respect to fitness can increase the fitness of a population (in Thoday's sense), any genetic contribution—proximate or ultimate—to that load also increases the population's fitness. Thus, genetic load (as an intellectual concept) has evolved from a burden that, of necessity, lowers the fitness of a population—even threatening its existence—to a property that, by facilitating self-thinning, facilitates the population's continued existence.

RANK-ORDER SELECTION

The ability of a population to cull excess individuals and to reduce the number of survivors to that corresponding to the carrrying capacity of the environment depends upon the ability of some individuals to eliminate others. The future of any population depends entirely upon the well-being of its surviving and reproducing individuals; the condition of the survivors depends in large measure upon the severity of the struggle between competing individuals (intraspecific competition): equals can inflict fatal harm upon one another whereas the superior individual of an unequal competing pair can survive unscathed.

The elimination of individuals during rank-order selection (Wills, 1978) proceeds (with some error, of course) from those of low competitive ability (fitness) toward those with ever higher abilities. As increasing numbers of individuals of poor fitness are eliminated, the probability of survival for the remaining ones increases: thus, rank-order selection is both frequency and density dependent—soft selection in Wallace's (1968*b*, 1975*a*) sense.

Soft selection calls for a quite different analysis of the data obtained by *ClB*-like procedures than that which was routinely carried out under earlier genetic load concepts. At that time, the genotypes of different individuals were thought to be virtually identical. True, all individuals (other than identical twins) were recognized as differing genetically, but these differences were thought to involve a miniscule fraction of all gene loci. The three genotypes generated by two alleles at each of 40 loci—perhaps 1% of the total genome—can generate nearly 10^{20} genotypes. Thus, although during the 1950s genetic variation could be stressed on the one hand, the essential homogeneity of populations and even of species could be emphasized on the other.

The assumed genetic uniformity of individuals and the emphasis placed on the average fitness of populations led to repeated attempts to determine the effect of deleterious mutations on their carriers; numerous studies of this sort have been reviewed by Wallace (1968*a*, 1981). That modified *ClB* techniques deal with entire chromosomes rather than with single gene loci was of small import to the conclusions based on their use. The inferior viability exhibited by flies homozygous for an entire chromosome was generally ascribed to mutant genes at one or at most a few loci; the remaining loci on the tested chromosome were assumed to be occupied by "normal" alleles, that is, by nonmutated wild-type alleles.

Among the studies that were made using the *ClB*-like techniques were those in which chromosomes were grouped according to the viability of their homozygous carriers; the average viabilities of wild-type flies heterozygous for chromosomes of various groups were then calculated. Such studies led to estimates of the dominance of deleterious genes; estimates of dominance *h* of approximately 2% (or more) for lethal mutations were common. Such studies have been reviewed in considerable detail by Wallace (1968*a*, pp. 171ff; 1981, pp. 228ff).

Rank-order selection seriously weakens the logic underlying these early studies. Selection in a population proceeds, according to the account presented earlier, from individuals with low fitness to those with ever higher fitnesses until the number of surviving individuals matches the carrying capacity of the environment. Now, the individual members of a population carry chromosomes that were passed on to them by parents who mated more or less at random (among those surviving to reproductive age). Individuals carrying random combinations of chromosomes are not identical in fitness. Consequently, a proper analysis of the relation between homozygous and heterozygous wild-type flies obtained through the use of *ClB*-like techniques should proceed first by grouping *heterozygotes* according to their relative viabilities. One should then ask: What sorts of chromosomes, defined in terms of their effects on homozygous carriers, are found in each of these classes of heterozygotes? What sorts, especially, occur in heterozygotes of high viability? The earlier studies dealt with questions that were badly posed.

AN EXCEPTIONAL STUDY

For several generations after new mutations arise in populations, they are exposed to natural selection only through their effects on heterozygous carriers. Such was the reasoning of Wallace (1963*a*) in claiming that previous analyses had been flawed: rather than asking about the average effect of a deleterious (when homozygous) mutation on its heterozygous carriers, one should ask what sorts of mutations (again, classified by their homozygous effects) are carried by heterozygotes of differing viabilities. The role of rank-order selection in populations, however, was not sufficiently appreciated in 1963. Granted that, as a rule, many individuals that possess low viabilities and low fitnesses must be eliminated from the populations each generation, the crucial question concerns the types of mutations that are carried by selectively favored heterozygotes.

Dobzhansky *et al.* (1960) presented a traditional analysis of 1000 second and third chromosomes of *D. pseudoobscura*. Their data were reanalyzed by Wallace (1963*a*). With increasing viability of heterozygotes, the homozygotes for chromosomes carried by these heterozygotes showed, over a considerable range, an increasing average viability as well. The two viabilities—that of heterozygotes and that of homozygotes—did not increase at equal rates, however; the regression of the average viability of homozygotes on that of heterozygotes was only 0.68. Even that regression was lost, however, in the case of heterozygotes with highest viabilities; the chromosomes which these superior heterozygotes carried exhibited unexpectedly large average detrimental effects on their homozygous carriers.

No subsequent analysis of the sort described by Wallace (1963*a*) seems to have been carried out. For one thing, an enormous amount of data are needed in order to subdivide the viabilities of heterozygous wild-type flies into small, but statistically reliable, increments. For another, the analysis was not presented as one at odds with conventional genetic load concepts. The data that are summarized in subsequent sections of this report involve over 16,000 cultures and more than 5,500,000 flies; additional data will be summarized in subsequent publications. The view underlying the present analysis is quite different from that upon which traditional genetic load concepts are based. Underlying the present analysis is the notion that natural selection is generally both density and frequency dependent, that in culling individuals from a population it proceeds from low to ever higher fitnesses (i.e., natural selection is rank-order selection), and that for such selection to be effective and efficient, variation in fitness (especially in competitive ability) must exist among the competing individuals of a population. In brief, the *average* effects of deleterious mutations on their heterozygous carriers are of little or no importance in understanding selective events occurring within populations. In natural populations, the proper question is: What genes give rise to the individuals that are most likely to survive and reproduce?

MATERIALS AND METHODS

The populations that have provided the data for the present analysis were experimental populations of *D. melanogaster* that were maintained at the Biological Laboratory, Cold Spring Harbor, New York for 7 or more years. Their study was intended to demonstrate the effect on fitness of genetic variation (Haldane, 1937). The fitness of a Mendelian population at equilibrium is reduced by an amount that equals its mutation rate. X- and γ-radiation were used to alter the mutation rate in these populations.

The early tests carried out on these populations quickly revealed that in their case obvious expectations were not being confirmed. The result was a series of additional analyses and studies that need not be reviewed here, references to pertinent papers can be fouund in Wallace (1968*a*, 1970, 1981).

Of more immediate concern than past findings is the experimental technique that provided the estimates of fitness of the various laboratory—control as well as experimental—populations. Even this technique, one of the modified *ClB* techniques, need not be outlined in detail, however, because numerous descriptions are already available (Wallace, 1968*a*, 1981, and references therein). Suffice it to say that, by the use of a genetically marked (*Curly Lobe*) chromosome that carries two inverted segments, wild-type second chromosomes could be recovered from male flies sampled (as eggs, not as adults) from the experimental populations. These chromosomes could then be manipulated through a series of crosses in order to obtain wild-type flies homozygous ($+_i/+_i$) for the sampled chromosomes or heterozygous ($+_i/+_j$) for chromosomes obtained from different males. By the use of systematic round-robin matings, these homozygotes and heterozygotes could be obtained in a series of cultures in which the wild-type flies can be represented as 1/1, 1/2, 2/2, 2/3, 3/3, 3/4, 4/4, . . . ; the numbers in this series refer to different wild-type second chromosomes.

The main interest in the early analyses concerned the relative fitnesses of the five populations that were under intensive study: population 1, whose progenitor flies had been exposed to a near-sterilizing level of x-radiation; population 3, a control population whose progenitors (except for the radiation exposure) were comparable to those of 1; populations 5 and 6, differing only in population size, which were exposed continuously to about 2000 r of γ radiation per generation; and population 7, a large population (similar in this respect to 1, 3, and 6) exposed continuously to about 700 r per generation. Although they were not started simultaneously with populations 1 and 3, the progenitor flies of populations 5–7 were derived from the same chromosomal stocks which had been retained for that purpose.

In brief, the early analyses showed that estimates of fitness (based on the frequency within the *ClB* test cultures of wild-type flies heterozygous for second chromosomes obtained from male flies sampled at intervals from the laboratory populations) for population 1 consistently exceeded those obtained for its nonirradiated control, population 3; these results were obtained despite the (at that time) higher frequencies of lethal, semilethal, and other deleterious chromosomes in the irradiated population. Equally discordant results were obtained for the continuously irradiated populations 5–7. The observed fitnesses of these three populations were lower than that of the control; however, as irradiation exposure continued through successive generations, the fitnesses of these populations increased, thus approaching that of the control (Wallace, 1956). This ostensible improvement in the fitnesses of these populations occurred despite an obvious increase (marked increase in the case of 5 and 6) in the frequency of lethal, semilethal, and deleterious chromosomes as revealed by concomitant homozygous tests.

The early results have been outlined here to emphasize that the present analysis is *not* concerned with the relative fitnesses of different populations. Here, we intend to rank the viabilities of wild-type flies (i.e., to rank these flies according to their frequencies within the F_3 test cultures))that were heterozygous for two different chromosomes ($+_i/+_j$) and then, proceeding increment by increment through that ranking, to determine what sorts of chromosomes (defined by their effect on homozygous carriers) these heterozygotes carried.

The chief technical problem confronting the present study is that of standardization. The laboratory populations were sampled at 2-week intervals over many years. The samples were taken and tested during all seasons of each year, a fact that caused small annual fluctuations in the observed frequencies of wild-type flies in both homozygous and heterozygous test cultures. Different technicians performed the matings and counted the flies in the final (F_3) cultures; several technicians were employed at any moment, and technicians departed and new ones were hired as the years went by. All such variables have small effects that must, if possible, be eliminated in the present study. In the older ones, it was enough that each technician contributed equally to the analysis of each population within each sample; that procedure permitted one population to be compared with another.

The data have been standardized for present purposes by analyzing them as a large collection of individual units. One unit consists of the cultures—homozygous and heterozygous—counted by *one* technician in *one* sample taken from *one* cage. The standard frequency of wild-type flies in the F_3 test cultures is the average frequency of wild-type flies observed in the quasinormal heterozygous cultures (that is, the few cultures in which the frequencies of wild-type flies fell below 50% of the expected frequency are not included in this average). Using the average frequency of quasinormal heterozygotes as the standard (= 1.00), we have recalculated the frequencies of wild-type flies—homozygous or heterozygous—in every test culture of that unit relative to the unit's standard.

The above recalculation has been carried out for every technician and every chromosome sample (where round-robin matings were used) for populations 1, 3, and 7. As one result, the average viability of the quasinormal heterozygotes of all populations is adjusted to 1.00; no interpopulation comparisons of averages are possible under the present standardization procedure. By means of this procedure, however, the small cyclic fluctuations in frequencies of wild-type flies are compensated for, as are the small idiosyncratic differences between technicians that cause one to observe slightly more and another slightly fewer wild type flies per culture.

RESULTS

The Data

The material to be reanalyzed here might best be introduced within the framework of previous studies. Specifically, the new analysis extends from sample 42 to 123 in the case of population 1, from 42 to 124 for population 3, and from 43 to 157 for population 7. These are the samples for which at least one technician analyzed homozygous and heterozygous F_3 cultures obtained by round-robin matings; the progeny emerging in these cultures were expected to consist of $66\frac{2}{3}\%$ (*Curly Lobe*) flies and $33\frac{1}{3}\%$ wild type. Because the sample interval was 2 weeks, the samples from populations 1 and 3 were taken over a 3-year period starting when the populations were somewhat less than 2 years old. Population 7 was started at sample 20 and therefore this population was less than 1 year old when the round-robin tests were started; this type of analysis continued in the case of population 7 for some 4 years.

A summary of the original data obtained during the stated intervals (and analyzed as it was originally) is given in Table I. The data in this table have not been standardized unit by unit; consequently, the average frequency of wild-type flies within normal heterozygous cultures reveals the relative fitnesses of these populations as originally estimated: 1.03 for population 1, 1.00 for 3 (the control), and 0.99 for 7. Included among the tests summarized in Table I are heterozygous cultures for which only one (or possibly neither) homozygous culture was successfully reared; conversely, homozygous cultures are included even though the corresponding heterozygous one(s) may have been lost. Of the 16,783 cultures included in Table I, only 16,196 were suitable for the standardized analysis of the present report. Nevertheless, Table I emphasizes the magnitude of the present analysis: 16,783 cultures containing nearly 5,600,000 flies were examined and counted in the original study.

TABLE I. Summary of the Total Numbers of Flies and Cultures Involved in the Present Analysis of Populations 1, 3, and 7[a]

	Wild-type flies	Total flies	Number of cultures	Average frequency of wild-type flies
Population 1				
All homozygotes	223,605	860,575	2,639	0.2474
Normal homozygotes	220,073	706,012	2,066	0.3096
Normal heterozygotes	329,577	948,744	2,673	0.3473
Population 3				
All homozygotes	216,845	938,110	2,835	0.2198
Normal homozygotes	213,766	693,316	1,997	0.3071
Normal heterozygotes	358,178	1,065,038	2,874	0.3365
Population 7				
All homozygotes	181,864	821,051	2,872	0.2084
Normal homozygotes	177,596	571,872	1,887	0.3089
Normal heterozygotes	313,237	937,394	2,890	0.3329
Total				
All homozygotes	—	2,619,736	8,346	—
Normal heterozygotes	—	2,951,176	8,437	—
Grand total	—	5,570,912	16,783	—

[a] "All homozygotes" include those cultures in which the tested chromosomes were lethal or semilethal when homozygous; "normal homozygotes" exclude lethals and semilethals. Heterozygous combinations that proved to be lethal or semilethal are not included among "normal heterozygotes." The numbers of cultures listed in this table are somewhat larger than those cited in subsequent ones because the latter do not include those instances where only one (or, rarely, neither) of the two homozygous cultures corresponding to a given heterozygous one was available.

Figures 1–4 present the data utilized in the present analysis: if the two wild-type chromosomes are represented as $+_i$ and $+_j$, only those sequences of three cultures ($+_i/+_i$, $+_i/+_j$, and $+_j/+_j$) where both homozygotes and their corresponding heterozygote were successfully tested have been included. Notice that in tabulating these data, each homozygous culture is counted twice: $+_i$, because of the round-robin matings, occurs in two heterozygotes, $+_i/+_j$, where $j = i - 1$ and $j = i + 1$. Thus, to use the total number of tests listed in Fig. 4 as an example, 8098 heterozygous combinations have been tested in all; the corresponding homozygous cultures, because they were counted twice each, amount to 16,196.

The numerical data illustrated in Figs. 1–4 are too complex to be grasped easily by mere inspection. An examination of the lower margin

HOMOZYGOTES

HETEROZYGOTES	0-.09	-.19	-.29	-.39	-.49	-.59	-.69	-.79	-.89	-.99	-1.09	-1.19	-1.29	-1.39	1.40+	TOTAL
1.15+	65	5	3	5	12	9	13	30	62	90	44	19	5	0	0	362
1.14	28	4	2	5	2	4	10	13	37	33	20	7	1	0	0	166
1.12	48	5	1	2	3	2	8	25	42	64	19	7	1	1	0	228
1.10	31	2	3	5	3	7	6	12	49	53	32	12	1	0	0	216
1.08	51	4	3	1	5	11	7	27	75	100	37	12	3	0	0	336
1.06	56	2	4	4	3	10	16	35	63	85	40	18	1	0	1	338
1.04	67	3	4	2	6	11	7	37	78	96	61	15	1	0	0	388
1.02	60	2	6	4	6	7	12	29	92	99	50	17	3	1	0	388
1.00	74	8	3	2	8	12	16	39	96	109	52	18	3	0	0	440
.98	91	5	7	6	4	13	24	31	99	110	45	14	3	0	0	452
.96	66	5	4	5	7	9	14	32	83	84	42	13	3	1	0	368
.94	61	4	3	5	8	7	15	51	67	81	38	12	1	0	1	354
.92	42	3	1	6	6	7	9	25	57	80	34	7	1	0	0	278
.90	44	3	0	4	4	9	12	24	56	69	28	4	1	0	0	258
.88	34	3	2	1	7	1	6	23	59	42	18	4	0	0	0	200
.86	25	1	1	1	3	0	6	8	30	35	6	6	0	0	0	122
.84-	39	0	3	2	3	10	7	28	61	60	26	12	0	1	0	252
TOTAL	882	59	50	60	90	129	188	469	1106	1290	592	197	28	4	2	5146

POPULATION I

FIG. 1. The distribution of cultures for population 1 according to (1) the relative viability of wild-type flies carrying two different second chromosomes (heterozygotes) and (2) that of flies homozygous for each of the chromosomes carried by the corresponding heterozygotes (homozygotes). Because each chromosome that was tested for its viability effects on homozygous carriers was tested in two different heterozygous combinations, the total number of tests listed in the figure is twice the number of homozygous tests actually made. Excluded from these data are those few heterozygous tests in which the wild-type flies had relative viabilities below 0.50.

(Total) reveals that chromosomes from almost any population, when rendered homozygous, result in a bimodal distribution of viabilities of homozygous carriers: many lethal and near-lethal chromosomes, few exhibiting intermediate viabilities, and many homozygotes with near-normal viabilities. Close examination reveals that the means of these near-normal viabilities are somewhat lower than 1.00, which, because of our standardization procedures, is "normal" viability, by definition.

Heterozygotes, in contrast to homozygotes, exhibit a unimodal distribution centered (as expected) almost precisely on 1.00. The spread of heterozygous viabilities illustrated in Figs. 1–4 is much less than that of homozygotes. The entire distribution of heterozygous viabilities shown in the four figures would fall within four segments (0.80–0.89 to 1.10–1.19) on the homozygous scale. In passing, it may be noted that each of

HOMOZYGOTES

HETEROZYGOTES	0-.09	-.19	-.29	-.39	-.49	-.59	-.69	-.79	-.89	-.99	-1.09	-1.19	-1.29	-1.39	1.40+	TOTAL
1.15 +	91	6	1	3	1	8	8	32	54	78	57	18	7	0	0	364
1.14	32	2	2	1	0	5	2	10	22	39	25	14	2	0	0	156
1.12	58	5	1	2	0	2	2	11	35	48	27	13	3	1	0	208
1.10	68	4	5	2	0	5	13	20	41	58	40	9	4	1	0	270
1.08	75	4	1	1	3	8	4	32	58	74	53	16	2	1	0	332
1.06	98	1	5	6	0	8	12	22	63	104	50	13	1	1	0	384
1.04	115	5	3	3	5	8	13	27	74	106	57	12	4	0	0	432
1.02	125	4	5	3	10	9	9	34	67	134	67	16	3	0	0	486
1.00	109	5	3	1	10	17	17	25	63	113	63	20	4	0	0	450
.98	112	5	3	5	6	10	16	23	79	120	51	21	1	0	0	452
.96	123	2	3	5	6	8	12	22	84	122	64	18	3	0	0	472
.94	111	4	3	4	2	10	12	23	50	81	49	18	3	0	0	370
.92	76	3	2	2	2	7	14	17	54	56	40	7	5	3	0	288
.90	67	2	1	2	3	5	9	12	50	73	39	6	3	0	0	272
.88	60	1	3	2	1	4	4	13	29	34	19	9	1	0	0	180
.86	40	1	1	1	2	1	3	12	30	24	17	5	3	0	0	140
.84-	75	1	5	2	2	5	9	23	36	54	31	11	3	1	0	258
TOTAL	1435	55	47	45	53	120	159	358	889	1318	749	226	52	8	0	5514

POPULATION 3

FIG. 2. The distribution of cultures for population 3 according to (1) the relative viability of wild-type flies carrying two different second chromosomes (heterozygotes) and (2) that of flies homozygous for each of the chromosomes carried by the corresponding heterozygote (homozygotes). See Fig. 1.

the 17 segments within the spread of heterozygous viabilities is represented by a number of tests (right-hand margin) comparable to entire experiments that have been reported by others (Dobzhansky and Queal, 1938; Dobzhansky *et al.*, 1942; Dobzhansky and Spassky, 1953, 1954; Pavan, *et al.*, 1951; Townsend, 1952; Cavalcanti, 1950, Hoenigsberg and DeNavas, 1965; Hoenigsberg *et al.*, 1977).

Traditional Analyses

Figures 5–8 present the traditional representation of homozygous and heterozygous viabilities; these figures are based on the marginal totals of Figs. 1–4. A careful examination of Figs. 5 and 6 reveals that more slightly deleterious homozygotes are to be found in population 1 than in 3. The same is not true for lethal chromosomes, however. Somewhat before

HOMOZYGOTES

HETEROZYGOTES	0-.09	-.19	-.29	-.39	-.49	-.59	-.69	-.79	-.89	-.99	-1.09	-1.19	-1.29	-1.39	1.40+	TOTAL
1.15+	95	1	7	3	9	5	10	37	52	74	57	25	5	0	0	380
1.14	49	3	2	2	3	5	3	10	26	40	21	10	2	1	1	178
1.12	61	3	1	2	2	4	3	22	26	49	43	9	1	0	0	226
1.10	74	3	4	5	4	5	6	11	23	67	22	13	1	0	0	238
1.08	63	7	0	9	8	3	5	19	49	86	52	16	1	0	0	318
1.06	115	9	6	3	3	7	4	25	65	108	64	24	2	0	1	436
1.04	133	10	1	5	3	7	10	26	73	106	54	24	6	0	0	458
1.02	131	8	3	4	3	6	13	18	66	90	68	26	2	1	1	440
1.00	151	11	4	6	8	6	8	31	64	116	60	19	4	0	0	488
.98	113	7	6	6	5	6	2	26	71	97	53	23	2	1	0	418
.96	138	4	3	4	8	7	8	20	50	96	44	18	3	0	1	404
.94	99	9	7	8	2	4	7	19	52	101	68	12	2	0	0	390
.92	85	5	5	0	7	6	4	15	50	74	44	11	4	0	0	310
.90	53	4	5	2	3	3	4	11	33	44	33	11	1	1	0	208
.88	48	2	4	3	1	4	1	9	36	38	21	9	0	0	0	176
.86	46	2	3	2	3	1	2	3	20	34	22	5	1	0	0	144
.84-	110	9	6	2	7	3	6	27	43	62	36	12	1	0	0	324
TOTAL	1564	97	67	66	79	82	96	329	799	1282	762	267	38	4	4	5536

POPULATION 7

FIG. 3. The distribution of cultures for population 7 according to (1) the relative viability of wild-type flies carrying two different second chromosomes (heterozygotes) and (2) that of flies homozygous for each of the chromosomes carried by the corresponding heterozygote (homozygotes). See Fig. 1.

sample 42, the frequency of lethals in population 1 began declining, while that in 3 did not (Wallace, 1956, Fig. 5); during the interval (sample 42–123/124) of the present study the nonirradiated control population possessed a higher average frequency of lethals than did population 1. Figure 8, which is based on the combined data of all three populations, probably illustrates the contrasting viability distributions of cultures containing wild-type flies (homozygous and heterozygous) better than most other published figures; it shows especially clearly the consistent difference between the viabilities of quasinormal homozygotes and heterozygotes. The averages of the latter invariably exceed those of the former.

The data presented so far are suitable for two traditional, though somewhat different, analyses. Table II, following a procedure described by Wallace and Madden (1953), shows the proportions of subvital and supervital chromosomes among the quasinormal homozygotes of populations 1, 3, and 7. In this analysis, a subvital chromosome is defined as

HOMOZYGOTES

HETEROZYGOTES	0-.09	-.19	-.29	-.39	-.49	-.59	-.69	-.79	-.89	-.99	-1.09	-1.19	-1.29	-1.39	1.40+	TOTAL
1.15+	251	12	11	11	22	22	31	99	168	242	158	62	17	0	0	1106
1.14	109	9	6	8	5	14	15	33	85	112	66	31	5	1	1	500
1.12	167	13	3	6	5	8	13	58	103	161	89	29	5	2	0	662
1.10	173	9	12	12	7	17	25	43	113	178	94	34	6	1	0	724
1.08	189	15	4	11	16	22	16	78	182	260	142	44	6	1	0	986
1.06	269	12	15	13	6	25	32	82	191	297	154	55	4	1	2	1158
1.04	315	18	8	10	14	26	30	90	225	308	172	51	11	0	0	1278
1.02	316	14	14	11	19	22	34	81	225	323	185	59	8	2	1	1314
1.00	334	24	10	9	26	35	41	95	223	338	175	57	11	0	0	1378
.98	316	17	16	17	15	29	42	80	249	327	149	58	6	1	0	1322
.96	327	11	10	14	21	24	34	74	217	302	150	49	9	1	1	1244
.94	271	17	13	17	12	21	34	93	169	263	155	42	6	0	1	1114
.92	203	11	8	8	15	20	27	57	161	210	118	25	10	3	0	876
.90	164	9	6	8	10	17	25	47	139	186	100	21	5	1	0	738
.88	142	6	9	6	9	9	11	45	124	114	58	22	1	0	0	556
.86	111	4	5	4	8	2	11	23	80	93	45	16	4	0	0	406
.84-	224	10	14	6	12	18	22	78	140	176	93	35	4	2	0	834
TOTAL	3881	211	164	171	222	331	443	1156	2794	3890	2103	690	118	16	6	16,196

POPULATIONS 1, 3, AND 7, COMBINED

FIG. 4. The distribution of cultures for populations 1, 3, and 7 combined, according to (1) the relative viability of wild-type flies carrying two different second chromosomes (heterozygotes) and (2) that of flies homozygous for each of the chromosomes carried by the corresponding heterozygote (homozygotes). See Fig. 1.

one whose homozygous carriers have lower viabilities than those heterozygotes lying two standard deviations (genetic variance only) below the mean of quasinormal heterozygotes. About half of all homozygotes qualify for the label "subvital."

Supervitals, on the other hand, are homozygotes whose viability equals or exceeds all heterozygotes except these lying two standard deviations (again, genetic variance only) above the mean of quasinormal heterozygotes. Ostensibly, the observed proportions of supervital homozygotes (0.05–0.09) exceed the 0.025 of heterozygotes that would normally fall above the two-standard-deviation limit; biologically this slight excess is probably meaningless.

Table III analyzes the same data in a manner described by Crow and Temin (1964). Using average viabilities alone and drawing upon the Poisson distribution, the Crow–Temin procedure leads to an estimation of the average number of lethal equivalents per chromosome. The calculation

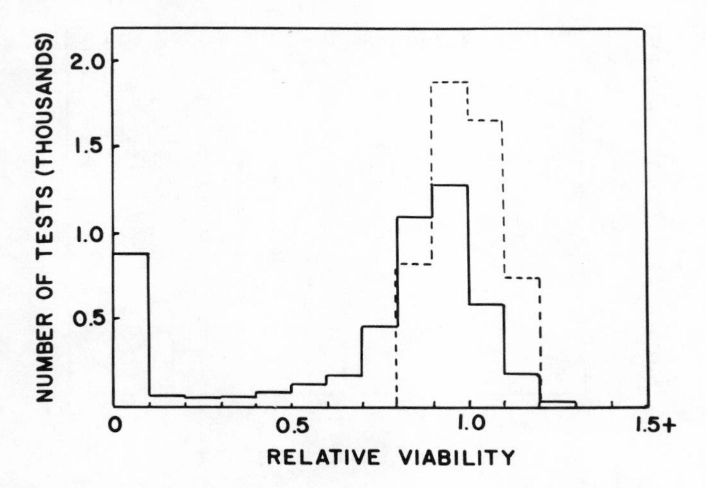

FIG. 5. The distributions of the relative viabilities of flies homozygous for various chromosomes from population 1 (solid line) and of flies heterozygous for random combinations of these same chromosomes (dashed line). The data on which this diagram is based are the marginal totals shown in Fig. 1.

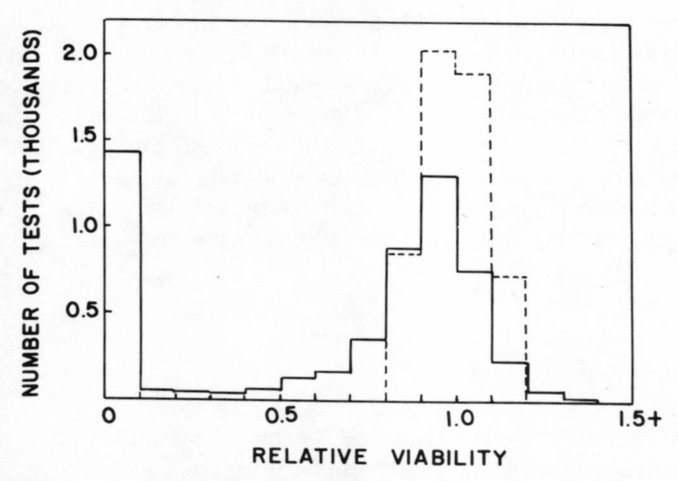

FIG. 6. The distributions of the relative viabilities of flies homozygous for various chromosomes from population 3 (solid line) and of flies heterozygous for random combinations of these same chromosomes (dashed line). The data on which this diagram is based are the marginal totals shown in Fig. 2.

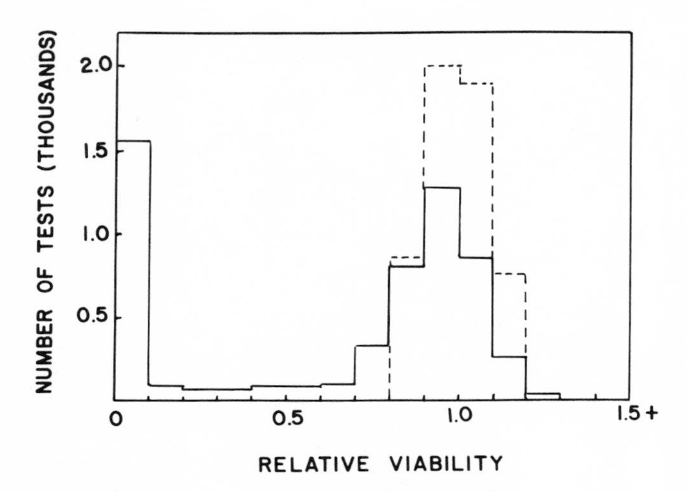

FIG. 7. The distributions of the relative viabilities of flies homozygous for various chromosomes from population 7 (solid line) and of flies heterozygous for random combinations of these same chromosomes (dashed line). The data on which this diagram is based are the marginal totals shown in Fig. 3.

permits one to partition the total load of lethal equivalents per chromosome into a lethal load and a detrimental load. Although, as was noted earlier, the lethal load of population 1 is less than that of the other two populations, the detrimental load is somewhat higher. Combining the two procedures illustrated in Tables II and III, one can calculate the average number of lethal equivalents per subvital chromosome. As in previous calculations of this sort (Wallace, 1968*a*, p. 53), the result is about 0.20—somewhat lower, perhaps than that (about 0.30) characteristic of tests involving *D. pseudoobscura*. Wallace (1984) has suggested why the calculated number of lethal equivalents per subvital chromosome should vary only slightly among tests involving different chromosomes and different species of *Drosophila*.

Distributions of Cultures

The data presented in Figs. 1–4, despite the complex pattern created by the intersection of two distributions (one unimodal and the other bimodal), can be examined to determine whether the number of cultures falling within certain regions of the figures exceed or fall short of expected numbers that can be estimated by means of marginal totals. Tables IV–

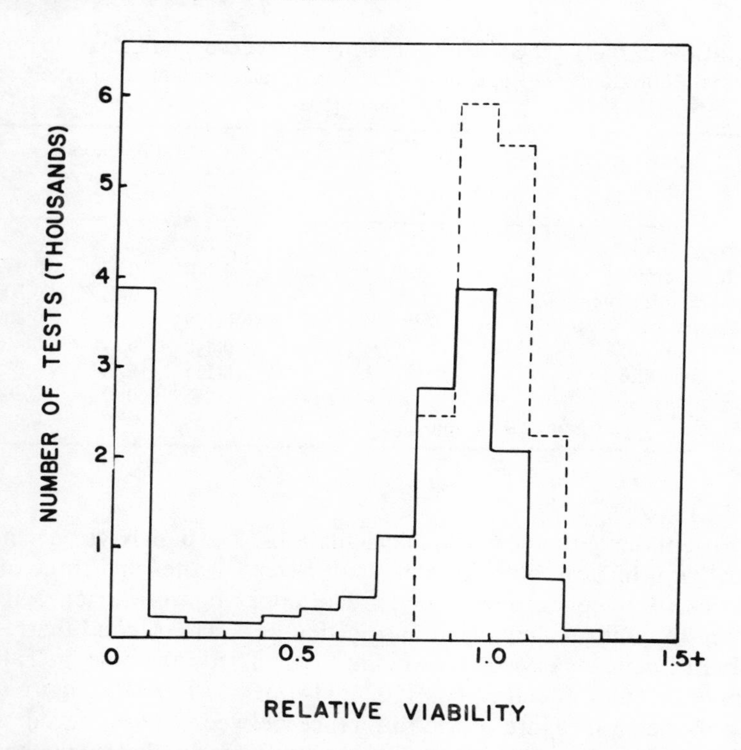

FIG. 8. The distributions of the relative viabilities of flies homozygous for various chromosomes from populations 1, 3, and 7 combined (solid line) and of flies heterozygous for random combinations of these same chromosomes (dashed line). The data on which this diagram is based are the marginal totals shown in Fig. 4.

TABLE II. The Proportions of Sub- and Supervitals in Populations 1, 3, and 7 Calculated According to Wallace and Madden (1953)

	Heterozygotes		Normal homozygotes		Proportions	
Population	$\bar{x}$	σ_{genet}	$\bar{x}$	σ_{genet}	Subvitals	Supervitals
1	0.995	0.050	0.890	0.123	0.52	0.05
3	0.995	0.050	0.913	0.122	0.44	0.07
7	0.995	0.045	0.923	0.119	0.44	0.09

TABLE III. The Total Genetic Load, Detrimental Load, and Lethal Load in Lethal Equivalents for Populations 1, 3, and 7 Calculated According to Crow and Temin (1964)

	Population		
	1	3	7
Average Viability			
Heterozygotes	0.3473	0.3365	0.3329
All homozygotes	0.2474	0.2198	0.2084
Normal homozygotes	0.3096	0.3071	0.3089
Total load	0.340	0.426	0.468
Detrimental load	0.115	0.091	0.075
Lethal load	0.225	0.335	0.393
Lethal equivalents/subvital chromosome	0.22	0.21	0.17

VII present the outcomes of such calculations. The data have been lumped somewhat arbitrarily into six classes of heterozygotes and three of homozygotes, thus generating 18 cells. The upper figure in each cell is the number of cultures actually found in that cell. The second figure is the number of cultures expected to fall in that cell (for example, in Table IV, 168 in the upper left cell equals 756 × 1141/5146). The third figure in each cell provides a measure of the difference between the observed and expected numbers: it equals the difference squared, divided by the expected number (or the contribution this cell would make to a χ^2 with ten degrees of freedom).

Table IV suggests for population 1 that considerably more lethal and near lethal chromosomes were carried by heterozygotes with high viability (1.11+) than were expected on the basis of marginal totals. Conversely, there are fewer chromosomes whose homozygous carriers exhibited high viability (1.00+) found among heterozygotes with relatively low viabilities (0.87–0.92, 0.93–0.98) than expected.

The pattern of excesses and deficiencies in the case of population 3 (Table V) agrees with what would be expected by most persons: deficiencies cluster about the upper left and lower right cells, while excesses tend to fall in cells lying on the lower left, upper right diagonal. In brief, chromosomes that produce homozygotes of high viability tend to be found in heterozygotes of high viability as well; conversely, chromosomes whose homozygotes have poor viabilities are found in heterozygotes of poor viability.

Table VI shows that the pattern of excesses and deficiencies in the case of population 7 resemble that of population 3; the largest excesses,

TABLE IV. Summary Analysis of the Distribution of Chromosomes with Given Effects on Their Homozygous Carriers among Heterozygotes of Various Viabilities for Population 1[a]

	Homozygous viability			
Heterozygous viability	0–49	50–99	100+	Total cultures
111+	190	442	124	756
	168	467	121	
	+2.9	−1.3	+0.1	
105–110	177	556	157	890
	197	550	142	
	−2.0	+0.1	+1.6	
99–104	255	740	221	1216
	270	752	194	
	−0.8	−0.2	+3.8	
93–98	281	720	173	1174
	260	726	188	
	+1.7	−0	−1.2	
87–92	160	479	97	736
	163	455	118	
	−0.1	+1.3	−3.7	
86–	78	245	51	374
	83	231	60	
	−0.3	+0.8	−1.4	
Total cultures	1141	3182	823	5146

[a] Within each cell of the table are listed (top), observed number, (middle) expected number, and (bottom) (difference)^2/(expected number). The sign of the difference (observed − expected) is indicated by a plus or minus sign.

at any rate, occur in the upper right and the lower left cells; deficiencies tend to fall in the opposite corners.

When the data from all three populations are combined, the result resembles the patterns described for populations 3 and 7: excesses tend to fall in cells lying at the lower left and the upper right and deficiencies in the cells at the upper left and lower right. Although marginal totals can be used in calculating expected numbers of cultures within each of the 18 cells of Tables IV–VII, any judgment concerning an excess or deficiency within an individual cell is somewhat risky because each row and column must balance precisely with no excess or deficiency overall. This matter will recur shortly.

TABLE V. Summary Analysis of the Distribution of Chromosomes with Given Effects on Their Homozygous Carriers among Heterozygotes of Various Viabilities for Population 3[a]

Heterozygous viability	Homozygous viability 0–49	50–99	100+	Total cultures
111+	205	356	167	728
	216	375	137	
	−0.6	−1.0	+5.6	
105–110	273	522	191	986
	292	509	185	
	−1.2	+0.3	+0.2	
99–104	406	716	246	1368
	406	706	257	
	0	+0.1	−0.5	
93–98	394	672	228	1294
	384	667	243	
	+0.3	+0	−0.9	
87–92	227	381	132	740
	219	382	139	
	+0.3	−0	−0.4	
86–	130	197	71	398
	118	205	75	
	+1.2	−0.3	−0.2	
Total cultures	1635	2844	1035	5514

[a] Within each cell of the table are listed (top) observed number, (middle) expected number, and (bottom) $(\text{difference})^2/(\text{expected number})$. The sign of the difference (observed − expected) is indicated by a plus or minus sign.

Correlations between Homozygous and Heterozygous Viabilities

Table VIII and Fig. 9 present data bearing on the main point of the present reanalysis. Here we have calculated the average viabilities of flies homozygous for chromosomes carried by heterozygotes of relative viabilities ranging from 0.84 or less to 1.15 or more. Population 1 presents a pattern that is clearly different from the other two. Because of the low average viability of homozygotes corresponding to high-viability heterozygotes and the converse, the high viability of homozygotes corresponding to low-viability heterozygotes, the regression of homozygous viabilities on those of heterozygotes is nearly zero (0.02). If the four aberrant

TABLE VI. Summary Analysis of the Distribution of Chromosomes with Given Effects on Their Homozygous Carriers among Heterozygotes of Various Viabilities for Population 7[a]

Heterozygous viability	Homozygous viability 0–49	50–99	100+	Total cultures
111+	243	366	175	784
	265	367	152	
	−1.8	−0	+3.5	
105–110	313	483	196	992
	336	464	193	
	−1.6	+0.8	−0	
99 − 104	481	640	265	1386
	469	648	269	
	+0.3	−0.1	−0.1	
93–98	419	566	227	1212
	410	567	235	
	+0.2	−0	−0.3	
87–92	227	332	135	694
	235	324	135	
	−0.3	+0.2	0	
86–	190	201	77	468
	158	219	91	
	+6.5	−1.5	−2.2	
Total cultures	1873	2588	1075	5536

[a] Within each cell of the table are listed (top) observed number, (middle) expected number, and (bottom) $(\text{difference})^2/(\text{expected number})$. The sign of the difference (observed − expected) is indicated by a plus or minus sign.

points are excluded from the data, the regressions for all three populations become remarkably similar: 0.20, 0.21, and 0.17.

In the above discussion of the analysis of Dobzhansky *et al.*'s data (Wallace, 1963*a*) under "An exceptional study," the regression of viabilities of homozygotes on those of heterozygotes was said to be 0.68. In the present case, the three regressions are nearer 0.20. This difference has an explanation which, although not obvious, is easily understood. The data provided by Dobzhansky and his colleagues consisted of replicate cultures; the present data consist of a single culture for each tested chromosome or chromosomal combination. The array of heterozygous viabilities in the earlier data was established *not* on the basis of the propor-

TABLE VII. Summary Analysis of the Distribution of Chromosomes with Given Effects on Their Homozygous Carriers among Heterozygotes of Various Viabilities for Populations 1, 3, and 7 Combined[a]

Heterozygous viability	Homozygous viability 0–49	50–99	100+	Total cultures
111+	638 651 −0.3	1164 1206 −1.5	466 411 +7.4	2,268
105–110	763 823 −4.4	1561 1525 +0.8	544 519 +1.2	2,868
99–104	1142 1140 +0	2096 2111 −0.1	732 719 +0.2	3,970
93–98	1094 1056 +1.4	1958 1957 0	628 666 −2.2	3,680
87–92	614 623 −0.1	1192 1154 +1.3	364 393 −2.1	2,170
86–	398 356 +5.0	643 660 −0.4	199 225 −3.0	1,240
Total cultures	4649	8614	2933	16,196

[a] Within each cell of the table are listed (top) observed number, (middle) expected number, and (bottom) $(\text{difference})^2/(\text{expected number})$. The sign of the difference (observed − expected) is indicated by a plus or minus sign.

tions of wild-type flies observed in particular cultures, but on the proportions of wild-type flies seen in the replicates of those cultures. The result, because replicate cultures tend to regress toward the mean, is a shorter range of heterozygous viabilities and consequently a greater slope to the regression of homozygotes on heterozygotes.

The data for population 1 presented in Table VIII and Fig. 9 mimic those reported by Wallace (1963*a*): the steady regression of the viabilities of homozygotes on those of heterozygotes disintegrates as one approaches heterozygotes exhibiting the highest viabilities. Population 3 does not show a similar pattern, at least not clearly; population 7 may or may not

TABLE VIII. The Average Viability of All Individuals Homozygous for Chromosomes Carried by Heterozygotes of Various Viabilities[a]

Viability of heterozygote	Population 1		Population 3		Population 7		Total (1 + 3 + 7)	
	Average	Number	Average	Number	Average	Number	Average	Number
0.84(–)	0.731[b]	252	0.613	258	0.556	324	0.627	834
0.85–0.86	0.684	122	0.630	140	0.595	144	0.634	406
0.87–0.88	0.694	200	0.582	180	0.631	176	0.638	556
0.89–0.90	0.705	258	0.669	272	0.652	208	0.677	738
0.91–0.92	0.728	278	0.647	288	0.641	310	0.671	876
0.93–0.94	0.703	354	0.618	370	0.651	390	0.657	1114
0.95–0.96	0.702	368	0.662	472	0.583	404	0.648	1244
0.97–0.98	0.678	452	0.656	452	0.644	418	0.660	1322
0.99–1.00	0.713	440	0.666	450	0.600	488	0.658	1378
1.01–1.02	0.742	388	0.655	486	0.629	440	0.672	1314
1.03–1.04	0.722	388	0.646	432	0.628	458	0.663	1278
1.05–1.06	0.724	338	0.661	384	0.653	436	0.676	1158
1.07–1.08	0.741	336	0.682	332	0.703	318	0.709	986
1.09–1.10	0.742	216	0.654	270	0.595	238	0.661[b]	724
1.11–1.12	0.677[b]	228	0.652	208	0.654	226	0.661[b]	662
1.13–1.14	0.689[b]	166	0.724	156	0.637	178	0.681[b]	500
1.15(+)	0.706[b]	362	0.670	364	0.665	380	0.680[b]	1106
Slope	0.02	(0.20)*	0.21*		0.17**		0.13	(0.21)
Correlation coefficient	0.08	(0.72)	0.68		0.47		0.66	(0.76)

[a] *, $p < 0.01$; **, $p \approx 0.05$.
[b] Data excluded in calculating the slope given in parentheses.

show a decline in the viabilities of homozygotes that correspond to high-viability heterozygotes.

In an attempt to generalize, data from all three populations have been combined in Table VIII; the corresponding diagram is shown in Fig. 10. Here we see that, if the four rightmost points are omitted from the calculations of the regression, they fall on or below the lower limit of the confidence interval for the regression (0.21) that is based on the remaining 13 points. When the original analysis (Wallace, 1963*a*) is considered together with the present one, it seems clear that, as a rule, the chromosomes carried by heterozygotes exhibiting high viabilities are not the chromosomes that also produce homozygotes with exceptionally high viabilities.

Table IX attempts to make the preceding point clear in the case of population 1. Here the distributions of homozygous viabilities that cor-

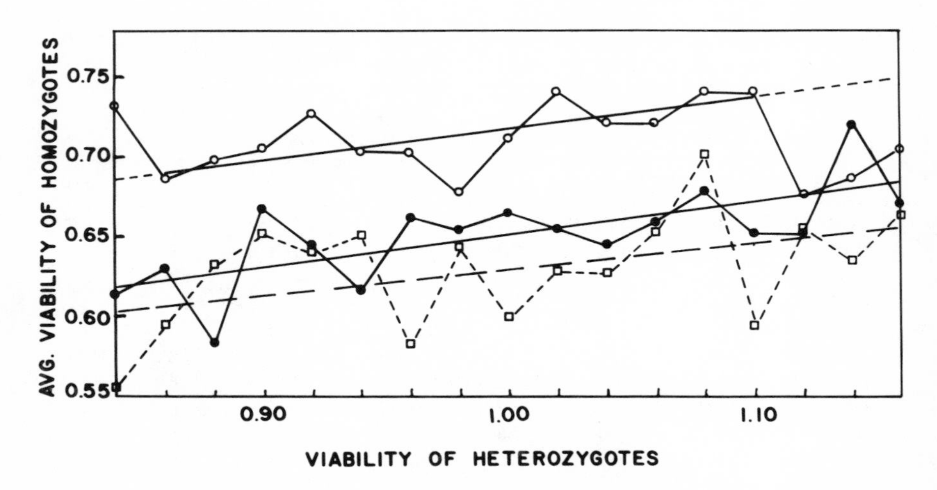

FIG. 9. Curves illustrating the relationship between the viabilities of flies heterozygous for pairs of chromosomes from populations 1, 3, and 7 and those of flies homozygous for these same chromosomes. (○) Population 1; (●) population 3; (□) population 7. The straight lines (dashed in the case of population 7) are the calculated regressions; in the case of population 1, the first and last three points were not included in the regression calculation.

respond to heterozygotes exhibiting viabilities 1.07–1.10 (the last two classes of heterozygotes that fit the linear regression) and to those exhibiting viabilities of 1.11 and above have been tabulated for comparison. The excess number of lethal and semilethal chromosomes among the homozygotes corresponding to the heterozygotes exhibiting the higher viabilities seems obvious by inspection alone; despite the large variances of these biomodal distributions, the *t*-test confirms what appears obvious to the eye. The disintegration of the regression of homozygous viabilities on those of heterozytoes mentioned above seems to be genuine.

Normalization of the Distributions Shown in Figures 1–4

Earlier, analyses were described in which marginal totals were used to calculate expected numbers of cultures per cell (18 in number; three per row and six per column). Alternative methods exist for examining the same data. Among these are techniques by which the observed data can be normalized; the normalization can be either "horizontal" or "vertical." By horizontal normalization is meant converting each of the 15

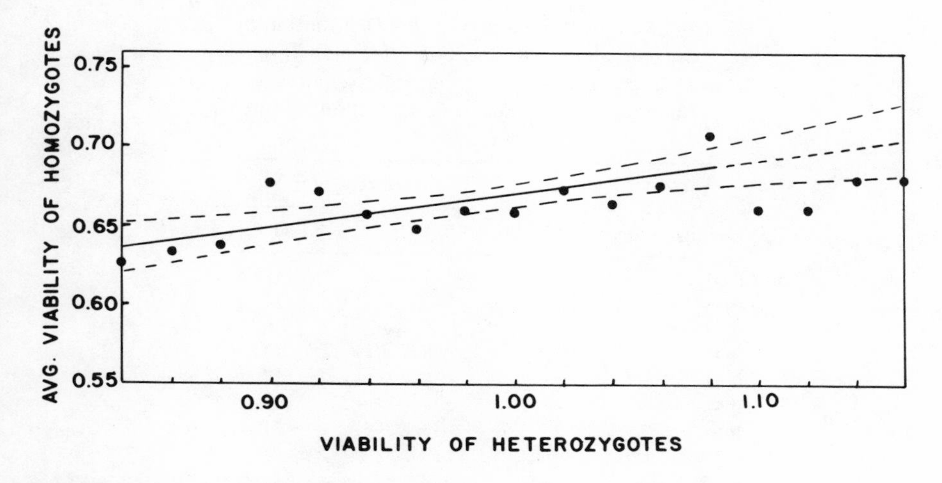

FIG. 10. The regression of the average viability of flies homozygous for different second chromosomes on the viability of individuals carrying these same chromosomes in heterozygous combinations. Data included in this figure are for populations 1, 3, and 7 combined. Note that the last four points on the right were not included in the calculation of the regression (additional information in text).

homozygous categories into a percentage based on the number of cultures counted within that increment of heterozygous viabilities. Thus, using Fig. 1 as an example, one can normalize the numbers 65, 5, 3, 5, 12, and others falling within the 1.15+ increment of heterozygous viabilities by dividing by 362. In the next lower increment, 28, 4, 2, 5, 2, 4, 10, and others are normalized by dividing by 166.

Normalization can, of course, be carried out vertically as well as horizontally. Referring again to Fig. 1, one sees that the numbers of lethal and near-lethal (0–0.09) homozygotes, which are given in successive increments of heterozygotes as 65, 28, 48, 31, and so on, can be converted to proportions by dividing by 882. In like manner, the entries in successive columns can be normalized by dividing by 59, 50, 60, and so on.

It is apparent by now that normalization of the data in Figs. 1–4 can be accomplished either horizontally or vertically; the outcomes of these standardizations are illustrated graphically in the eight three-dimensional diagrams in Figs. 11–18. A visual interpretation of these diagrams is performed as follows: If the standardization has been carried out horizontally (1H, 3H, 7H, and 11H), the heights of bars in the vertical columns can be compared; if the standardization has been carried out vertically (1V, 3V, 7V, and 11V), the heights of bars in horizontal rows can be compared.

TABLE IX. A Comparison of the Distribution of Chromosomes with Various Effects on Their Homozygous Carriers among Heterozygotes with Viabilities 1.07–1.10 and Greater than 1.11 in Population 1

	Heterozygotes	
Homozygotes	1.07–1.10	1.11+
0.05	82	141
0.15	6	14
0.25	6	6
0.35	6	12
0.45	8	17
0.55	18	15
0.65	13	31
0.75	39	68
0.85	124	141
0.95	153	187
1.05	69	83
1.15	24	33
1.25	4	7
1.35	0	1
Total	522	756
$\bar{x}$	0.7529	0.7071
$\sigma^2_{\bar{x}}$	0.0002130	0.0001793
Difference	0.0458	
σ difference	0.0198	
	$t = 2.31$; $p \approx 0.02$	

A minor exception to this account occurs in the rightmost column of Fig. 12 (1V); the two bars in this column should represent 50% each; the computer has made them no taller than the four bars in the adjoining column, which represent heights of 25% each. The interpretation of these eight diagrams need not depend, however, on the graphic representation of six culture bottles.

An examination of Fig. 11 reveals that the heights of bars in the upper left corner exceed those of their neighbors; the greater heights of these bars reflect the higher than expected frequencies of lethal and semilethal chromosomes in these high-viability heterozygotes. The same is shown less clearly in Fig. 12, where comparisons are to be made horizontally along rows. Ignoring the six chromosomes mentioned above that lie in the two rightmost columns, one sees an indication that heterozygotes of

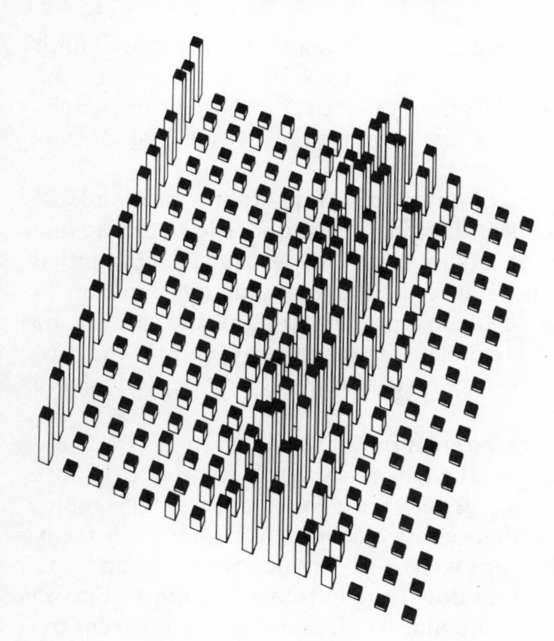

FIG. 11. A three-dimensional diagram representing the data presented in Fig. 1 (population 1), but normalized so that the heights of the bars reflect proportions rather than numbers. In this figure as in Figs. 13, 15, and 17 the sum of proportions in horizontal rows (*X* axis) equals 1.00; thus, visual comparisons are to be made along columns (*Y* axis). Further explanations in text. Low viabilities for both heterozygous and homozygous tests are in the lower left-hand corner of the diagram.

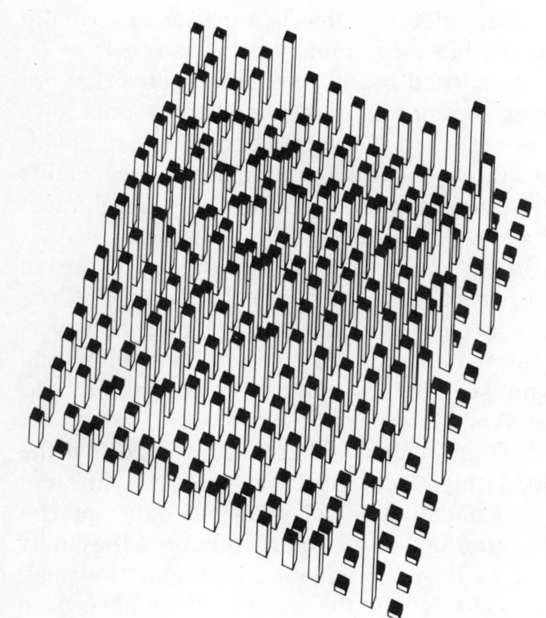

FIG. 12. A three-dimensional diagram representing the data presented in Fig. 1 (population 1), but normalized so that the heights of the bars reflect proportions rather than numbers. In this figure as in Figs. 14, 16, and 18 the sum of proportions in columns (*Y* axis) equals 1.00; thus, visual comparisons are to be made across rows (*X* axis). Further explanation in text. Low viabilities for both heterozygous and homozygous tests are in the lower left-hand corner of the diagram.

high viability can involve chromosomes whose homozygotes also exhibit high viability. Otherwise, however, one is struck by the heights of the bars in the upper left corner of the diagram; these represent chromosomes whose homozygotes exhibit poor viability but whose heterozygous carriers exhibit high viability.

Figure 13 illustrates the situation in population 3 following standardization by horizontal rows. At best, looking at the upper left corner, one can only claim that there is no obvious lack of lethal and near-lethal chromosomes. The same can be said of the diagram illustrated in Fig. 14 (3V): Here comparisons must be made from left to right and, again, the bars in the upper left corner are not all exceptionally short. The two rightmost columns involve very few tests and need not be considered seriously.

The standardized results for population 7 are shown in Figs. 15 and 16. This population, it will be recalled, was exposed continuously to a low level of radiation (700 r per generation); in a sense, it resembles population 3 except that (1) mutations occurred at a higher than spontaneous rate and (2) these mutations were γ-ray-induced rather than "natural." Nevertheless, upon sighting down the leftmost column in Fig. 15 where horizontal rows have been standardized, one sees an apparent excess of lethals among the chromosomes carried by high-viability heterozygotes. An excess number can also be seen among low-viability heterozygotes, but, under rank-order selection, that is a matter of virtually no importance. The question we have continually emphasized here is: What sorts of chromosomes are carried by those heterozygotes that are likely to represent the surviving and reproducing segment of a population?

Figure 16 illustrates the result when the vertical columns are standardized for population 7. Here the legitimate comparisons of the heights of bars is to be made across horizontal rows. Ignoring the two rightmost columns, where numbers of cultures are always low, we again see an apparent excess of tall bars in the upper right corner. We also see an excess in the lower left corner, an excess that corresponds to the conventional expectation. There is no clear evidence in this figure, however, that there is a deficiency of tall bars in the upper left corner; lethal and semilethal chromosomes are not strikingly absent from high-viability heterozygotes in population 7.

There is reason to believe from evidence already presented that the three populations we are considering represent three somewhat different situations. Nevertheless, a consolidation of all available data appears worthwhile; the total data are enormous and when combined they may suggest generalities overlooked during the population-by-population analyses. In Fig. 17, the combined data for the three populations have been

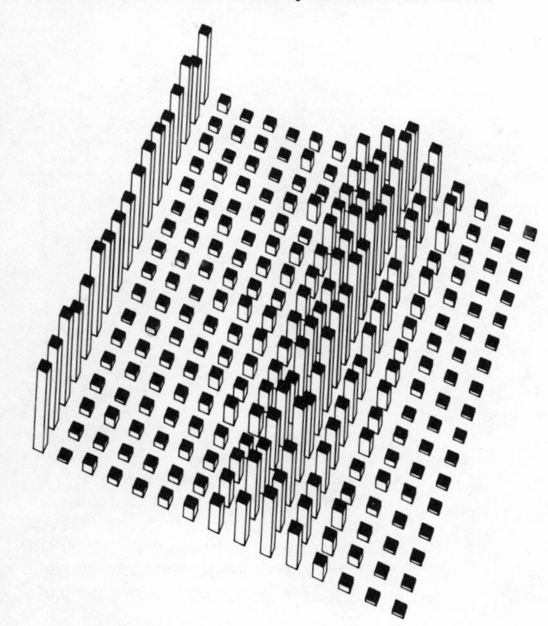

FIG. 13. A three-dimensional diagram representing the data presented in Fig. 2 (population 3), but normalized so that the heights of the bars reflect proportions rather than numbers. See Fig. 11 and text for further explanation.

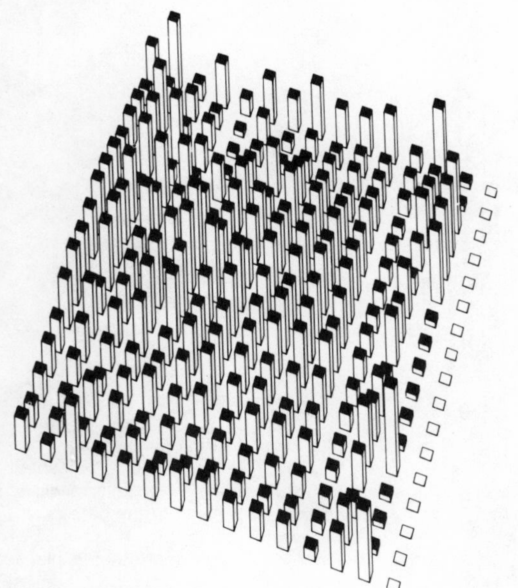

FIG. 14. A three-dimensional diagram representing the data presented in Fig. 2 (population 3), but normalized so that the heights of the bars reflect proportions rather than numbers. See Fig. 12 and text for further explanation.

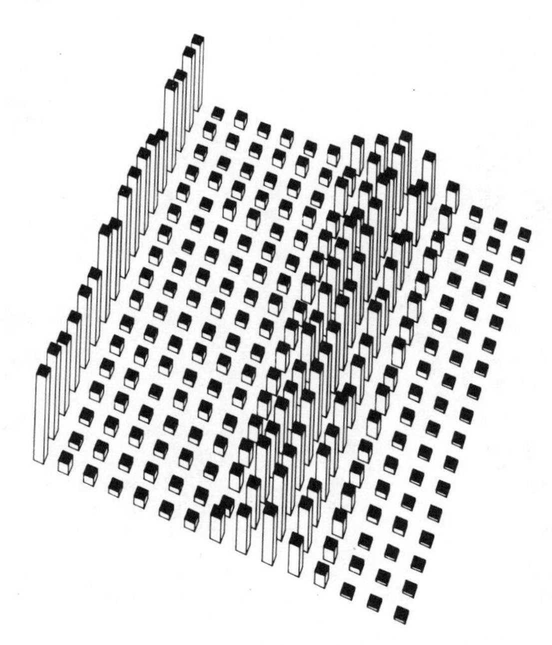

FIG. 15. A three-dimensional diagram representing the data presented in Fig. 3 (population 7), but normalized so that the heights of the bars reflect proportions rather than numbers. See Fig. 11 and text for further explanation.

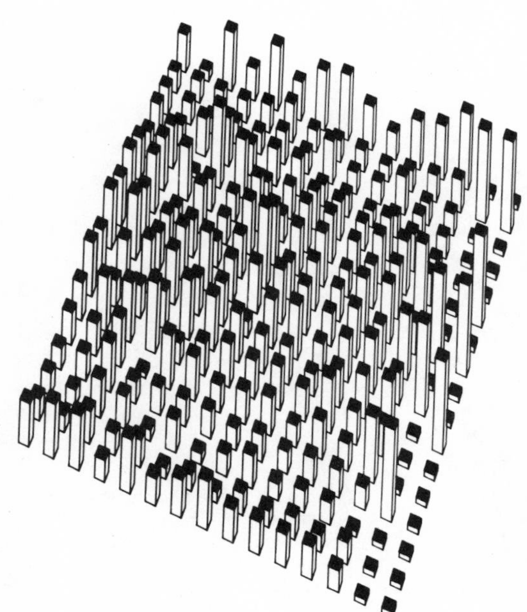

FIG. 16. A three-dimensional diagram representing the data presented in Fig. 3 (population 7), but normalized so that the heights of the bars reflect proportions rather than numbers. See Fig. 12 and text for further explanation.

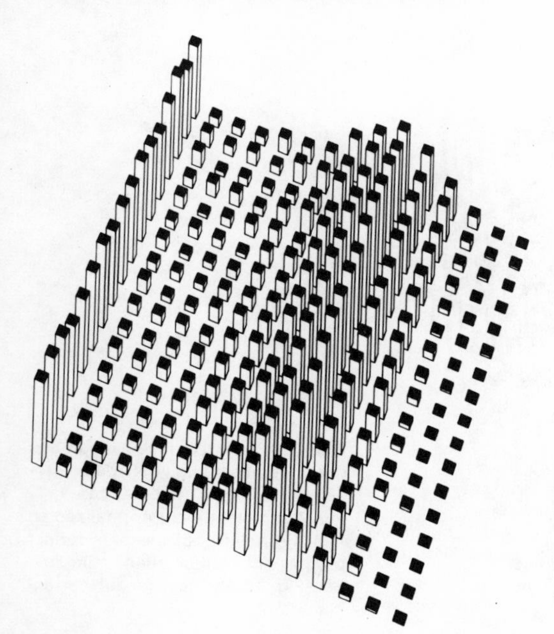

FIG. 17. A three-dimensional diagram representing the data presented in Fig. 4 (populations 1, 3, and 7 combined), but normalized so that the heights of the bars reflect proportions rather than numbers. See Fig. 11 and text for further explanation.

standardized horizontally, row by row; consequently, comparisons can be made vertically along each column. The striking feature is the series of tall bars representing lethal chromosomes among high-viability heterozygotes (upper left corner). Upon turning to Fig. 18, no feature stands out as striking, but there is no compelling evidence in this diagram that the heights of the bars systematically decrease as one approaches the left-hand ends of the upper rows.

Table X represents an attempt to reduce the normalized data presented in Figs. 11–18 to easily comprehended numbers. It will have been noted that the standardization of rows or columns as illustrated in Figs. 11–18 removed one of the distribution patterns (unimodal or bimodal) but left the other one. Specifically, the standardization of horizontal rows in Figs. 11, 13, 15, and 17 removed the bell-shaped distribution of heterozygous viabilities but left the bimodal distribution of homozygous viabilities: many lethals, few semilethals, and a bell-shaped distribution of quasinormal homozygotes. Standardization of the vertical columns (Figs. 12 and 14–16) smoothed out the bimodal distribution of homozygous viabilities, but left the unimodal distribution of heterozygous viabilities (disguised visually, unfortunately, by the sporadic tall bars in the two right-most columns).

The numbers summarized in Table X were compiled in an effort to

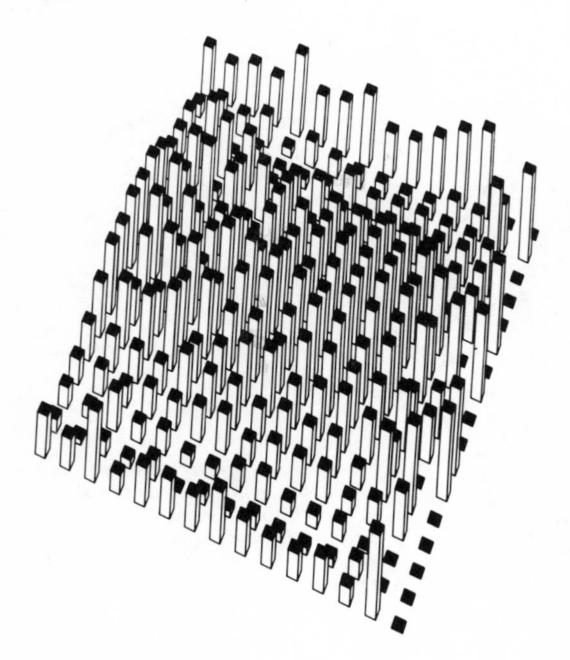

FIG. 18. A three-dimensional diagram representing the data presented in Fig. 4 (populations 1, 3, and 7 combined), but normalized so that the heights of the bars reflect proportions rather than numbers. See Fig. 12 and text for further explanation.

remove both distribution patterns, both the bell-shaped distribution of heterozygotes *and* the bimodal distribution of homozygotes. The procedure consisted in calculating the mean proportion of cultures in each cell of a standardized row or column (column when rows had been standardized; row when columns had been standardized), followed by calculating a ratio of proportion observed in each cell to the average proportion. Because the numbers of cultures in many cells of low (but not lethal) homozygous viability are not large, running averages of three adjacent cells were used; this procedure gave excessive weight to lethal and near-lethal chromosomes, but, because of their large numbers, they probably deserve this extra weight. Finally, both horizontal and vertical standardizations were combined because no compelling reason exists for using one in preference to the other.

The results of these rather complex calculations are summarized in Table X. Once more, the cells have been grouped for analyses: homozygous viabilities have been grouped into three categories, heterozygous viabilities into six. The conclusion arrived at by an examination of the four sets of data in Table X is that lethals do not occur in consistently lower than expected numbers among heterozygotes with high viability; only in population 7, which was continuously exposed to radiation, is

TABLE X. The Relative Frequency with Which Chromosomes with Various Effects on Their Homozygous Carriers Are Found in Heterozygous Combinations of Differing Viabilities[a]

Heterozygotes, population 1	Homozygotes 0–0.39	Homozygotes 0.40–0.79	Homozygotes 0.80–1.19	Heterozygotes, population 3	Homozygotes 0–0.39	Homozygotes 0.40–0.79	Homozygotes 0.80–1.19
1.11+	1.20	0.99	0.97	1.11+	1.06	0.80	1.13
1.05–1.10	0.93	0.94	1.06	1.05–1.10	1.00	0.93	1.03
0.99–1.04	0.95	0.93	1.08	0.99–1.04	0.95	1.14	0.97
0.93–0.98	1.07	1.04	0.94	0.93–0.98	0.99	1.05	0.99
0.87–0.92	0.95	1.09	0.94	0.87–0.92	0.97	1.02	0.95
0.86–	0.88	1.01	1.01	0.86–	1.10	1.01	0.93

Heterozygotes, population 7	Homozygotes 0–0.39	Homozygotes 0.40–0.79	Homozygotes 0.80–1.19	Heterozygotes, 1 + 3 + 7	Homozygotes 0–0.39	Homozygotes 0.40–0.79	Homozygotes 0.80–1.19
1.11+	0.86	1.20	1.01	1.11+	1.01	1.00	1.02
1.05–1.10	1.00	0.99	1.03	1.05–1.10	0.98	0.96	1.04
0.99–1.04	0.94	0.96	1.02	0.99–1.04	0.95	0.99	1.02
0.93–0.98	1.03	0.94	0.99	0.93–0.98	1.02	1.02	0.98
0.87–0.92	1.06	0.95	1.01	0.87–0.92	1.01	1.04	0.98
0.86–	1.19	0.94	0.90	0.86–	1.08	0.98	0.94

[a] The frequencies shown in these tables involve running averages and a "normalization" procedure that makes the expected frequency in each cell equal to 1.00 (explanation in text).

there an indication that high-viability heterozygotes carry fewer than expected lethal and near-lethal chromosomes. In the case of population 7, low-viability heterozygotes appear to carry more lethal and near-lethal chromosomes than might be expected; these might be regarded as truly deleterious chromosomes.

Frequency of Lethals following Rank-Order Truncation

Figure 19 presents still another view of the expected role of lethal chromosomes in *Drosophila* populations. The analysis underlying Fig. 19 is one introduced by Milkman (1978). Suppose that heterozygous individuals that are free of lethals have distributions of viabilities such as those represented by the dashed curves in Figs. 5–7 (the three populations are considered individually here); suppose, too, that individuals heterozygous for lethal chromosomes have a similar distribution, but one in

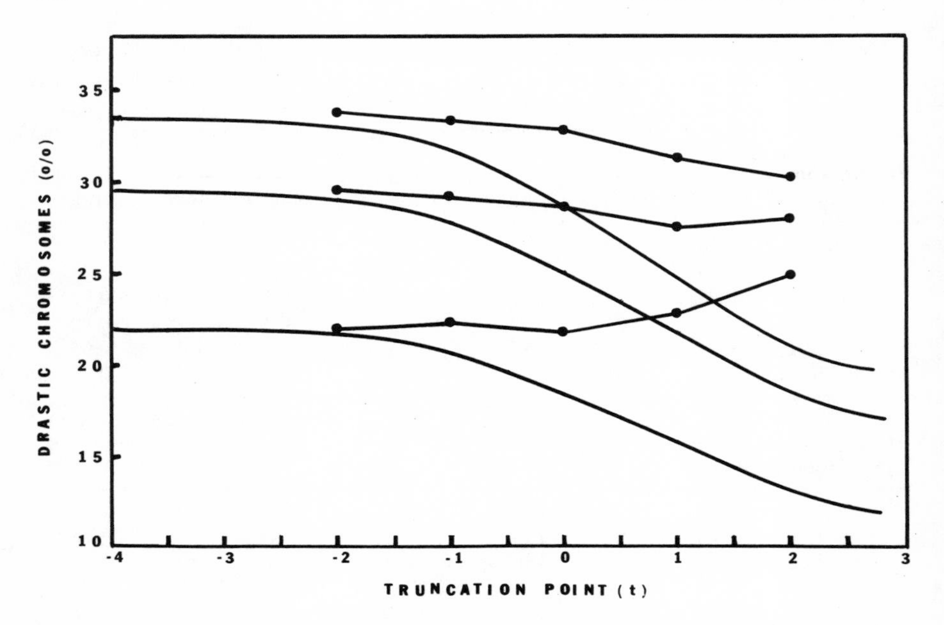

FIG. 19. The observed (dots) and expected (smooth curves) frequencies of lethal and semilethal (drastic) chromosomes among successive truncated segments of the distributions of heterozygous viabilities observed in population 7 (top), 3 (center), and 1 (bottom). The expected frequencies are based on a 2% reduction in the mean viabilities of carriers of drastic chromosomes (h = 0.02).

which the mean viability is lowered (in theory) because of the partially dominant deleterious effect of the lethal mutations. Now, attempts to determine experimentally the partial dominance h of supposedly recessive lethals have provided considerable employment to *Drosophila* geneticists since such studies were first undertaken by Stern *et al.* (1952); an extensive review of this and later studies has been provided by Wallace (1968*a*, p. 172ff). Although various studies have led to somewhat different estimates of the partial dominance of recessive lethals, a modest estimate would be h = 0.02. This value has been used in the following analysis.

Using Milkman's procedure, and using the frequency of lethal chromosomes observed in each of the three populations, one can truncate the two overlapping distributions of viabilities (those of lethal-free and lethal-carrying heterozygotes) at various points starting at the far left; with each truncation, the expected proportion of lethal chromosomes among all remaining ones can be estimated by means of standard statistical parameters. These proportions for the three populations are shown by the smooth curves in Fig. 19. It should be noted [as Haldane (1932) pointed out] that truncation at the far right of the distribution curves (that is, intense selection overall) does not result in increasingly severe selection against the lethal chromosomes: the smooth curves become more nearly horizontal toward the right.

The segmented lines connecting the dots in Fig. 19 show the actual proportions of drastic (lethal and semilethal) chromosomes among those remaining following truncation of the heterozygous distribution curves at various points. The observed line does not follow the expected curve for any one of the three populations. Only population 7 looks as if it could be fitted to a hypothetical curve resembling the one shown in the figure; the value of h needed to generate the smooth curve would be considerably less than 2%, a value already admitted to be modestly low.

Viability Effects of "Common" versus "Sporadic" Lethals

The concluding part of this discussion of experimental results will be devoted to subsidiary, but nevertheless important, issues. Throughout, the frequency of wild-type flies in the F_3 test cultures has been equated with egg-to-adult viability under somewhat crowded conditions; furthermore, viability has been equated with fitness because viability—the ability to survive—is surely an important component of fitness. If the discussion were limited to tests of flies homozygous for this or that chromosome, there would be no hesitation in equating frequency with viability: lethal chromosomes, by definition, kill their homozygous carriers. Furthermore, as Timofeeff-Ressovsky (1935) and Dobzhansky and Queal (1938) demonstrated many years ago, the slight deviations from expected frequencies commonly observed in *ClB*-like tests are correlated from generation to generation.

Events involving lethal chromosomes in population 7 over a period of 2 or more years can be used to demonstrate that the fitness of lethal-bearing heterozygotes within that population and the experimental assessment of viability (i.e., frequency of wild-type flies in heterozygous test cultures) are related.

Among the heterozygous combinations studied for any population, the wild-type flies $(+_i/+_j)$ occasionally prove to be lethal. If the matings have been arranged in round-robin fashion, one finds that lethal homozygotes precede and follow the culture lacking heterozygous wild-type flies. A lethal heterozygous combination involves two chromosomes seemingly carrying the same lethal gene.

Figure 20 shows the cumulative number of lethal heterozygous combinations observed through time in population 7. Because the number of

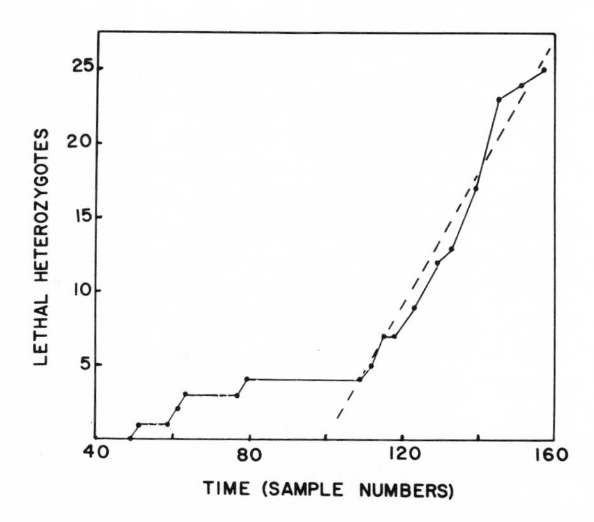

FIG. 20. The cumulative number of lethal heterozygotes arising from the chance allelism of lethal chromosomes observed in population 7. The sample interval in these studies was 2 weeks; consequently, these data span 240 weeks or approximately 5 years. The dashed line indicates the greater rate at which allelic combinations of lethal chromosomes occurred following (approximately) sample 100.

tests per sample throughout the period covered in the diagram was roughly constant, a change in the rate at which the numbers of lethal heterozygotes increased also represents a change in the frequency with which they occurred among all tests.

Figure 20 shows clearly that, beginning at about sample 100 and lasting throughout the remaining 60 samples of the figure, lethal heterozygotes occurred at a rate higher than they had in the preceding 30 or more samples. The higher rate at which lethal heterozygotes occurred suggests that one (or at best a few) lethal chromosomes increased in frequency within population 7. Furthermore, the persistence of the higher rate of allelism (that is, of the lethality seen among heterozygous combinations) suggests that the lethal chromosome(s) involved increased in frequency because it or they were favored by selection; chance events are not often consistent events.

If the lethal(s) involved in the increased rate of allelism were favored by selection, and if this advantage resided in the relative viability of its carriers, then a corresponding effect should appear in the F_3 test cultures. Suppose (to use an example) that among the round-robin tests, the wild-type flies designated 4/5 are lethal. It will be found that cultures 4/4 and 5/5 contain no wild-type homozygotes. Two other cultures can be identified in which chromosomes 4 and 5 are carried in heterozygous condition: 3/4 and 5/6. The frequency of wild-type flies in cultures 3/4 and 5/6 can be compared, then, with either the standardized frequency (1.00) or the frequency of wild-type flies found in, for example, cultures 10/11 and 11/12, where the homozygote 11/11 represents the next isolated homozygous lethal culture to occur within the round-robin crosses of the particular unit.

Forty-four heterozygous cultures in which the wild-type flies carried a "common" lethal (e.g., 3/4 and 5/6, where it is known that 4/4, 4/5, and 5/5 are lethal), and a similar number of heterozygous cultures in which the wild-type flies carried a "sporadic" lethal (e.g., 10/11 and 11/12, where 11 is the lethal chromosome) were available for study in population 7. The standardized frequency of wild-type flies that carried a common lethal proved to be 1.01 compared with the true standard, 1.00; the standardized frequency of wild-type flies heterozygous for a "sporadic" lethal was 0.98. These results suggest that the increased rate at which lethal heterozygotes occurred in population 7 after sample 100 was a consequence of the selective superiority of other, *non*lethal heterozygous carriers of these lethals. The 2% disadvantage of wild-type flies heterozygous for "sporadic" lethals, by coincidence, equals the disadvantage that was chosen for illustrative purposes in Fig. 19.

Still another comparison of the standardized viabilities of common and sporadic lethal heterozygotes can be obtained by ranking the 44 values of each in succession, from highest to lowest. A pairwise comparison of corresponding members of the two sequences reveals that the viabilities of carriers of common lethals exceed those of sporadic ones in 34 instances, equal sporadic ones in five instances, and are smaller in five instances. Like the average viabilities cited above, these comparisons suggest that as a rule carriers of common lethals had the greater viabilities. Still more is revealed by this analysis: the five instances in which common lethal heterozygotes were lower in viability than those carrying sporadic lethals were clustered together at the low end of the viability ranking. First, that the five negative signs should be clustered together anywhere in the array has a probability of only 0.002; thus, the observed pattern departs from an expectation based on random events. Second, the distribution reveals that common lethals enter into low-viability combinations with chromosomes other than lethal ones, thus suggesting commonality between these lethals and at least some other nonlethal chromosomes (thus casting doubt on the convenient term, lethal "gene"). Third, the distribution of plus and minus signs reveals that the variation among the viabilities of heterozygotes carrying common lethals exceeds that of heterozygotes carrying sporadic lethals. This fact demands that considerable caution be exercised in discussing the fate of lethals in populations. Haldane (1932, p. 177), somewhat to his own surprise, one suspects, showed that intense selection favors not the race with the higher average, but that with the higher spread. Intense selection would, in the case of population 7, favor common lethals because of the superior viabilities of some of their carriers.

Ruling Out "Segregation Distortion" As a Confounding Factor

The assumption that frequencies of wild-type flies in the F_3 cultures reflect actual viabilities (and hence selection patterns) in populations has been advanced in the preceding paragraphs. There remains a second matter, however, that should be disposed of. When cultures are encountered within which the frequency of wild-type flies (homozygous or heterozygous) is exceptionally high, segregation distortion, as well as viability, might be advanced as an explanation. To state the bleakest case, might not high frequencies of wild-type flies reflect a breakdown of Mendelian expectations rather than evidence of high egg-to-adult viability of wild-type flies?

The common name for a breakdown of Mendelian expectations is "segregation distortion." Of all instances known to us (e.g., Hiraizumi and Thomas, 1984; Temin and Marthas, 1984), only heterozygous male *D. melanogaster* exhibit gross departures from the 1:1 segregation ratios expected under Mendelian genetics. Because this is the case, the round-robin mating scheme can reveal the role segregation distortion might play in determining the frequencies of wild-type flies in the F_3 test cultures, frequencies that we have consistently assumed to reflect egg-to-adult viability.

In setting up the round-robin crosses when studying the Cold Spring Harbor Laboratory populations, the males of the *i*th culture were always mated with the females of the *j*th ($=i + 1$) culture in order to obtain $+_i/+_j$ wild-type heterozygotes (Wallace, 1952). Let us suppose that $CyL/+_i$ males produce only $+_i$ functional sperm. Instead of the expected two $CyL/+_i$:one $+_i/+_i$ ratios in the *i*th homozygous culture, one would expect one $CyL/+_i$:one $+_i/+_i$; these ratios would be determined solely by the 1:1 ratio of *CyL* and $+_i$ gametes produced by $CyL/+_i$ females.

When $CyL/+_i$ males were mated with $CyL/+_j$ females, because of the postulated extreme segregation distortion of $CyL/+_i$ males (only $+_i$ functional sperm produced), the F_3 cultures of this cross would also exhibit the 1:1 ratio: one $CyL/+_i$:one $+_i/+_j$. The next homozygous culture

TABLE XI. Differences in the Viabilities of Individuals Homozygous for Each of the Two Chromosomes Carried by Heterozygotes of Superior Viability (1.13+)[a]

	Population				Plus			
	Population				Population			
Difference	1	3	7	Total	1	3	7	Total
0–4	16	14	20	50	19	17	21	57
5–8	17	12	13	42	15	10	16	41
9–12	8	7	3	18	14	11	10	35
13–16	11	15	7	33	8	10	7	25
17–20	5	9	5	19	7	7	7	21
21–24	9	4	7	20	5	4	3	12
25–28	3	2	2	7	5	5	5	15
29–32	4	2	10	16	1	3	5	9
33–36	2	4	1	7	5	1	1	7
37–40	4	4	3	11	3	1	4	8
41–44	2	3	2	7	2	2	1	5
45–48	3	0	4	7	3	0	1	4
49–52	3	1	1	5	2	1	2	5
53–56	2	1	0	3	2	1	1	4
57–60	2	1	2	5	1	2	0	3
61–64	4	3	2	9	3	2	3	8
65–68	5	1	4	10	1	2	2	5
69–72	3	1	0	4	0	1	3	4
73–76	7	4	2	13	3	4	3	10
77–80	2	5	3	10	4	7	6	17
81–84	4	3	4	11	4	7	5	16
85–88	2	5	5	12	2	4	3	9
89–92	4	8	3	15	4	7	8	19
93–96	5	5	5	15	4	5	7	16
97–100	5	3	12	20	2	5	2	9
101–104	2	3	8	13	0	4	6	10
105–108	1	2	3	6	2	4	1	7
109–112	1	3	0	4	2	0	4	6
113–116	0	2	5	7	1	1	2	4
117–120	3	1	0	4	1	1	0	2
121–124	0	0	1	1	0	0	1	1
125–128	0	0	0	0	0	0	0	0
129–132	0	0	0	0	0	0	1	1
133–136	0	0	0	0	0	0	0	0
137–140	0	0	0	0	0	0	1	1

Population 1: $\bar{d} = -0.049$, $\sigma_{\bar{d}} = 0.032$, $p \approx 0.15$
Population 3: $\bar{d} = 0.009$, $\sigma_{\bar{d}} = 0.037$, $p > 0.50$
Population 7: $\bar{d} = -0.008$, $\sigma_{\bar{d}} = 0.037$, $p > 0.50$

[a] If "superior viability" were really a reflection of abnormal segregation, plus and minus differences should not occur with equal frequencies, because segregation distortion affects males, rather than females (further explanation in text).

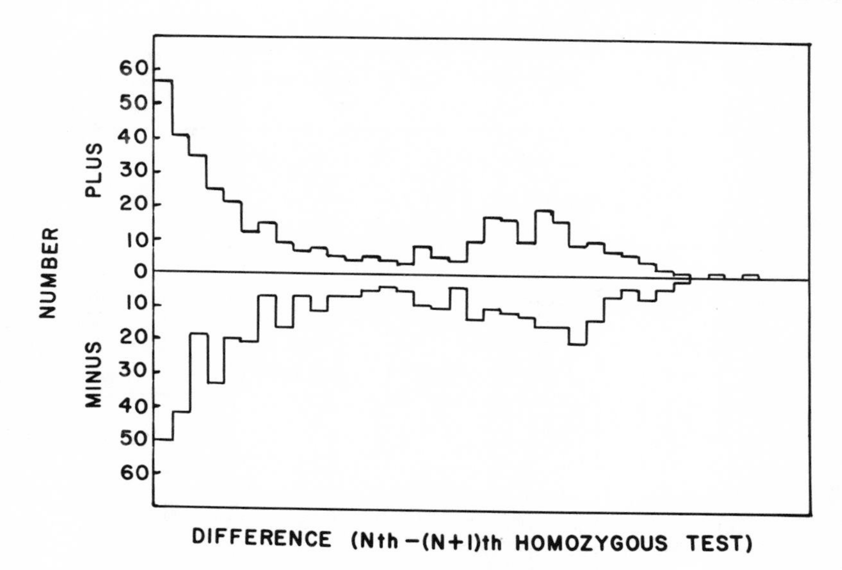

FIG. 21. The distribution of differences between the viabilities of the homozygous carriers of chromosomes that in random combination produced heterozygous flies with exceptionally high relative viabilities (1.13 or higher). If the measure of "viability" used in the tests described in this report were influenced by distorted segregation ratios, the frequencies of plus and minus differences should not form a symmetric pattern as they do in this figure. The divisions of the horizontal axis are the increments of 0.04 listed in Table XI.

in the series, $CyL/+_j \times CyL/+_j$, would exhibit the expected 2:1 ratio: two $CyL/+_j$:one $+_j/+_j$.

Because segretation distortion affects primarily (if not always) male flies, and because the round-robin crosses were set up consistently with males from the lower numbered homozygote ($CyL/+_i$) mated to females from the next higher numbered homozygote ($CyL/+_j$), a common occurrence of segregation distortion within the laboratory populations would lead, among seemingly high-viability heterozygotes, to dissimilar distributions of plus and minus "viability" differences if these were consistently calculated by subtracting the standardized viability of jth ($=i + 1$) homozygote from that of the ith homozygote. Table XI and Fig. 21 fail to reveal any evidence for dissimilar distributions. For every difference in viability (from 0–4% to 137–140%) between the ith and the $(i + 1)$th homozygous cultures, the number of positive differences and the number of negative differences are virtually identical. No evidence exists upon which one could argue that the supposed egg-to-adult viabilities reflect instead a distortion of the segregation ratios of $CyL/+$ male flies.

DISCUSSION

The topic discussed in this chapter recently has receded into near obscurity as molecular analyses of evolutionary problems have expanded. Not only has Muller's *ClB* technique receded from view, but so, unfortunately, has the memory of Muller himself, as Pontocorvo (1982) has ruefully reported.

In many respects, the questions of selective change, relative fitnesses, and fitness components such as viability and fertility resemble those concerned with longevity. Despite the technical sophistication of molecular investigations into gerontological problems, final decisions are, of necessity, generally based on the survivorship curves of aging individuals. Longevity ultimately depends upon, and is measured by observing, the rate at which individuals die. Similarly with fitness and viability: despite the beauty of current and future molecular studies, someone eventually must measure the means by which, and the rate at which, genetic novelties spread through populations. This measure will ultimately concern the relative fates of those individuals carrying and of others lacking the novelity under investigation. Only if and when molecular biologists have shown that an individual's soma "contaminates" its germ cells with adaptively preprogrammed bits of DNA will "gene" frequencies and changes in these frequencies through differential survival and reproduction cease being the focus of the evolutionist's attention.

Although the topic of the present chapter has been largely resurrected from the past, its treatment has been new. Indeed, the use of *ClB*-like techniques in population analyses has not entirely ceased: Mukai and Nagano (1983) analyzed a Florida population of *D. melanogaster*; their paper contains references to numerous previous publications by Mukai and others.

The paper by Mukai and Nagano (1983) can be used as a convenient foil for contrasting the traditional interpretation of data obtained by *ClB*-like techniques and the interpretation attempted in the present chapter.

As in Tables I and II, Mukai and Nagano have used their data to calculate homozygous loads: the total load and its two components, the lethal and detrimental loads. Next, the allelism of lethal chromosomes in random heterozygous combinations has been used to estimate the size of the population from which the original wild flies were sampled. Their estimate of an infinitely large population must be accepted with caution in the absence of detailed information on the means by which the wild flies were captured and the length of time over which trapping occurred. This point reemerges in their discussion when, despite the negative evidence obtained in viability tests, they argue that lethals must be partially dominant; otherwise, lethal chromosomes would occur in higher frequencies than those observed. This argument relies on the postulated "infinite" size of the sampled population. If that population, in fact, had an effective size of 10,000 [a size which Prout (1954) was unable to distinguish from infinity], the expected equilibrium frequency of each recessive lethal could be halved.

The above discussion has been presented in order to provide an example of similar ones commonly encountered during the 1950s and 1960s when an understanding of the nature and cause of genetic loads was the aim of much of population genetic research. The arguments advanced in the present account render many of the traditional discussions irrelevant. The frequency of lethals in a population, according to the present account, is determined by their frequency among surviving and reproducing individuals. The *average* viability of lethal heterozygotes has little or no bearing on the matter: many heterozygous chromosomal combinations in population 7 that were characterized by low viability carried lethal chromosomes; that was shown rather clearly in Figs. 15 and 16. Many heterozygotes that possessed the highest viabilities also carried lethal chromosomes. To the extent that an individual's survival or nonsurvival reflects rank-order selection, the average effect of lethal mutations on their carriers would play a minor role in determining the frequency of lethals among the reproducing adult individuals of a population.

Thanks largely to the persistence of Milkman (1967, 1973, 1978), the role of rank-order selection in maintaining numerous genetic polymorphisms at low "cost" to the population is now generally acknowledged. Milkman's calculations have been confirmed and extended by Kimura and Crow (1978) and Crow and Kimura (1979). We would only offer two additional suggestions.

1. In representing the relative fitnesses of (statistical) populations of two genetically different individuals (*aa* as compared with *Aa*, for example), we would not insist that this difference be density and frequency independent. Given that the two types differ in fitness by a certain amount when both are equally numerous (and, hence, equally frequent), we can visualize this difference vanishing (a) if both types remained equally frequent but both became less numerous and (b) if the total number of the two types remained constant but the fitter genotype approached either 0 or 100%. An improvement in the fitness of individuals of the poorer genotype when the number of the better fit decreases merely acknowledges that the presence of one depresses the fitness of the other. On the other hand, as the "fitter" type approaches 100% in frequency, intragenotype

competition may outweigh intergenotype competition, thus giving an edge to the otherwise poorer genotype (Wallace, 1981, pp. 455, 459).

2. The second suggestion concerns a sentence that appears in Crow and Kimura's (1979) abstract: "Whether nature ranks and truncates, or approximates this behavior, is an empirical question, yet to be answered." Confusion often arises from the reification of processes. For example, to say that natural selection *causes* the differential survival and reproduction of individuals is to become embroiled in the circularity of (a) the fittest survive and (b) those that survive are the fittest. The correct statement (e.g., Lerner, 1958, p. 10) is that natural selection *is* the differential survival and reproduction of individuals. Although their statement is not altered so simply, I would suggest that Crow and Kimura's sentence would be more appropriate if it read, "With respect to fitness and major components of fitness such as viability and fecundity, ranking (and perhaps truncating) *is* nature," because, with respect to fitness and its major components, natural selection *is* rank-order selection.

Mukai and Nagano (1983) raise an interesting point when they ask, "Does excess genetic variance originate from polymorphic isozyme loci?" They provide a tentative answer by suggesting that most of the genetic variance in viability is likely to be due to nonstructural genes. We suspect that this suggestion is correct. In recent studies on the fate of *sepia* and allozyme variants at the *Est-6* and *Adh* loci when introduced into laboratory populations by sporadic migrant flies (Wallace 1979, 1982*b*), it appeared that both *sepia* and *Est-6* heterozygotes had a twofold selective advantage over the "homozygous" wild-type (Riverside, California) flies of the experimental populations. This advantage appeared to reside, however, at neither the *sepia* nor *Est-6* locus, but in the chromosomal region lying between them (the two loci are separated by ten map units).

An admission that traditional structural genes may be of less importance in determining selective forces than are the presently, largely unknown chromosomal "control regions" has ramifications of some importance to population genetics and population geneticists. The traditional view of gene action with its emphasis of "gene products" was subject to satirical restatement in pharmaceutical terms: "If one teaspoonful of a medicine is good, two must be better." An attempt to divert attention from the traditional product formation of genes to the timing of gene action was made by Wallace (1963*b*) and later by Wallace and Kass (1974) and Wallace (1975*b*, 1976); MacIntyre (1982) has reviewed the problem of regulatory genes and adaptation.

The importance of gene regulation relative to concepts of survival and reproduction is that alternative alleles can be better seen as complementing each other in an advantageous manner through temporal or spatial environmental variation, or through challenges arising during the individual's development (Wallace, 1959*b*). If the surviving individual is viewed as one that relies first on one allele and then on its homologue rather than one that relies on the total "product" made by two alleles functioning in synchrony, a constructive role for seemingly deleterious alleles can be easily visualized. A lethal allele (by "allele" we mean both the structural gene and its controlling regions) can be lethal because of a single, even trivial, malfunction during the individual's development; when heterozygous with an allele that covers that one, lethal-causing lesion, such a lethal need not be harmful to any marked degree. Such lethals may be those that make the observed curves in Fig. 19 depart so markedly from the theoretical ones.

Genetic regulatory devices are now known to come in a variety of forms: flip-flop switches in DNA (Zieg *et al.*, 1977), transposable elements that move about the genome (McClintock 1951, 1956, 1965; Fedoroff 1983), and DNA segments (introns) that lie within the gene locus itself. Confronted with such exotica, many persons have joined with molecular biologists in claiming that the new molecular discoveries require that much (most, or even all) of population genetics must be revamped or discarded. Most of these calls for change come, we believe, from persons who fail to appreciate the power of abstract symbols. In discussing genetic diversity, genetic uniformity, and heterosis, Wallace (1963*b*) was fully aware of the systems of genetic control that had been proposed by Jacob and Monod (1961), Maas (1961), and McClintock (1961 and earlier publications). Nevertheless, as Wallace pointed out then, if allele A acts only under the direction of a controlling element C, and if C varies (C_1, C_2, C_3, . . .) while A remains constant, the variants C_1A, C_2A, C_3A, . . . can be represented symbolically as A_1, A_2, A_3, ... If both vary, the notation can be changed to A_{11}, A_{12}, A_{13}, . . . , A_{22}, A_{23}, . . . , A_{ij} if one wants; such a change is unnecessary, however, since a single subscript can be identified with any paired combination.

The symbolism of population genetics did not develop as a means for representing a particular genetic situation; on the contrary, the symbolism of population genetics is general. Individual population geneticists may superimpose their views on this symbolism just as they, of necessity, formulate assumptions in order to perform various calculations. These views and these assumptions may require alteration (many erroneous views and assumptions in need of change may reside, of course, in the minds of those who are *not* population geneticists), but that branch of genetics known as population genetics need not undergo any drastic revision because of discoveries made so far by molecular biologists. The contributions of Hardy and Weinberg, we may recall, were not affected when DNA, rather than protein, was recognized as the hereditary material.

SUMMARY

During the past 30 years, genetic variation with respect to fitness or the major components of fitness has been viewed as a cost to the population (Haldane, 1937), a load borne by the population (Muller, 1950), as a portion of a larger ecological load (Turner and Williamson, 1968), as relatively unimportant in the ecology of a population (Wallace 1968*b*, 1970), and as an aid to the self-thinning of grossly overcrowded populations (Wallace, 1977, 1982*a*). Under rank-order selection, the elimination of (generally) excess individuals from a population proceeds from the least competitive (least viable, least fit) ones to those that are more and more competitive; with each individual that is eliminated, the probability of ultimate survival of each remaining one increases. In the present chapter, *D. melanogaster* flies carrying random combinations of second chromosomes (obtained by the use of a *ClB*-like technique) have been ranked according to their relative viabilities. It appears that, even though lethal chromosomes may have an *average* deleterious effect on viability, heterozygotes of superior viability may carry at least as high a proportion of lethals as that occurring in the entire population. It is suggested that the fate of lethals in populations is determined by their presence among individuals that survive and reproduce, not by their average effects. The discussion touches on the limited impact molecular discoveries have had on the symbolism of population genetics; the mental images of individual population geneticists (and of their critics) may, of course, need considerable revision.

ACKNOWLEDGMENTS

We wish to thank Profs. Ross MacIntyre and Marvin Wasserman for their critical reading of an early version of this manuscript. One of us (B. W.) would like to extend his thanks to the many persons at the Biological Laboratory, Cold Spring Harbor, New York who from 1949 to 1958 were connected with these studies; special acknowledgment must be made to J. C. King and C. V. Madden. The original work was done under contract with the Atomic Energy Commission; the present analysis was made while the senior author's research was supported by grants GM29810 and GM31687, National Institute of General Medical Sciences, U. S. Public Health Services.

REFERENCES

Cavalcanti, A. G. L., 1950, Contribuicão a genetica das populacões naturais: Analise dos gens autosomicos recessivos dos chromosomos II e III de *Drosophila prosaltans* Duda, Tese da Facul. Nac. Filosofia, Rio de Janeiro Univ., Brazil.

Crow, J. F., 1958, Some possibilities for measuring selection intensities in man, *Hum. Biol.* **30:**1–13.

Crow, J. F., and Kimura, M., 1979, Efficiency of truncation selection, *Proc. Natl. Acad. Sci. USA* **76:**396–399.

Crow, J. F., and Temin, R. G., 1964, Evidence for the partial dominance of recessive genes in natural populations of *Drosophila*, *Am. Nat.* **98:**21–33.

Dobzhansky, Th., and Queal, M. L., 1938, Genetics of natural populations. II. Genic variation in populations of *Drosophila pseudoobscura* inhabiting isolated mountain ranges, *Genetics* **23:**463–484.

Dobzhansky, Th., and Spassky, B., 1953, Genetics of natural populations. XXI. Concealed variability in two sympatric species of *Drosophila*, *Genetics* **38:**471–484.

Dbozhansky, Th., and Spassky, B., 1954, Genetics of natural populations. XXII. A comparison of the concealed variability in *Drosophila prosaltans* with that in other species, *Genetics* **39:**472–487.

Dobzhansky, Th., and Wallace, B., 1953, The genetics of homeostasis in *Drosophila*, *Proc. Natl. Acad. Sci. USA* **39:**162–171.

Dobzhansky, Th., Holz, A. M., and Spassky, B., 1942, Genetics of natural populations. VIII. Concealed variability in the second and fourth chromosomes of *Drosophila pseudoobscura* and its bearing on the problem of heterosis, *Genetics* **27:**463–490.

Dobzhansky, Th., Krimbas, C., and Krimbas, M. G., 1960, Genetics of natural populations. XXX. Is the genetic load in *Drosophila pseudoobscura* mutational or balanced?, *Genetics* **45:**741–753.

Fedoroff, N. V., 1983, Controlling elements in maize, in: *Mobile Genetic Elements* (J. A. Shaprio, ed.), pp. 1–63, Academic Press, New York.

Haldane, J. B. S., 1932, *The Causes of Evolution*, Harper & Row, New York.

Haldane, J. B. S., 1937, The effect of variation on fitness, *Am. Nat.* **71:**337–349.

Harris, H., 1966, Enzyme polymorphisms in man, *Proc. Roy. Soc. Lond. B* **164:**298–310.

Hiraizumi, Y., and Thomas, A. M., 1984, Suppressor systems of Segregation Distorter (SD) chromosomes in natural populations of *Drosophila melanogaster*, *Genetics* **106:**279–292.

Hoenigsberg, H. F., and DeNavas, Y. G., 1965, Population genetics in the American Tropics. I. Concealed recessives in different bioclimatic regions, *Evolution* **19:**506–513.

Hoenigsberg, H. F., Palomino, J. J., Hayes, M. J., Zandstra, I. Z., and Rojas, G. G., 1977, Population genetics in the American Tropics. X. Genetic load differences in *Drosophila willistoni* from Colombia, *Evolution* **31:**805–811.

Jacob, F., and Monod, J., 1961, On the regulation of gene activity, *Cold Spring Harbor Symp. Quant. Biol.* **26:**193–211.

Kimura, M., and Crow, J. F., 1978, Effect of overall phenotypic selection on genetic change at individual loci, *Proc. Natl. Acad. Sci. USA* **75**:6168–6171.

Lerner, I. M., 1958, *The Genetic Basis of Selection*, Wiley, New York.

Lewontin, R. C., 1974, *The Genetic Basis of Evolutionary Change*, Columbia University Press, New York.

Lewontin, R. C., and Hubby, J. L., 1966, A molecular approach to the study of genic heterozygosity in natural populations. 2. Amount of variation and degree of heterozygosity in natural populations of *Drosophila pseudoobscura*, *Genetics* **54**:595–609.

Maas, W. K., 1961, Studies on repression of argenine biosynthesis in *Escherichia coli*, *Cold Spring Harbor Symp. Quant. Biol.* **26**:183–191.

MacIntyre, R. J., 1982, Regulatory genes and adaptation: Past, present, and future, *Evol. Biol.* **15**:247–285.

McClintock, B., 1951, Chromosome organization and genic expression, *Cold Spring Harbor Symp. Quant. Biol.* **16**:13–47.

McClintock, B., 1956, Controlling elements and the gene, *Cold Spring Harbor Symp. Quant. Biol.* **21**:197–216.

McClintock, B., 1961, Some parallels between gene control systems in maize and bacteria, *Am. Nat.* **95**:265–277.

McClintock, B., 1965, The control of gene action in maize, *Brookhaven Symp. Biol.* **18**:162–184.

Milkman, R. D., 1967, Heterosis as a major cause of heterozygosity in nature, *Genetics* **55**:493–495.

Milkman, R. D., 1973, A competitive selection model, *Genetics* **74**:727–732.

Milkman, R. D., 1978, Selection differentials and selection coefficients, *Genetics* **88**:391–403.

Mukai, T., and Nagano, S., 1983, The genetic structure of natural populations of *Drosophila melanogaster*. XVI. Excess of additive genetic variance of viability, *Genetics* **105**:115–134.

Muller, H. J., 1928, The measurement of gene mutation rate in *Drosophila*, its high variability, and its dependence upon temperature, *Genetics* **13**:279–357.

Muller, H. J., 1950, Our load of mutations, *Am. J. Hum. Genet.* **2**:111–176.

Pavan, C., Cordeiro, A. R., Dobzhansky, Th., Dobzhansky, N., Malogolowkin, C., Spassky, B., and Wedel, M., 1951, Concealed genic variability in Brazilian populations of *Drosophila willistoni*, *Genetics* **36**:13–30.

Pontecorvo, G., 1982, Who was H. J. Muller (1890–1967)? (A review of *Genes, Radiation, and Society*, by Elof Axel Carlson. 1982. Cornell University Press, Ithaca, N.Y.), *Nature* **298**:203–204.

Prout, T., 1954, Genetic drift in irradiated experimental populations of *Drosophila melanogaster*, *Genetics* **39**:529–545.

Stern, C., Carson, G., Kinst, M., Novitski, E., and Uphoff, D., 1952, The viability of heterozygotes for lethals, *Genetics* **37**:413–449.

Sturtevant, A. H., 1937, Autosomal lethals in wild populations of *Drosophila pseudoobscura*, *Biol. Bull.* **73**:542–551.

Temin, R. G., and Marthas, M., 1984, Factors influencing the effect of segregation distortion in natural populations of *Drosophila melanogaster*, *Genetics* **107**:375–393.

Thoday, J. M., 1953, Components of fitness, *Symp. Soc. Exp. Biol.* **7**:96–113.

Timofeeff-Ressovsky, N. W., 1935, Auslosung von Vitalitatmutationen durch Rontgenbestralung bei *Drosophila melanogaster*, *Nachr, Ges. Wiss. Gott. Biol. N. F.* **1**:163–180.

Townsend, J. I., 1952, Genetics of marginal populations of *Drosophila willistoni*, *Evolution* **6**:428–442.

Turner, J. R. G., and Williamson, M. H., 1968, Population size, natural selection, and genetic load, *Nature* **218:**700.

Wallace, B., 1950, Autosomal lethals in experimental populations of *Drosophila melanogaster*, *Evolution* **4:**172–174.

Wallace, B., 1951, Genetic changes within populations after X-irradiation, *Genetics* **36:**612–628.

Wallace, B., 1952, The estimation of adaptive values of experimental populations, *Evolution* **6:**333–341.

Wallace, B., 1956, Studies on irradiated populations of *Drosophila melanogaster*, *J. Genet.* **54:**280–293.

Wallace, B., 1959*a*, Studies on the relative fitnesses of experimental populations of *Drosophila melanogaster*, *Am. Nat.* **93:**295–314.

Wallace, B., 1959*b*, The role of heterozygosity in *Drosophila* populations, *Proc. 10th Int. Cong. Genet.* **1:**408–419.

Wallace, B., 1963*a*, A comparison of the viability effects of chromosomes in heterozygous and homozygous condition, *Proc. Natl. Acad. Sci. USA* **49:**801–806.

Wallace, B., 1963*b*, The annual invitation lecture. Genetic diversity, genetic uniformity, and heterosis, *Can. J. Genet. Cytol.* **5:**239–253.

Wallace, B., 1968*a*, *Topics in Population Genetics*, Norton, New York.

Wallace, B., 1968*b*, Polymorphism, population size, and genetic load, in: *Population Biology and Evolution* (R. C. Lewontin, ed.), pp. 87–108, Syracuse University Press, Syracuse, New York.

Wallace, B., 1970, *Genetic Load: Its Biological and Conceptual Aspects*, Prentice-Hall, Englewood Cliffs, New Jersey.

Wallace, B., 1975*a*, Hard and soft selection revisited, *Evolution* **29:**465–473.

Wallace, B., 1975*b*, Gene control mechanisms and their possible bearing on the neutralist–selectionist controversy, *Evolution* **29:**193–202.

Wallace, B., 1976, The structure of gene control regions and its bearing on diverse aspects of population genetics, in: *Population Genetics and Ecology* (S. Karlin and E. Nevo, eds.), pp. 499–521, Academic Press, New York.

Wallace, B., 1977, Automatic culling and population fitness, *Evol. Biol.* **10:**265–276.

Wallace, B., 1979, The migration of a mutant gene into isolated populations of *Drosophila melanogaster*, *Genetica* **50:**67–72.

Wallace, B., 1981, *Basic Population Genetics*, Columbia University Press, New York.

Wallace, B., 1982*a*, Phenotypic variation with respect to fitness: The basis for rank-order selection, *Biol. J. Linn. Soc.* **17:**269–274.

Wallace, B., 1982*b*, The fate of several migrant genes in isolated populations of *Drosophila melanogaster*, *Genetica* **58:**141–151.

Wallace, B., 1984, Quantifying the concealed, detrimental genetic variation of Mendelian populations, *Genetika* (*Yugoslavia*) **16:**1–12.

Wallace, B., and Kass, T. L., 1974, On the structure of gene control regions, *Genetics* **77:**541–558.

Wallace, B., and King, J. C., 1951, Genetic changes in populations under irradiation, *Am. Nat.* **85:**209–222.

Wallace, B., and King, J. C., 1952, A genetic analysis of the adaptive values of populations, *Proc. Natl. Acad. Sci. USA* **38:**706–715.

Wallace, B., and Madden, C., 1953, The frequencies of sub- and supervitals in experimental populations of *Drosophila melanogaster*, *Genetics* **38:**456–470.

Wills, C., 1978, Rank-order selection is capable of maintaining all genetic polymorphisms, *Genetics* **89:**403–417.

Zieg, J., Silverman, M., Hilmen, M., and Simon, M., 1977, Recombinational switch for gene expression, *Science* **196:**170–172.

4

Evolutionary Origin of Ethological Reproductive Isolation in Cricket Frogs, *Acris*

EVIATAR NEVO

and

ROBERT R. CAPRANICA

INTRODUCTION

Theories of Reproductive Isolating Mechanisms

Reproductive isolating mechanisms, either premating or postmating, are the key to speciation; hence, they are among the most important biological properties of species (Mayr, 1970; Dobzhansky *et al.*, 1977) or of sympatric individuals of different adaptive systems (Littlejohn, 1980). Yet, whereas their nature is quite well known, their mode of origin, particularly of premating as opposed to postmating mechanisms, still remains largely obscure and controversial. Two opposing theories have attempted to resolve the question regarding their origin: the *sympatric* and *allopatric theories* (Mayr, 1970, pp. 325–330; Littlejohn, 1980). According to the *sympatric theory*, reproductive isolating mechanisms are directly selected when two incipient species become sympatric, due to hybrid inferiority

EVIATAR NEVO • Institute of Evolution, University of Haifa, Haifa, Israel.
ROBERT R. CAPRANICA • Section of Neurobiology and Behavior, Division of Biological Sciences, Cornell University, Ithaca, New York 14853.

in their overlap zone (Wallace, 1889; Fisher, 1930; Dobzhansky, 1940, 1951; V. Grant, 1971; Blair, 1974) [see Littlejohn (1980) for review and a critical discussion; for mathematical analysis see Sawyer and Hartl (1981)]. According to the *allopatric theory*, isolating mechanisms arise as incidental by-products of adaptive genetic divergence in allopatry (Darwin, 1859; Muller, 1942; Mayr, 1942; Patterson and Stone, 1952). Recent attempts to reconcile these two opposing theories have suggested that, while primary isolating mechanisms could develop before contact is encountered, they may be perfected and/or reinforced by a secondary isolating mechanism in sympatry (Levin, 1971, 1978, and references therein). Yet this reconciliation does not provide a clear distinction or direct proof of the actual origin of primary isolating mechanisms. Are they due to selection in sympatry or are they incidential by-products of adaptive differentiation in allopatry? Alternatively, do they arise because of direct selection in accordance with factors in the original abiotic and biotic allopatric environment?

Role of Anuran Mating Calls in Reproductive Isolation

Anurans offer an excellent opportunity to test the mode of origin of premating reproductive isolating mechanisms. In most sympatric anurans, vocalization by males is the primary and by far the most important reproductive isolating barrier: females respond selectively to conspecific males on the basis of the distinct signal characteristics in their calls (Blair, 1958*a*, 1964*a,b*; Littlejohn, 1969; Schiøtz, 1973; Straughan, 1973; Otte, 1974). By means of systematic analysis of field-recorded calls, these signal features can be measured precisely in terms of their spectral (frequency) composition and their temporal parameters. Behavioral studies of female phonotaxis to playback of recorded mating calls provides a reliable measure of the discriminatory abilities of different species (Martof and Thompson, 1958; Littlejohn and Michaud, 1959; Straughan 1966; Littlejohn and Loftus-Hills, 1968; Littlejohn and Martin, 1969; Gerhardt, 1974; Oldham and Gerhardt, 1975). The underlying sensory basis for this selectivity can be assessed through appropriate electrophysiological studies of the peripheral and central auditory nervous system (Frishkopf and Goldstein, 1963; Frishkopf *et al.*, 1968; Capranica *et al.*, 1973; Capranica and Moffat, 1975; Capranica 1976, 1977; Mudry *et al.*, 1977). Finally, the mating call can be synthesized electronically and the relative importance of its spectral and temporal parameters in eliciting a species-selective response can be ascertained in subsequent behavioral studies (Awbrey, 1965; Capranica, 1965; Gerhardt, 1974).

Cricket Frogs As a Model System

Cricket frogs, genus *Acris*, are semiaquatic members of the treefrog family Hylidae. They represent ideal specimens for studying the origin of isolating mechanisms because they have extensive allopatric and sympatric ranges, are abundant, and have a long breeding season. Currently, two closely related sibling species are recognized in *Acris*. *Acris crepitans* is distributed widely over the central and eastern U.S. and comprises two subspecies: *A. crepitans blanchardi* ranges chiefly in the semiarid and subhumid grassland biome, and the smaller, *A. crepitans crepitans* ranges primarily in the humid deciduous forest and secondarily in the southern pineries and northern hardwood Atlantic forest (Conant, 1975, pp. 316–318, map 291) (Fig. 1). *Acris gryllus* is restricted to Florida and the lowlands from eastern Louisiana to southeastern Virginia and also comprises two subspecies: A. *gryllus gryllus*, which ranges from southeast Virginia to the Gulf Coast and Mississippi River, and the similar but smaller *A. gryllus dorsalis*, which extends from southeast Georgia to the tip of Florida (Conant, 1975, pp. 316–317, map 292) (Fig. 1). Both species share a broad overlap zone in the southeastern U. S. (Fig. 1), where *A. c. crepitans* and *A. g. gryllus* overlap without interbreeding despite interfertility (Neill, 1954; Mecham, 1964). Since both species share the same breeding season without pronounced spatial separation where they coexist, Mecham (1964) suggested that mating call discrimination by females is the critical isolating mechanism involved. Earlier, Blair (1958*a*) indicated, based on evidence then available, that call differences are distinct in sympatry but are slight in allopatry, suggesting that reinforcement of isolating mechanisms by selection occurs in the overlap zone. Thus, *Acris* appeared to support the sympatric reinforcement selection theory (Blair, 1958*a*, 1964*a,b*, 1974).

Our objective was to test the mode of origin of mating call differentiation and consequent reproductive isolation in cricket frogs. Specifically, we compared the sympatric and allopatric theories concerning the origin of isolating mechanisms. We employed a systematic, multidisciplinary approach in our overall experimental study (Nevo, 1969). First, we recorded and then analyzed the signal (both spectral and temporal) characteristics of their mating calls across the entire range of *Acris*. Second, we conducted female (phonotaxis) discrimination tests in both sympatric and allopatric populations, utilizing natural (field-recorded) mating calls as well as electronically generated synthetic calls. Finally, electrophysiological studies were pursued to gain a measure of the frequency selectivity of the auditory nervous system of females from different populations (Capranica *et al.*, 1973). It was hoped that such a multifaceted

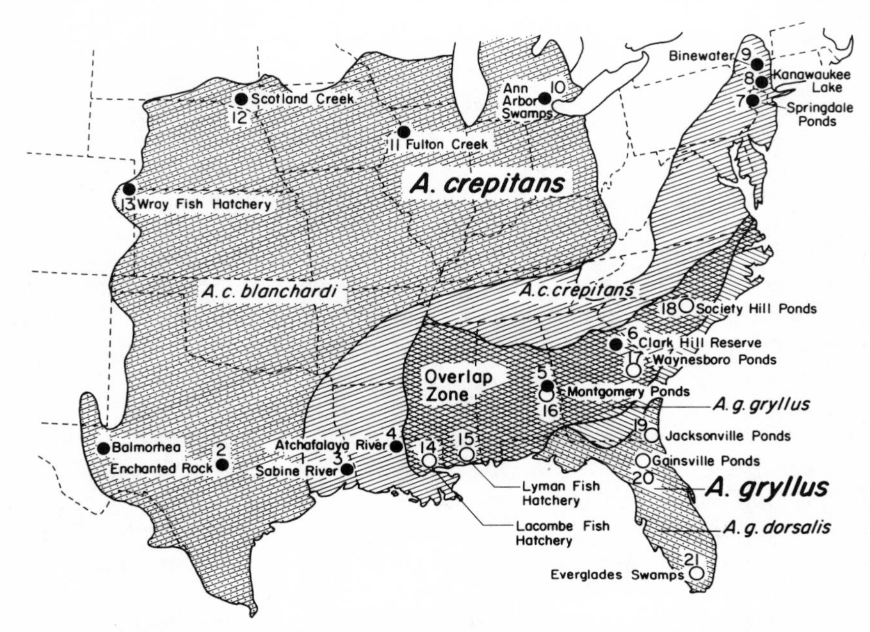

FIG. 1. Localities of allopatric and sympatric populations of *Acris crepitans* and *Acris gryllus*.

approach might provide insight into the alternative theories regarding the origins of ethological reproductive isolation in cricket frogs, particularly since our analysis encompassed the entire range of this genus.

MATERIALS AND METHODS

Localities

A number of mating calls from each of 393 males were tape-recorded in the field in 46 different localities across the entire extent of the range of *Acris* between May and July of 1965, 1966, and 1967. This extensive sample involved 318 *A. crepitans* males from 37 different geographic localities and 75 *A. gryllus* males from nine localities. Of this set of recordings, due to the very time-consuming analyses involved, a representative group of 112 males was selected: 74 *A. crepitans* from 13 localities (11 allopatric, two sympatric) and 38 *A. gryllus* from eight localities (three allopatric, five sympatric), which are indicated in Fig. 1 and tabulated,

TABLE I. Localities and Geographic, Climatic, and Recording Data for the 23 Populations Studied of *Acris crepitans* and *Acris gryllus*

Locality[a]			Collecting date	Sample size analyzed N	Latitude Lt	Longitude Ln	Altitude Al	Rainfall Rn,[b] in.	Temperature Ta,[b] °F	Mean recording temperature, °C
No.		Area								
Acris crepitans										
1	A	Balmorhea, TX	6-30-67	5	30.99	103.75	3200	15	65	19.70
2	A	Enchanted Rock, TX	7-1-67	5	30.55	98.70	2000	25	65	23.70
3	A	Sabine River, TX	7-2-67	5	30.13	93.70	30	50	70	26.00
4	A	Atchafalaya River, LA	7-3-67	5	30.54	91.74	25	60	70	28.60
5	S	Montgomery ponds, AL	7-4-67	5	32.33	85.59	350	55	65	25.66
6a	S	Clark Hill Reserve, GA	5-25-66	5	33.58	82.53	500	50	65	21.00
6b	S	Clark Hill Reserve, GA	6-15-66	5	33.58	82.53	500	50	65	24.50
7a	A	Springdale ponds, NJ	5-21-66	5	41.02	74.77	800	45	60	16.90
7b	A	Springdale ponds, NJ	6-25-66	5	41.02	74.77	800	45	60	23.00
8	A	Kanawaukee Lake, NY	7-9-66	4	41.22	74.22	1100	45	50	23.00
9	A	Binnewater Lake, NY	6-31-66	5	41.87	74.08	400	40	50	22.20
10	A	Ann Arbor swamps, MI	6-17-67	5	42.29	83.79	800	35	45	20.70
11	A	Fulton Creek, IL	6-17-67	5	41.89	90.19	600	35	50	24.60
12	A	Scotland Creek, SD	6-20-67	5	43.15	97.67	1200	25	45	21.00
13	A	Wray Fish Hatchery, CO	6-20-67	5	40.29	102.27	3650	45	50	18.00
			Total	74						
Acris gryllus										
14	S	Lacombe Fish Hatchery, LA	7-11-65	5	30.32	89.95	25	60	70	23.90
15	S	Lyman Fish Hatchery, MS	7-12-65	5	30.50	89.11	50	60	70	26.30
16	S	Montgomery ponds, AL	7-4-67	5	32.33	85.59	350	55	65	24.80
17	S	Waynesboro ponds, GA	5-25-66	3	33.10	82.15	250	50	65	22.30
18	S	Society Hill ponds, SC	5-30-66	5	34.29	79.14	250	50	60	24.00
19	A	Jacksonville ponds, FL	5-26-66	5	30.63	81.61	75	55	70	21.40
20	A	Gainesville ponds, FL	5-27-66	5	29.30	82.18	125	55	70	22.80
21	A	Everglades swamps, FL	5-28-66	5	26.00	81.58	5	60	75	24.20
			Total	38						

[a] A, Allopatric; S, sympatric.
[b] Mean annual value.

along with geographic and climatic data, in Table I. During our field studies, body (snout–vent) lengths for 5286 cricket frogs from 108 localities were measured, including those whose calls were recorded, in a separate study of the variation in size of specimens within the genus *Acris* (Nevo, 1973*a*).

Mating Call Analysis

Individual mating calls were recorded in the field with an AKG model MBH dynamic microphone and a Nagra IIIB portable tape recorder operated at 7½ in./sec. The microphone and tape recorder had a frequency response of ±4 dB from 100 to 5000 Hz and all distortion products were less than 36 dB below the fundamental frequency. A minimum of five mating calls for each male was analyzed in our selected sample of each locality.

A typical mating call consists of a number of groups of clicks (Fig. 2). The groups are separated from each other by silent intervals and are readily identifiable. Within a group, individual clicks are generally separated by short intervals from other clicks. Thus, each click is a distinct auditory event. Furthermore, each click is comprised of a number of pulses, but individual pulses are not resolvable to the (human) ear.

Our analysis of each call was expedited by first recording it on a continuous tape loop (Cousino cartridge) so that it repeated itself once every 30 sec (playback via a Magnecord tape recorder, 7½ in./sec). This repetitive playback enabled a systematic measurement of spectral and temporal features throughout the call. A Kay 7029A spectrum analyzer was used for generating normal (narrowband and wideband) sonagrams as well as section displays (45-Hz narrowband) at a variety of instants throughout each male's calls. The spectral peak (i.e., dominant frequency) in each click of a male's call was very stereotyped and exhibited little variation (in fact, the spectral peaks in the clicks of different males within a given population are all quite similar; see Table III). Therefore, the frequencies corresponding to the spectral peaks in the sequential analysis through each call were averaged and then taken as a representative measure of the calls for that particular male.

Analysis of the gross and fine temporal structure within each call was achieved by combining two basic approaches: one involved direct measurement and the other involved an automated instrumentation measure. The results of these two approaches were compared to verify our conclusions regarding the temporal features in the mating calls of *Acris*. For direct measurements, each call was displayed on various time scales in

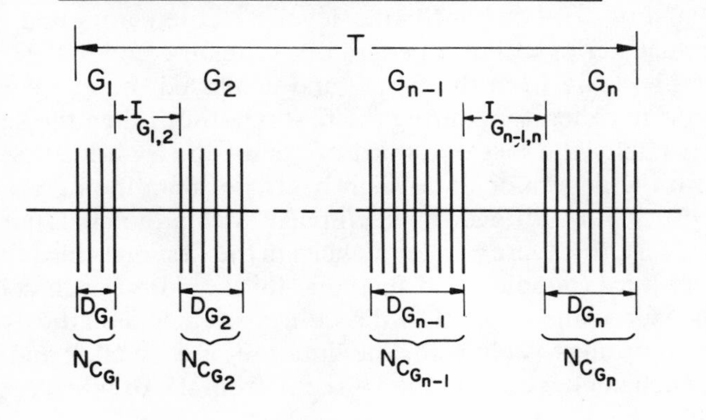

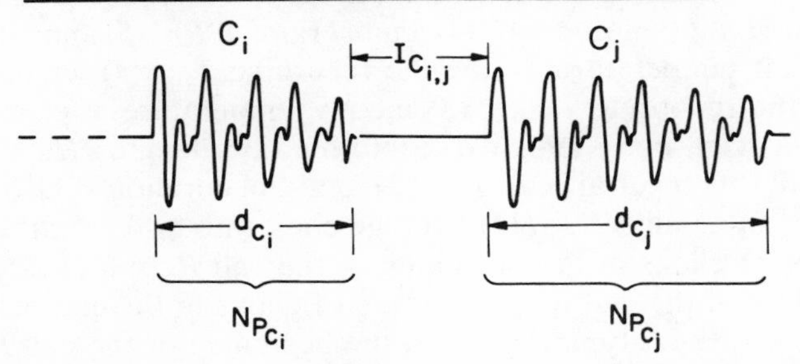

FIG. 2. Diagrammatic gross and fine temporal properties of *Acris* mating calls. The total duration of the call is T seconds and consists of a sequence of groups G_1, G_2, . . . , G_{12} of clicks. $I_{G1,2}$, . . . , $I_{Gn-1,n}$ represent the time intervals between successive groups; D_{G1}, D_{G2}, . . . , D_{Gn} indicate the durations of each group; and the N_C's are the numbers of clicks in each successive group. Each of the clicks C_1, C_2, . . . , C_i, C_j within a single group is separated from its neighbors by interclick intervals $I_{C1,2}$ $I_{C2,3}$, . . . , $I_{Ci,j}$. The duration of each click in an individual group is denoted by d_{c1}, d_{c2}, . . . , d_{ci}, d_{cj}, and the N_p's represent the number of pulses in each click.

systematic increments of 500-msec steps from the start to the end on a storage oscilloscope (Tektronix 564) equipped with a calibrated delay plug-in. Later in our study we used a high-speed Honeywell Visicorder to obtain a continuous waveform trace (recorded on light-sensitive strip recording paper). This recorded trace was then measured by means of a calibrated time scale to quantify the temporal rates of various parameters in the calls of different males. To automate and verify these measurements and conclusions, an electronic gated counter (Hewlett-Packard 5632B) was used to display the rate of clicks at the beginning, then middle, and finally the end of each male's call. The instrumentation for this analysis involved a series of Grason-Stadler level zone detectors and timers and an Ortec 4650 series of pulse generators. Thus the counter was gated by an adjustable delay from the timers and displayed the number of pulses from the pulse generators during the first one-third, then the second one-third, and finally the last one-third of each call. By adjusting the pulse width from the generator to be slightly greater than the duration of each click (in order not to trigger on each pulse within the click but rather on the click itself), the average rate of clicks in the first one-third (beginning), middle one-third (middle), and final one-third (end) of each call could be readily measured and verified by direct measurements on the oscilloscope and strip recordings. Values for the duration of each click and pattern of pulses in each click were obtained directly from the oscilloscope and strip recordings.

Twenty-six variables, ten background variables and 16 call variables, entered the analysis. *Background variables* involved: (1) latitude *Lt*; (2) longitude *Ln*; (3) altitude *Al*; (4) annual rainfall *Rn*; (5) annual temperature *Ta*; (6) air temperature during the recording T_{air}; (7) water temperature during the recording T_{water}; (8) mouth temperature after calling T_{mouth}; (9) snout–vent body length *BL*; (10) hindfoot length *HL*. *Call variables* included: (11) spectral peak *SP*; (12) total call duration *T*; (13) total number of clicks in a call *NCT*; (14) average click rate in the entire call *NCT/T*; (15) rate of clicks in the beginning of the call *Rcb*; (16) rate of clicks in the middle of the call *Rcm*; (17) rate of clicks in the end of the call *Rce*; (18) duration of a typical click in the beginning of the call *Dcb*; (19) duration of a typical click in the middle of the call *Dcm*; (20) duration of a typical click at the end of the call *Dce*; (21) number of pulses in a typical click at the beginning of the call *Npcb*; (22) number of pulses in a typical click in the middle of the call *Npcm*; (23) number of pulses in a typical click at the end of the call *Npce*; (24) number of groups of pulses in a typical click at the beginning of the call *Ngpcb*; (25) number of groups of pulses in a typical click in the middle of the call *Ngpcm*; (26) number of groups of pulses in a typical click at the end of the call *Ngpce*. The dimensions of the variables are: *SP* in hertz (Hz); *T* in seconds, rates in clicks per second, durations in milliseconds, and number of pulses and number of groups of pulses per click. The variables have been broken into three parts, beginning, middle, and end of call, because they differ throughout the call of an individual male as well as for males of different populations. Usually click rates are slower and click durations are less at the beginning of a call than at the end. Since these variables change systematically in a call, we calculated them accordingly rather than just averaging them over the total call. The values for durations and number of pulses and groups of pulses involve averages of several samples from each of the three portions of each call. Finally, population values for each of the 16 call variables are based on five calls from each of five males. Air, water, and mouth temperatures of calling males were taken with a rapid reading thermometer. The temperatures in Table I are population averages of the mouth temperatures of the five recorded males in a particular station. Recorded males were preserved and deposited in the Department of Herpetology, Museum of Comparative Zoology, Harvard University.

Statistics

The data were analyzed first by univariate and then by multivariate analyses. The univariate analysis included means and standard errors for each call variable standardized to a mouth temperature of 22.79°C for each of the 23 populations. Multiple group differences between means of call variables were analyzed by an *a posteriori* Contrast Test. The latter is a systematic procedure for comparing possible pairs of group means [Scheffe's Multiple Range Test (Nie *et al.*, 1975)]. Pearsonian correlations were computed among environmental variables and call variables and between call variables and environmental factors. The multivariate analysis included first a stepwise multiple regression (Draper and Smith, 1966) to determine whether environmental factors influence or are associated with each call variable. Then two multivariate analyses of variance (MANOVA), metric and nonmetric, were conducted in order to plot the differential call patterns based on *all* 16 call variables.

The metric technique—Canonical decomposition (Candec) and biplots (Gabriel, 1971, 1973)—is complementary to principal component analysis. The biplot is a graphical display of a two-dimensional approximation to the data matrix. The approximation is obtained by least squares, using the first and second singular value components of the matrix. The display consists of vectors plotted from a common origin, a vector for each row (representing populations) and a vector for each column (representing variables). The vectors are chosen so that any element in the matrix is exactly the inner product of the vectors corresponding to its row and its column. The biplot shows statistical distances (i.e., standardized measures of dissimilarity) between the populations studied, and indicates clustering of variables characterizing specific populations, their variances, and correlations. [For a detailed description, mathematical background, and computer program of Candec and biplots see Gabriel (1973).]

Our nonmetric analysis involved Partial Order Scalogram Analysis (POSA) and is a specific version of Multidimensional Scalogram Analysis (MSA), which belongs to the Guttman–Lingoes families of computer programs (Lingoes, 1973). MSA-1, the most general of the MSA nommetric analyses, represents a set of quantitative variables by a set of partitions in a Euclidean space. A qualitative variable is a set of discrete categories such that each subject is placed in exactly one category. In our analysis each of the 16 call variables was represented by 3–5 categories. A variable with three categories was represented by a partition into three regions. Each point in the space diagram represented the call structupule, or profile, based on nine variables singled out due to their clear partitioning of the space. In our case, each structupule was composed of a category from each of the nine call variables selected. The POSA generated a partial ordering of the structupules in a Cartesian space. It portrayed the similarities and differences of the call profiles, ranking them within the perfect scale of the partial order. The order generated by the POSA is thus a result of a preliminary multidimensional scalogram analysis (GL-MSA-1) which requires no *a priori* specification of order (Guttman, 1966).

Mating Call Discrimination and Phonotaxis by Females

The effectiveness of the mating call as an isolating mechanism was tested through female phonotaxis discrimination tests. Testing was conducted in the field rather than in laboratory tanks to avoid acoustical distortion and other technical pitfalls, and to permit a populational approach. We conducted 486 call discrimination tests involving 123 female *Acris crepitans* from three populations, one sympatric (Clark Hill Reserve, Georgia) and two allopatric (Springdale, New Jersey and Palmetto Fish Hatcheries, Texas; see localities in Fig. 1). Females in each of the above populations were tested for their preference among paired, broadcasted calls from the following: a local call, other calls of increasingly distant allopatric *A. crepitans* populations, and a call of *A. gryllus*. Each representative call was transcribed onto a repeating tape loop. The duration of the call was first measured and then the silent interval on the loop was made equal to one-half the call's duration (see Table VIII). Thus, when calls of different durations were broadcast, the average energy (i.e., the overall "duty cycle") presented to the female was the same for all of the calls. This procedure was adopted as a precaution to avoid possible bias of a female's choice by presenting more energy through one loudspeaker compared to the other paired loudspeaker.

Our field discrimination setup consisted of two battery-powered

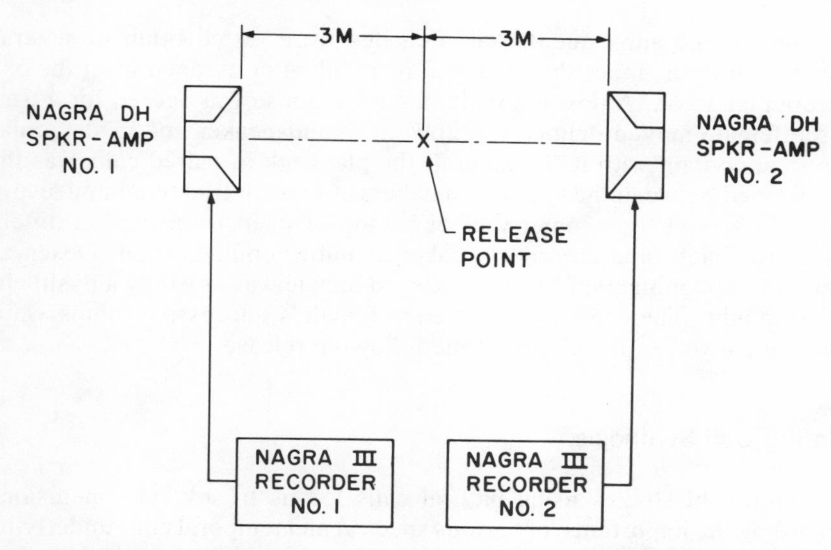

FIG. 3. Field setup for behavioral studies of call discrimination by female cricket frogs. Recorded natural and synthetic mating calls were presented simultaneously through the two loudspeakers and each female was released midway between them. These phonotactic trials therefore involved a two-choice paradigm and resulted in a very clear, positive response (i.e., physical contact with one of the loudspeakers) to the preferred call.

Nagra III tape recorders with their line outputs connected to two battery-powered Nagra DH speaker-amplifiers facing each other and separated by a distance of 6 m (Fig. 3). Playback levels of all calls were standardized to 92 dB SPL (sound pressure level, peak relative to 20 $\mu N/m^2$) at a distance of 1 m in front of each speaker-amplifier by means of a Bruel and Kjaer 2203 precision sound level meter equipped with an octave filter set. Readings on the sound level meter for each tape loop were calibrated at Bell Telephone Laboratories (by R. R. Capranica) prior to the discrimination tests in order to accurately adjust the playback level of each stimulus (in the field) to our standardized peak sound pressure reference.

Gravid female cricket frogs were collected during amplexus in breeding choruses after dusk in early evening, placed in separate plastic bags, and tested within 10–120 min after capture. The discrimination trials were conducted at nearby sites in open areas, but out of audible range of the chorus, on a damp substrate in free-field conditions. Each female to be tested was placed under a glass container equidistant between the two loudspeakers (see Fig. 3), allowed to rest for a few minutes, and then carefully released by slowly raising the container. Repetitive playback of each of the paired calls commenced just prior to release and continued throughout the subsequent trial. Females were tested often in several successive trials under they stopped responding or escaped from the experimental arena. A positive phonotactic response was scored for a trial if the female moved deliberately toward a loudspeaker and made actual physical contact with it. In general, the playback of paired calls was interchanged between the two loudspeakers after each trial to minimize any possible bias in the geometry of our setup or of instrumentation differences for each tape recorder/speaker-amplifier configuration. Observation of the phonotactic trajectory of each female was aided by a flashlight or headlight. The movements of each female's successive jumps were charted as well as the elapsed time following release.

Mating Call Synthesis

Our field studies using natural calls led us to several conclusions regarding the importance of various spectral and temporal cues underlying a female's discriminatory ability. In order to test these conclusions further, we decided to include electronically synthesized mating calls in our studies in 1967 in central Texas and Georgia. Since those studies will be the subject of a separate report, we therefore simply summarize the techniques involved; a brief description of some of our findings will be presented at the end of the next section.

Synthetic calls were generated by specially designed electronic circuits. The temporal pattern of clicks in any given natural call was used as a guide. That is, the natural call was played from a tape recorder into a series of Schmitt triggers and pulse generators (similar to our automated call analysis approach) so that the output consisted of a sequence of standard pulses identical in pattern to the clicks in the natural call. Again the output pulses had a pulse width slightly greater than the duration of the clicks in the natural call in order to preserve the individual click pattern. These rather wide output pulses were then fed into another pulse generator module, which would trigger on the onset of each wide pulse and produce, in return, a standard narrow pulse (0.1 msec) coincident with each click's onset. This transformed pulse, because of its narrow width, contained spectral energy over a much broader frequency band. Thus, the output of our pulse generator modules consisted of a train of narrow pulses, each of which occurred at the onset of each click in the natural call.

The train of standard pulses was then passed through a series of adjustable low-pass and high-pass filters: the cutoff frequency and the amount of damping of each filter could be controlled independently. This effectively provided a band-pass filter whose center frequency and bandwidth could be varied, as could the slopes on the low- and high-frequency sides. Thus, we could shape the spectral profile of each filtered click (whose waveform represented a damped sinusoidal oscillation). This scheme enabled us to generate a synthetic call having the same temporal pattern of clicks in any selected field-recorded call but with a spectral peak equal to that in any other call.

RESULTS

Representative samples of temporal (oscillographic) and spectral (narrowband section) displays are shown in Figs. 4–9. Our quantitative analyses are presented numerically in Tables II–IX and graphically in Figs. 10 and 11.

Univariate Analysis

Geographic Variation in Call Characteristics

The means of the 16 call variables standardized to a mouth temperature of 22.79°C for all 23 populations are given in Table II, and the recalculated means and standard errors for *A. crepitans* and its two subspecies, *A. c. blanchardi* and *A. c. crepitans*, as well as for *A. gryllus* and its two subspecies, *A. g. dorsalis* and *A. g. gryllus*, are shown in Table III. Note that sympatric populations are represented by *A. c. crepitans* and *A. g. gryllus*, whereas allopatric populations are represented by *A. c. blanchardi* and *A. g. dorsalis*.

The mean spectral peak *SP* for the populations of *A. gryllus* collectively was 3546 Hz and for *A. crepitans* was 3679 Hz. In general, the mean spectral peak decreased northward and westward in both species: values decreased from *A. g. dorsalis* to *A. g. gryllus* and from *A. c. crepitans* to *A. c. blanchardi* (3790 − 3357 Hz = 433 Hz) and these spectral differences were similar to those between the sympatric subspecies (i.e., *A. c. crepitans* and *A. gryllus*, 3893 − 3407 Hz = 486 Hz). The mean peak decreased westward in *A. crepitans* in its two subspecies not only in the southern, combined allopatric–sympatric transect (from 4023 Hz at sympatric Clark Hill, Georgia to 3217 Hz at allopatric Balmorhea, west Texas) but, more importantly, also in the northern, solely allopatric transect (from 3853 Hz at Kanawaukee Lake, New York to 3163 Hz at

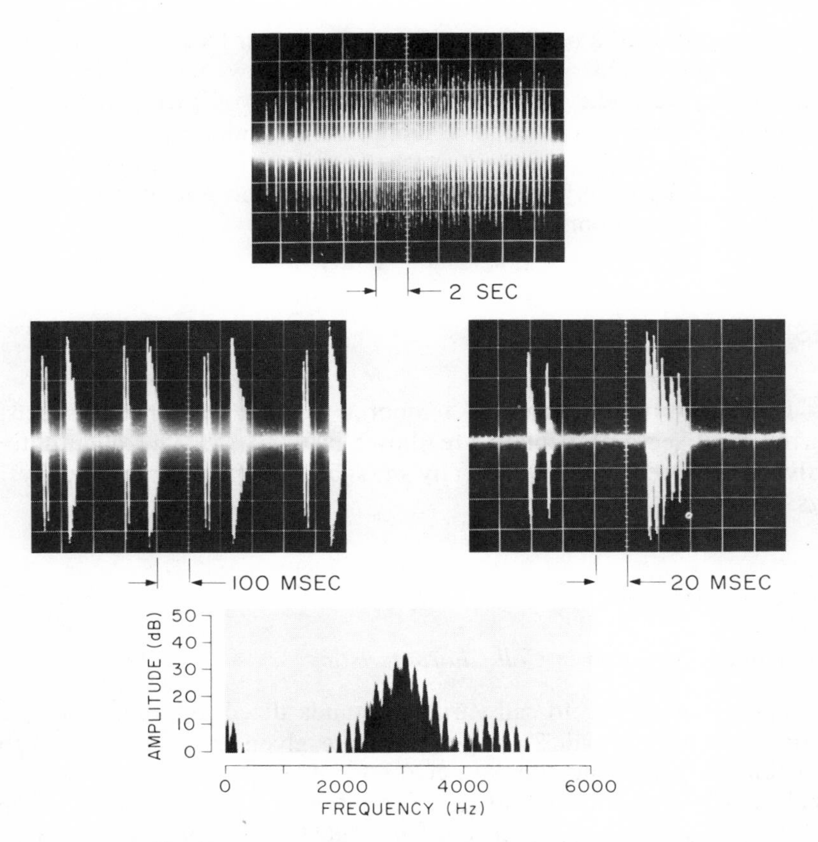

FIG. 4. Signal characteristics of a representative mating call from a male *Acris crepitans blanchardi* in west Texas. The top photograph displays the gross temporal structure, whereas the two oscillograms in the middle row show the fine-temporal pattern of clicks in the call. The bottom display is a spectrographic section of the energy distribution in the call, which was a spectral peak around 3000 Hz. The frequency of the spectral peak increases with temperature, whereas the time scale of the fine-temporal structure is inversely correlated with temperature. Thus, as temperature increases, the time scale decreases as if it were on an elastic band, but the temporal *pattern* of the clicks remains invariant. [R. R. Capranica and E. Nevo (unpublished results).]

Scotland Creek, South Dakota). The maximal decrease in mean spectral peaks westward for the southern and northern transects was 806 and 690 Hz, respectively. In *A. gryllus* the mean spectral peak decreased northward from 3856 Hz for *A. g. dorsalis* at Everglades, Florida to 3466 Hz for *A. g. gryllus* at Society Hill, South Carolina, and westward to 3173 and 3507 Hz at Lyman, Mississippi and Lacombe, Louisiana, respectively. Note that the lowest frequencies of the spectral peaks across the

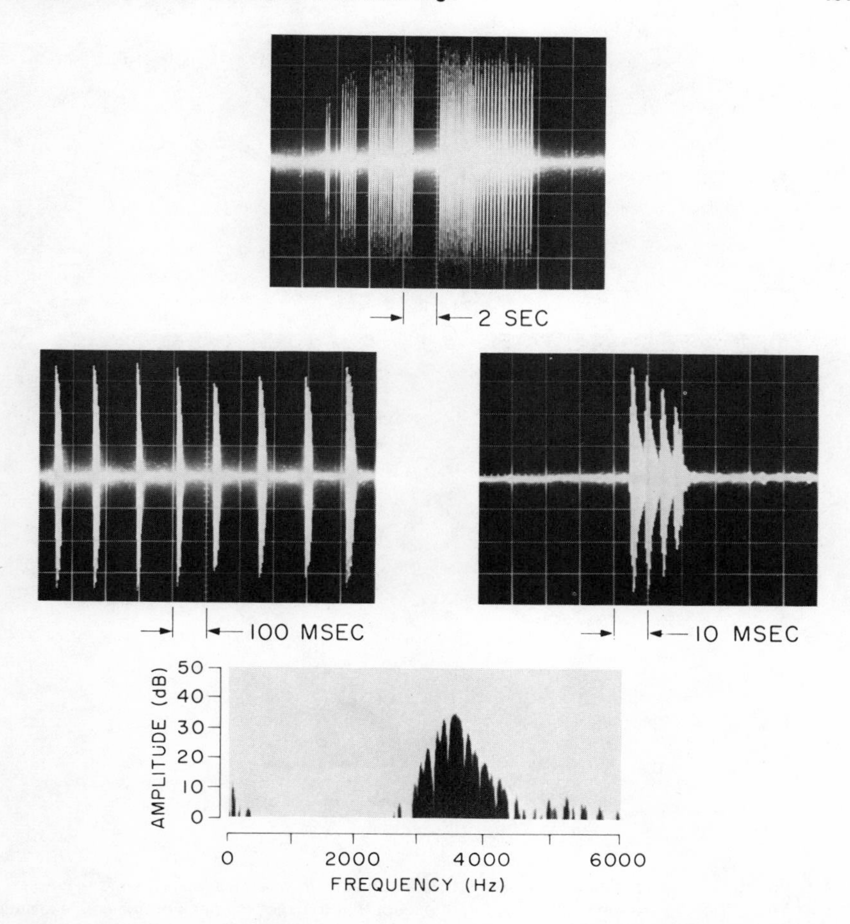

FIG. 5. Signal characteristics of a representative mating call from a male *Acris crepitans blanchardi* in central Texas. Details are the same as in Fig. 4 except that the spectral peak in this call is at 3650 Hz and the temporal pattern of clicks is clearly different from males in west Texas.

genus range were found in the xeric midwest (populations 1 and 11–13 in Table II), whereas the highest values characterized the mesic southeast.

Our analysis involved measurements first of overall gross call patterns and then of fine-temporal click and pulse structure at the beginning, middle, and end of the call (Table II). The overall average total call duration *T* was similar for both species (16.5 vs. 14.9 sec in *A. gryllus* and *A. crepitans*, respectively) but within each species it increased in a westward transect from *A. g. dorsalis* to *A. g. gryllus* and from *A. c. crepitans*

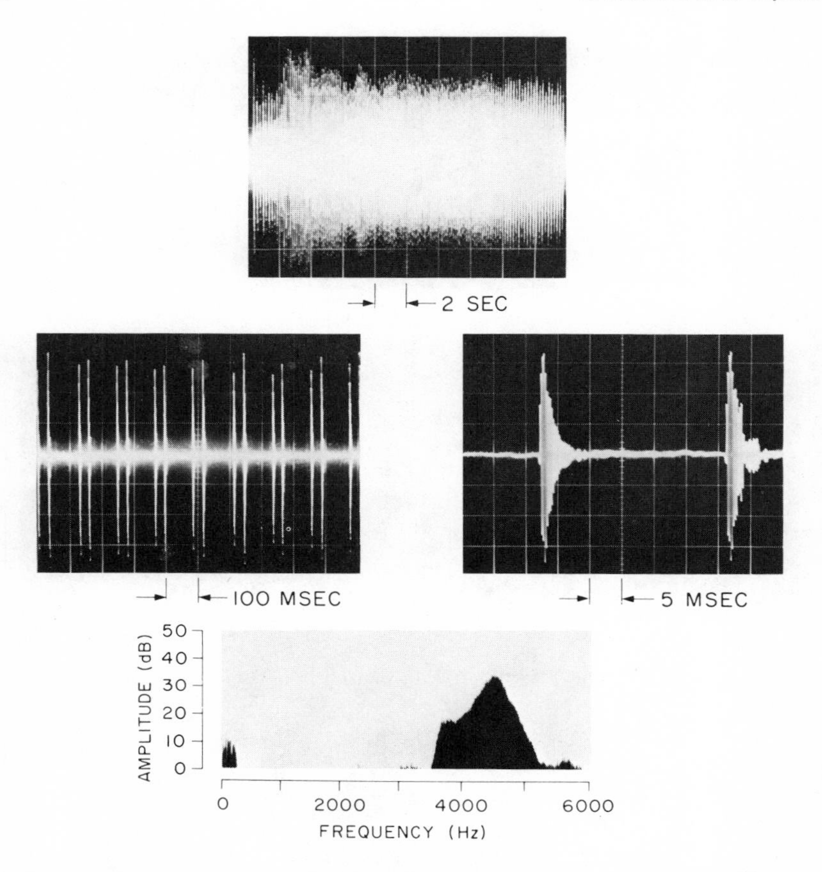

FIG. 6. Signal characteristics of a representative mating call from a male *Acris crepitans crepitans* in east Texas. Details are the same as in Fig. 4 except that the spectral peak in this call is around 4500 Hz and the temporal pattern of clicks is different from calls in west and central Texas.

to *A. c. blanchardi* (15.5 to 17.1 and 14.0 to 16.2 sec, or 10 and 16%, respectively). This trend was obvious in both the southern and northern transects, regardless of allopatry and sympatry. Again, as in measures of the spectral peak, the result of these parallel trends in both species was that the largest difference in call duration was found in the overlap zone between *A. g. gryllus* and *A. c. crepitans* (17.1 vs. 14.0 sec, namely a difference of 3.1 sec compared to a difference of 0.7 sec for the allopatric subspecies).

The total number of clicks in a call *NCT* was distinctly different in both species, 27.8 vs. 56.3 in *A. gryllus* compared to *A. crepitans*, but it

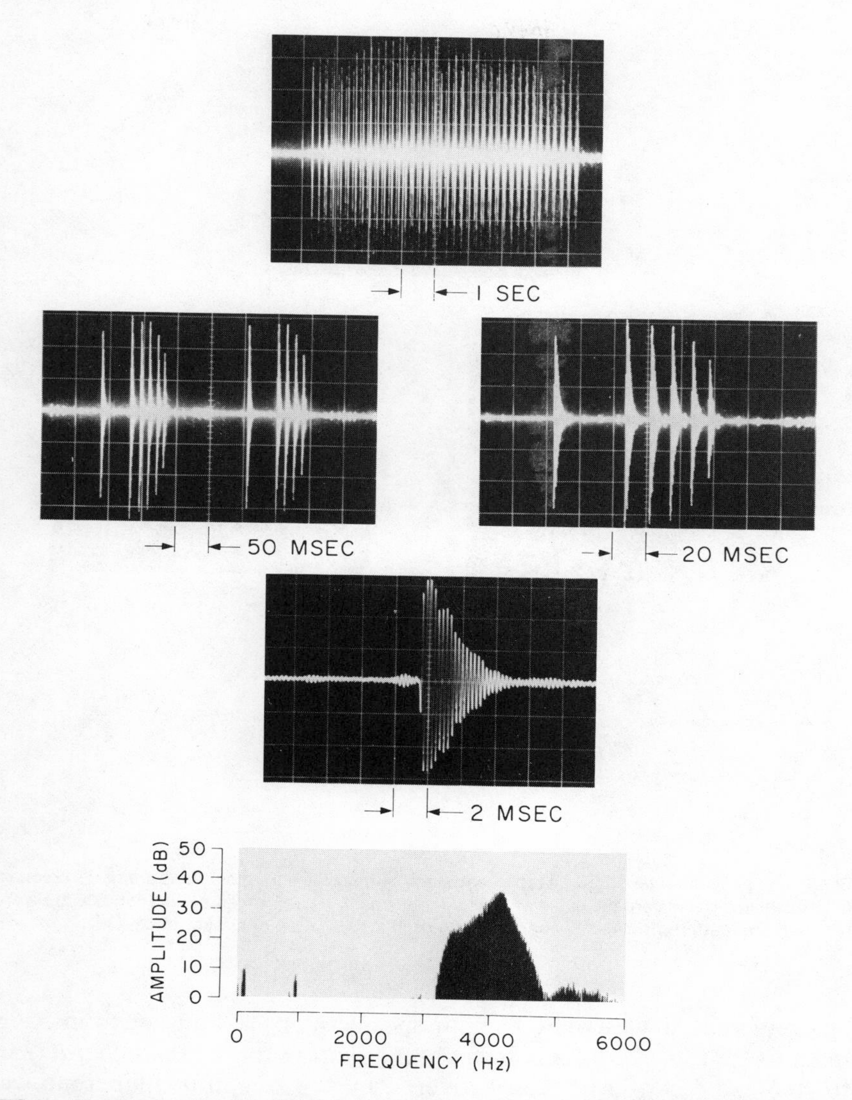

FIG. 7. Signal characteristics of a representative mating call from a male *Acris crepitans crepitans* in Georgia. Details are the same as in Fig. 4; the temporal pattern of clicks is unique to this population and the spectral energy is centered around 4000 Hz.

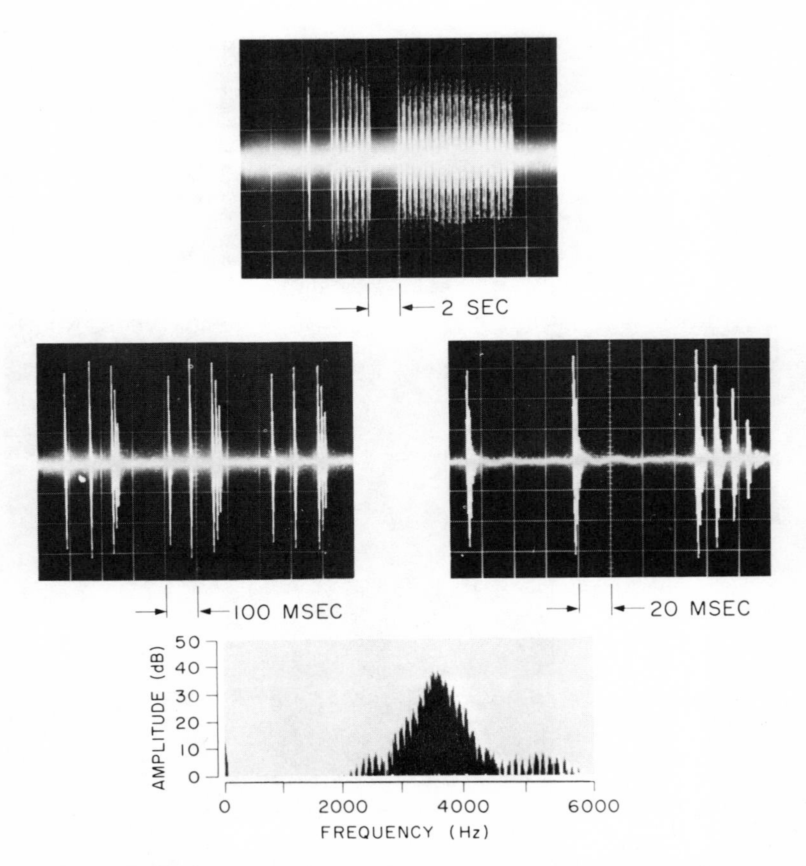

FIG. 8. Signal characteristics of a representative mating call from a male *Acris crepitans crepitans* in New Jersey. Details are the same as in Fig. 4; the spectral peak is centered around 3600 Hz and the temporal pattern of clicks is characteristic of this population of cricket frogs.

again increased westward in both species along a transect from *A. g. dorsalis* to *A. g. gryllus* and from *A. c. crepitans* to *A. c. blanchardi* (23.1 to 30.6 and 52.8 to 61.5, respectively). In *A. c. crepitans* this trend was similar in both the southern and northern allopatric transects. The differences in the number of clicks between *A. g. gryllus* and *A. c. crepitans* were substantial and nonoverlapping in their sympatric zone, 30.6 vs. 52.8, respectively, but these differences were less than those of *A. g. dorsalis* and *A. c. blanchardi* (23.1 vs. 61.5).

The rate of clicks in an entire call *NCT/T* was twice as high in *A. crepitans* as compared to *A. gryllus* (3.9 vs. 2.0). It also increased west-

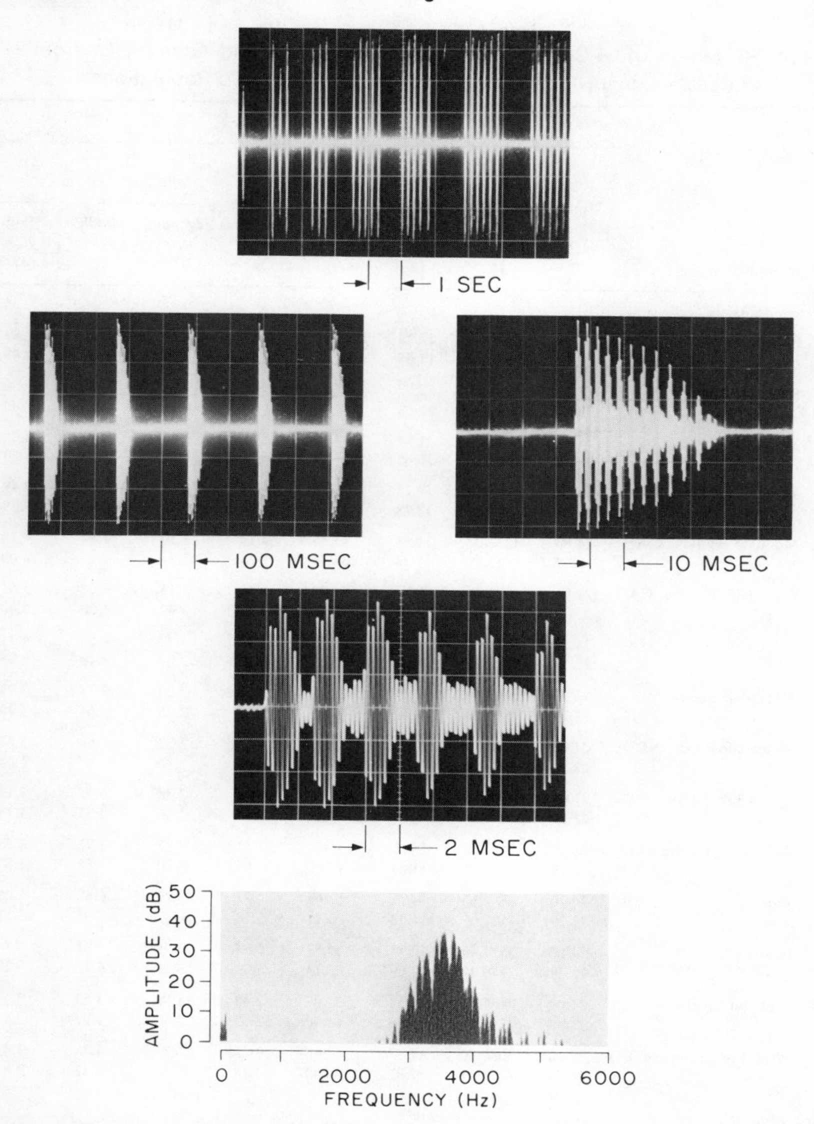

FIG. 9. Signal characteristics of a representative mating call from a male *Acris gryllus gryllus* in Georgia. Details are the same as in Fig. 4. Note that each individual click is more diffuse (i.e., contains a prolonged series of pulses of diminishing amplitude) than the clicks in calls of *Acris crepitans*.

TABLE II. Means of 16 Call Variables of *Acris crepitans* and *Acris gryllus* from 21 Localities, Involving 15 Allopatric and Six Sympatric Populations[a]

		Gross call structure							
							Rate of clicks in calls		
		Mouth temperature,	Spectral peak *SP*, Hz	Total call duration *T*, sec	Total number clicks in a call *NCT*	Average click rate *NCT/T*	Beginning *Rcb*	Middle *Rcm*	End *Rce*
	Variable number:	°C	11	12	13	14	15	16	17
Acris crepitans									
1	Balmorhea, TX	19.70	2994.80	17.20	44.92	2.81	2.21	3.68	3.19
		22.79	3217.18	17.65	55.91	3.41	2.50	4.35	3.83
2	Enchanted Rock, TX	23.70	3680.20	12.92	51.42	4.46	4.24	7.07	5.08
		22.79	3614.90	12.78	48.20	4.29	4.15	6.88	4.89
3	Sabine River, TX	26.00	4484.40	21.94	114.63	5.18	3.60	6.40	4.00
		22.79	4253.68	21.46	103.23	4.56	3.29	5.71	3.33
4	Atchafalaya River, LA	28.60	4166.80	19.13	98.50	5.35	3.98	6.21	4.49
		22.79	3749.08	18.28	77.86	4.23	3.43	4.96	3.29
5	Montgomery ponds, AL	24.80	4232.20	11.48	45.58	4.13	3.50	4.54	4.50
		22.79	4087.78	11.18	38.44	3.74	3.31	4.11	4.09
6a	Clark Hill Reserve, GA	21.00	4070.40	9.83	31.20	3.45	3.13	3.61	3.63
		22.79	4199.28	10.09	37.57	3.79	3.30	4.00	4.00
6b	Clark Hill Reserve, GA	24.50	3971.60	8.74	35.92	4.23	3.38	4.85	4.31
		22.79	3848.76	8.49	29.85	3.90	3.22	4.48	3.96
7a	Springdale ponds, NJ	16.90	3469.40	10.55	23.68	2.31	2.20	2.34	2.27
		22.79	3893.16	11.42	44.62	3.45	2.76	3.61	3.49
7b	Springdale ponds, NJ	23.00	3524.20	16.03	57.56	3.73	3.32	3.69	3.67
		22.79	3509.24	16.00	56.82	3.69	3.30	3.65	3.63
8	Kanawaukee Lake, NY	23.00	3868.50	15.64	40.03	2.86	2.38	3.55	3.19
		22.79	3853.54	15.61	39.29	2.82	2.36	3.51	3.15
9	Binnenwater Lake, NY	22.20	3602.80	13.56	45.31	3.51	3.15	3.75	3.53
		22.79	3645.38	13.65	47.41	3.63	3.20	3.88	3.65
10	Ann Arbor swamps, MI	20.70	3371.60	14.21	62.68	4.77	4.22	6.60	5.08
		22.79	3522.06	14.52	70.11	5.18	4.42	7.05	5.51
11	Fulton Creek, IL	24.60	3378.40	12.95	69.57	5.74	3.74	6.81	5.56
		22.79	3248.37	12.69	63.14	5.39	3.57	6.42	5.19
12	Scotland Creek, SD	21.00	3035.00	21.58	55.03	2.62	1.96	3.53	2.42
		22.79	3163.88	21.85	61.40	2.97	2.13	3.92	2.79
13	Wray Fish Hatchery, CO	18.00	3030.80	17.17	53.45	3.23	3.21	4.38	3.43
		22.79	3375.44	17.87	70.48	4.15	3.67	5.42	4.42
Acris gryllus									
14	Lacombe Fish Hatchery, LA	23.90	3587.40	9.51	31.76	3.72	3.30	5.43	6.21
		22.79	3507.71	9.34	27.82	3.51	3.20	5.19	5.98
15	Lyman Fish Hatchery, MS	26.30	3426.00	16.63	40.07	2.89	2.08	3.05	4.67
		22.79	3173.70	16.11	27.60	2.21	1.74	2.30	3.94
16	Montgomery ponds, AL	25.66	3464.80	17.14	42.50	2.62	1.91	3.09	4.05
		22.79	3258.53	16.72	32.31	2.06	1.64	2.47	3.46
17	Waynesboro ponds, GA	22.33	3564.00	24.13	37.05	1.52	0.91	1.63	3.78
		22.79	3596.99	24.19	38.68	1.61	0.96	1.73	3.87

Fine call structure								
Duration of clicks in cells			Number of pulses in a click in calls			Number of groups of pulses in clicks in calls		
Beginning *Dcb* 18	Middle *Dcm* 19	End *Dce* 20	Beginning *Npcb* 21	Middle *Npcm* 22	End *Npce* 23	Beginning *Ngpcb* 24	Middle *Ngpcm* 25	End *Ngpce* 26
22.63	76.01	119.38	5.20	6.81	9.41	1.00	1.76	2.41
14.05	60.38	92.15	5.40	7.30	9.95	1.00	1.66	2.12
14.75	19.62	62.45	4.16	5.19	10.52	1.00	1.05	1.63
17.27	24.21	70.45	4.11	5.05	10.37	1.04	1.08	1.72
13.42	71.21	103.63	1.83	3.84	8.37	1.32	2.80	2.83
39.64	107.93	139.17	1.62	3.33	7.82	1.45	2.91	3.14
28.75	78.73	108.57	2.73	5.02	8.59	1.65	2.84	2.70
76.22	145.20	172.93	2.37	4.10	7.58	1.88	3.04	3.26
62.68	118.17	111.77	3.31	4.84	5.69	2.39	3.42	3.20
79.09	141.15	134.02	3.18	4.52	5.34	2.46	3.49	3.39
94.82	137.56	133.80	3.38	5.09	6.55	2.24	3.22	2.84
80.17	117.05	113.94	3.50	5.37	6.86	2.17	3.16	2.67
48.68	99.44	111.00	2.80	4.44	6.16	1.72	3.00	3.00
62.64	118.99	129.93	2.69	4.17	5.86	1.79	3.06	3.16
120.75	193.81	218.32	3.68	4.85	5.99	2.43	3.09	3.81
72.59	126.39	153.03	4.05	5.79	7.01	2.20	2.89	3.25
62.12	137.68	141.20	2.48	5.36	6.28	2.12	3.44	3.48
63.82	140.06	143.50	2.47	5.33	6.24	2.13	3.45	3.50
64.49	144.80	159.10	2.30	4.04	5.61	1.61	3.15	3.35
66.19	147.18	161.40	2.29	4.00	5.58	1.62	3.16	3.37
55.87	141.78	149.26	2.51	4.67	5.50	1.82	3.15	3.14
51.03	135.01	142.70	2.55	4.76	5.60	1.80	3.13	3.08
15.80	30.22	67.34	3.40	4.64	7.54	1.04	1.20	1.40
10.00	19.65	48.92	3.53	4.97	7.90	1.00	1.13	1.20
12.36	33.89	49.74	2.98	3.98	6.40	1.05	1.58	1.47
17.38	43.03	65.67	2.87	3.69	6.08	1.12	1.64	1.64
27.45	92.45	127.47	4.33	5.78	13.37	1.27	2.95	2.47
22.48	83.39	111.69	4.45	6.07	13.68	1.20	2.89	2.29
32.69	44.08	92.37	6.35	6.84	9.96	1.00	1.16	1.52
19.40	19.86	50.17	6.65	7.60	10.79	1.00	1.00	1.06
31.04	35.84	39.08	12.04	13.48	14.08	1.00	1.00	1.00
30.59	36.49	40.94	11.97	13.30	13.89	1.00	1.00	1.00
25.30	31.96	38.20	9.79	12.30	14.14	1.00	1.00	1.00
23.89	34.00	44.10	9.57	11.74	13.53	1.00	1.00	1.00
23.90	30.61	35.37	8.25	10.04	11.12	1.00	1.00	1.00
22.74	32.28	40.19	8.07	9.59	10.62	1.00	1.00	1.00
14.98	27.55	41.47	4.76	9.29	12.28	1.00	1.00	1.00
15.16	27.28	40.70	4.79	9.36	12.36	1.00	1.00	1.00

(*Continued*)

TABLE II. (Continued)

		Gross call structure						
						Rate of clicks in calls		
	Variable number:	Mouth temperature, °C	Spectral peak SP, Hz 11	Total call duration T, sec 12	Total number clicks in a call NCT 13	Average click rate NCT/T 14	Beginning Rcb 15	Middle Rcm 16	End Rce 17
18	Society Hill ponds, GA	24.00	3553.00	19.34	30.98	1.77	1.31	2.37	3.56
		22.79	3466.12	19.17	26.69	1.53	1.20	2.11	3.31
19	Jacksonville ponds, FL	21.40	3542.40	17.22	27.21	1.72	1.24	2.54	3.64
		22.79	3642.51	17.43	32.15	1.99	1.38	2.84	3.92
20	Gainesville, FL	22.80	3872.40	16.31	24.08	1.54	1.31	2.11	2.88
		22.79	3871.82	16.30	24.05	1.54	1.31	2.11	2.87
21	Everglades swamps, FL	24.20	3957.60	13.02	18.16	1.63	1.36	2.24	2.68
		22.79	3856.34	12.81	13.16	1.36	1.23	1.93	2.39

[a] Upper row: analysis of raw recorded data; lower row: transformed data for standardized mouth temperature of 22.79°C.

ward in both species, from *A. g. dorsalis* to *A. g. gryllus* and from *A. c. crepitans* to *A. c. blanchardi* (1.6 to 2.2 and 3.9 to 4.2, or 37 and 11%, respectively). The increase in the average rate of clicks *NCT/T* westward was distinct in *A. g. gryllus* and in *A. c. crepitans*, but involved an opposite change within *A. c. blanchardi*. The differences in *NCT/T* between *A. g. gryllus* and *A. c. crepitans* were still substantial and nonoverlapping in the sympatric zone, but these differences were smaller in the sympatric than in the allopatric zones (1.6 vs. 2.6, respectively).

Rates of clicks in the beginning, middle, and end of call *Rcb, Rcm,* and *Rce* all had distinctly higher average values in *A. crepitans* compared with *A. gryllus* (3.2 vs. 1.6, 4.8 vs. 2.6, and 3.9 vs. 3.7). All three variables exhibited similar spatial patterns, increasing generally westward in both species from *A. g. dorsalis* to *A. g. gryllus* and from *A. c. crepitans* to *A. c. blanchardi* (for *Rcb*, 1.3 to 1.7 and 3.1 to 3.4; for *Rcm*, 2.3 to 2.8 and 4.2 to 5.7; and for *Rce*, 3.1 to 4.1 and 3.6 to 4.4, respectively). The differences in these values between *A. crepitans* and *A. gryllus* were again smaller in the sympatric than in the allopatric zones (*Rcb* = 1.4 vs. 2.1, *Rcm* = 1.4 vs. 3.4, and *Rce* = 0.5 vs. 1.3, and at least for *Rce* the values of both species overlapped in the sympatric zone).

The durations of clicks in the beginning, middle, and end of a call *Dcb, Dcm,* and *Dce* all had distinctly higher average values in *A. crepitans* compared with *A. gryllus* (for *Dcb, Dcm,* and *Dce* the means were 46.1 vs. 23.6, 95.3 vs. 32.8, and 115.3 vs. 41.1). All three click duration variables exhibited similar spatial patterns, increasing in a generally westward direction in *A. gryllus* and decreasing in *A. crepitans*, from *A. g.*

Fine call structure								
Duration of clicks in cells			Number of pulses in a click in calls			Number of groups of pulses in clicks in calls		
Beginning *Dcb* 18	Middle *Dcm* 19	End *Dce* 20	Beginning *Npcb* 21	Middle *Npcm* 22	End *Npce* 23	Beginning *Ngpcb* 24	Middle *Ngpcm* 25	End *Ngpce* 26
28.36	41.10	48.68	9.52	12.50	13.50	1.00	1.00	1.00
27.87	41.80	50.71	9.44	12.31	13.29	1.00	1.00	1.00
26.31	35.01	42.19	7.97	10.47	11.91	1.00	1.00	1.00
26.87	34.20	39.85	8.06	10.70	12.15	1.00	1.00	1.00
24.22	31.46	40.50	6.82	8.70	10.12	1.00	1.00	1.00
24.22	31.46	40.51	6.82	8.70	10.12	1.00	1.00	1.00
18.18	24.39	29.51	5.68	8.03	8.67	1.00	1.00	1.00
17.62	25.21	31.87	5.59	7.81	8.43	1.00	1.00	1.00

dorsalis to *A. g. gryllus* (22.9 vs. 24.1, 30.3 vs. 34.4, 37.5 vs. 43.3, respectively) and from *A. c. crepitans* to *A. c. blanchardi* (65.7 vs. 16.8, 131.0 vs. 41.8, and 143.4 vs. 73.2, respectively). The differences in these values between *A. crepitans* and *A. gryllus* were larger in sympatry than in allopatry (*Dcb* = 41.6 vs. 6.1, *Dcm* = 96.6 vs. 11.5, and *Dce* = 100.1 vs. 35.3).

The number of pulses in a click in the beginning, middle, and end of a call *Npcb, Npcm*, and *Npce* had higher values for *A. gryllus* compared with *A. crepitans* (for *Npcb, Npcm*, and *Npce* the means were 8.0 vs. 3.4, 10.4 vs. 5.0, and 11.8 vs. 7.8, respectively). All three variables displayed similar spatial patterns, increasing westward in both species from *A. g. dorsalis* to *A. g. gryllus* and from *A. c. crepitans* to *A. c. blanchardi* (for *Npcb*, 6.8 to 8.8 and 2.7 to 4.5; for *Npcm*, 9.1 to 11.3 and 4.6 to 5.8; and for *Npce*, 10.2 to 12.7 and 6.4 to 9.8, respectively) with no overlap in values between *A. gryllus* and *A. crepitans*. The differences in these values between *A. crepitans* and *A. gryllus* were larger in the sympatric than in the allopatric zones (*Npcb* = 6.1 vs. 2.3, *Npcm* = 6.7 vs. 3.3, and *Npce* = 6.3 vs. 0.4).

With regard to the number of groups of pulses in a click at the beginning, middle, and end of a call *Ngpcb, Ngpcm*, and *Ngpce*, *A. gryllus* always produced one group of pulses in each click throughout a call and across the entire range of its two subspecies. On the other hand, for *A. crepitans* the values of these three variables decreased westward from *A. c. crepitans* to *A. c. blanchardi* (*Ngpcb*, 1.9 to 1.1, *Ngpcm*, 3.1 to 1.6,

TABLE III. Mean and Standard Errors of 16 Call Variables, Standardized to Mouth Temperature of 22.79°C, of *Acris crepitans* and *Acris gryllus* and Their Subspecies[a]

	Acris crepitans						*Acris gryllus*					
	A.c.b. + *A.c.c.* $N = 15$		*A. c. blanchardi* $N = 6$		*A. c. crepitans* $N = 9$		*A.g.g.* + *A.g.d.* $N = 8$		*A. g. gryllus* $N = 5$		*A. g. dorsalis* $N = 3$	
Call variable	Mean	± SE	Mean	± SE	Mean	± SE	Mean	± SE	Mean	± SE	Mean	± SE
SP	3679	89.8	3357	73.7	3893	82.9	3546	88.99	3401	79.4	3790	74.0
T	14.9	1.03	16.2	1.46	14.0	1.40	16.5	1.54	17.1	2.41	15.5	1.39
NCT	56.3	4.91	61.5	3.49	52.8	7.84	27.8	2.63	30.6	2.24	23.1	5.50
NCT/T	3.9	0.19	4.2	0.39	3.8	0.16	2.0	0.24	2.2	0.36	1.6	0.19
Rcb	3.2	0.16	3.4	0.37	3.1	0.11	1.6	0.25	1.7	0.39	1.3	0.04
Rcm	4.8	0.32	5.7	0.54	4.2	0.24	2.6	0.39	2.8	0.62	2.3	0.28
Rce	3.9	0.20	4.4	0.41	3.6	0.11	3.7	0.38	4.1	0.48	3.1	0.45
Dcb	46.1	6.96	16.8	1.76	65.7	4.49	23.6	1.82	24.1	2.63	22.9	2.75
Dcm	95.3	12.65	41.8	10.58	131.0	4.65	32.8	1.83	34.4	2.39	30.3	2.66
Dce	115.3	10.52	73.2	10.04	143.4	5.83	41.1	1.84	43.3	1.97	37.5	2.78
Npcb	3.4	0.34	4.5	0.55	2.7	0.24	8.0	0.82	8.8	1.18	6.8	0.71
Npcm	5.0	0.32	5.8	0.61	4.6	0.26	10.4	0.67	11.3	0.77	9.1	0.85
Npce	7.8	0.62	9.8	1.06	6.4	0.31	11.8	0.68	12.7	0.59	10.2	1.08
Ngpcb	1.6	0.13	1.1	0.03	1.9	0.11	1.00		1.00		1.00	
Ngpcm	2.5	0.24	1.6	0.29	3.1	0.07	1.00		1.00		1.00	
Ngpce	2.6	0.22	1.7	0.20	3.2	0.08	1.00		1.00		1.00	

[a] Standardization to 22.79°C was conducted for all variables except "duration" for both species. "Duration" means were standardized for each subspecies separately since their regressions varied drastically for *Dcb*, *Dcm*, and *Dce* among subspecies. *N*, number of populations. *SP*, Spectral peak (dominant frequency); *T*, total call duration; *NCT*, total number of clicks in a call; *NCT/T*, rate of clicks in entire call; *Rcb*, rate of clicks in beginning of call; *Rcm*, rate of clicks in middle of call; *Rce*, rate of clicks in end of call; *Dcb*, duration of a click in the beginning of the call; *Dcm*, duration of a click in the middle of the call; *Dce*, duration of a click in the end of the call; *Npcb*, number of pulses in a click in the beginning of the call; *Npcm*, number of pulses in a click in the middle of the call; *Npce*, number of pulses in a click in the end of the call; *Ngpcb*, number of groups of pulses in the beginning of the call; *Ngpcm*, number of groups of pulses in the middle of the call; *Ngpce*, number of groups of pulses in the end of the call.

and *Ngpce*, 3.2 to 1.7). For all three variables there was no overlap in values between *A. gryllus* and *A. crepitans*.

In summary, for 12 of the call variables (*NCT, NCT/T, Rcb, Rcm, Dcm, Dce, Npcb, Npcm, Npce, Ngpcb, Ngpcm, Ngpce*), *A. gryllus* distinctly differed from *A. crepitans* regardless of their spatial relationships, whether allopatric or sympatric. For the remaining four call variables the values of both species overlapped and showed parallel changes with geographic direction in three variables (*SP, T, Rce*) and opposed changes in *Dcb*. In five of the temporal call variables (*NCT, NCT/T, Rcb, Rcm,* and *Rce*) the differences in values for the two species were smaller in their sympatric than in their allopatric zones; for the remaining ten temporal call variables, this was reversed: the differences in values were greater in the sympatric than in the allopatric zones. Only for the spectral peak *SP* was the difference similar in both zones.

Multiple comparisons of the paired means of various call parameters for different geographic populations of the two species (Table II) were performed with the Scheffe Test, which is the most conservative of the seven *a posteriori* contrast tests available in the SPSS computer programs. This discrimination analysis leads to a very clear distinction between the call characteristics of the two species and furthermore to the distinction of the two subspecies of *A. crepitans*, namely *A. c. crepitans* and *A. c. blanchardi*. Most of the population comparisons between *A. c. crepitans* and *A. gryllus* were significant with regard to the variables *Dcm, Dce, Npcb, Npcm, Npce, Ngpcm*, and *Ngpce*. On the other hand,*A. c. blanchardi* differs significantly in most comparisons with *A. gryllus* in *Npcb* and *Npcm*, and with *A. c. crepitans* in *SP* and *Dcm*.

Environmental and Morphological Correlates of Call Variation

In our search for correlates of call variation, we considered several environmental and morphological variables (Pearsonian correlations; see Materials and Methods). The intercorrelations of these variables either in *A. crepitans, A. gryllus*, or in both species combined are given in Table IV. The outstanding high correlations were between latitude and annual temperature ($r = -0.95$ for each species and both combined); annual rainfall and altitude ($r = -0.8$ for all three); body length and rainfall ($r = -0.72$ for *A. crepitans* and -0.67 for both species).

The partial correlation matrix among all 16 call variables for each species separately and for both species combined is given in Table V. The following significant correlations are outstanding: First, concerning gross call structure, spectral peaks are significantly correlated with several temporal call variables (click duration, number of pulses in a click,

TABLE IV. Correlation Coefficients *r* among Nine Environmental and Morphological Background Variables for *Acris crepitans*, *Acris gryllus*, and Both Species Combined[a]

	Latitude *Lt*	Longitude *Ln*	Altitude *Al*	Annual rainfall *Rn*	Annual temperature *Ta*	Recording temperature standardized on mouth temperature (partial *r*) T_{air}	T_{water}	Body length *BL*	Hindfoot length *FL*
Lt									
Ln	−0.427								
	−0.118								
	−0.234								
Al	0.033	−0.594							
	0.763	−0.365							
	0.295	0.573							
Rn	−0.235	−0.620	−0.877						
	−0.753	0.696	−0.799						
	−0.499	−0.545	−0.897						
Ta	−0.961	0.274	−0.237	0.441					
	−0.952	0.304	−0.837	0.823					
	−0.973	0.123	−0.444	0.639					
T_{air}	−0.288	0.195	−0.131	0.051	0.342				
	−0.115	0.245	−0.139	0.185	0.222				
	−0.346	0.106	−0.239	0.161	0.397				
T_{water}	0.271	−0.099	0.218	−0.135	−0.324	−0.895			
	−0.544	0.189	−0.318	0.501	0.516	−0.265			
	0.322	−0.003	0.322	−0.227	−0.373	−0.890			
BL	0.248	0.534	0.595	−0.719	−0.404	0.105	−0.083		
	0.688	0.170	0.425	−0.285	−0.536	−0.146	−0.193		
	0.339	0.524	0.595	−0.672	0.446	−0.003	−0.002		
FL	0.050	0.567	0.395	−0.543	−0.175	0.143	−0.121	0.893	
	0.528	0.404	0.195	−0.060	−0.369	−0.048	−0.262	0.860	
	0.007	0.492	0.235	−0.295	−0.070	0.083	−0.107	0.808	

[a] *A. crepitans*, 13 localities, top value; *A. gryllus*, eight localities, middle value; both species combined, 20 localities, bottom value.

and number of groups of pulses in a click). Similarly, gross call variables are partly intercorrelated. Call duration *T* is highly and significantly correlated with total number of clicks *NCT*, as well as with other temporal call variables, though less strongly; and click rate in the entire call *NCT/T* is correlated with call duration *T* and number of clicks, *NCT* in both species. Second, concerning fine call structure, each of the four classes of call variables—click rates (*Rcb, Rcm, Rce*), click duration (*Dcb, Dcm, Dce*), number of pulses in a click (*Npcb, Npcm, Npce*), and number of groups of pulses in a click (*Ngpcb, Ngpcm, Ngpce*)—displayed significant correlations within its three variables at the beginning, middle, and end of the call ($r > 0.50$). Similarly, click rates in the three portions of the call (*Rcb, Rcm, Rce*) were highly correlated with the total rates of clicks in the entire call ($r = 0.763, 0.866, 0.511$); click durations (*Dcb, Dcm, Dce*) in *A. gryllus* with number of pulses in a click ($r = 0.46$–0.89); and number of pulses within a click ($r \leq 0.65$). It is important to appreciate these intercorrelations of call variables in order to interpret our subsequent discussion concerning call structure and geographic differentiation.

Partial correlation coefficients between standardized call variables and morphological, ecological, and geographic measures are given in Table VI. In each species and within the entire genus, spectral peaks are primarily correlated negatively and highly significantly with body length *BL* and secondarily with several environmental variables: annual rainfall *Rn* (excepting *A. gryllus*), annual temperature *Ta*, altitude *Al*, longitude *Ln*, and latitude *Lt*. Call duration *T* displayed a rather low but nevertheless significant correlation in *A. crepitans* with *BL, Rn,* and *Ln*; and in *A. gryllus* this call variable was significantly correlated with *Rn, Ta, Al, Ln,* and *Lt*. The total number of clicks in a call *NCT*, particularly in *A. gryllus*, was highly correlated with *BL, FL, Ta, Al,* and *Lt*; and in *Acris* overall this call variable was especially correlated with annual rainfall *Rn*. Average click rate *NCT/T* in *A. gryllus* exhibited highly significant correlation with longitude *Ln*, and in the genus overall this call variable was also correlated primarily with *Rn, Ta,* and *Lt*.

The rate of clicks within a call displayed medium correlations for *Rn* and *Ln* (with *Rcb* and *Rcm*) and for foot length *FL* (with *Rce*). Click durations displayed several high correlations with *Rn, T,* and *Lt* and medium correlations with *BL*. The number of pulses per click was highly correlated with *Ln* and *Rn* (particularly in *A. crepitans* for *Npcb* and *Npce*) and with *Lt* in *A. gryllus*. Finally, the number of groups of pulses in a click was invariant in *A. gryllus* but showed very high correlations with *Rn, Ln,* and *BL* within *A. crepitans*.

In summary, the spectral and temporal characteristics of the mating calls of both species were significantly correlated with the environment.

TABLE V. Partial Correlation Coefficients *r* among All 16 Call Variables in *Acris crepitans, Acris gryllus,* and Both Species Combined on Standardized Mouth Temperature[a]

	Gross call structure				Rates of clicks in call		
Variable	Spectral peak *Sp*, Hz	Total call duration *T*, sec	Total number of clicks in a call *NCT*	Total rates of clicks *NCT/T*	Beginning *Rcb*	Middle *Rcm*	End *Rce*
11 *SP*		0.174	−0.092	−0.114	−0.065	−0.216	−0.216
		−0.108	−0.463**	−0.356*	−0.111	−0.115	−0.439**
		−0.137	0.035	0.049	0.056	0.008	−0.217*
12 *T*			0.801***	−0.340**	−0.430***	−0.034	−0.311**
			0.656***	−0.566***	−0.428**	−0.497**	−0.238
			0.545***	−0.346***	−0.402***	−0.197*	−0.289**
13 *NCT*				0.201	−0.158	0.360***	−0.035
				0.124	0.019	0.053	0.379*
				0.520***	0.260**	0.543***	0.083
14 *NCT/T*					0.547***	0.774***	0.605***
					0.697***	0.798***	0.805***
					0.763***	0.866***	0.511***
15 *Rcb*						0.421***	0.636***
						0.871***	0.668***
						0.706***	0.556***
16 *Rcm*							0.652***
							0.744***
							0.593***
17 *Rce*							
18 *Dcb*							
19 *Dcm*							
20 *Dce*							
21 *Npcb*							
22 *Npcm*							
25 *Npce*							
24 *Ngpcb*							
25 *Ngpcm*							
26 *Ngpce*							

[a] *A. crepitans*, $N = 74$, top value; *A. gryllus*, $N = 38$, middle value; both species combined, $N = 112$, bottom value. Levels of significance, *: $p < 0.05$; **: $p < 0.01$; ***: $p < 0.001$.

Fine call structure								
Duration of clicks in call			Number of pulses in a click			Number of groups of pulses in click		
Beginning *Dcb*	Middle *Dcm*	End *Dce*	Beginning *Npcb*	Middle *Npcm*	End *Npce*	Beginning *Ngpcb*	Middle *Ngpcm*	End *Ngpce*
0.532***	0.509***	0.482***	-0.377***	-0.256*	-0.390***	0.487***	0.430***	0.484***
-0.245	-0.387*	-0.391*	-0.382*	-0.476**	-0.573***	—	—	—
0.429***	0.437***	0.397***	-0.386***	0.344***	-0.457***	0.455***	0.417***	0.434***
-0.304**	-0.106	0.107	0.032	0.084	0.480***	-0.372***	-0.029	0.075
-0.184	0.041	0.240	-0.298	-0.138	-0.015	—	—	—
-0.288**	-0.134	-0.001	-0.001	0.070	0.312***	-0.320***	-0.085	-0.029
-0.384***	-0.328**	-0.092	-0.052	-0.084	0.362***	-0.404***	-0.265*	-0.111
0.038	0.171	0.190	0.082	0.198	0.207	—	—	—
-0.159	0.046	0.296***	-0.476***	-0.495***	-0.159	-0.022	0.217*	0.317***
-0.279*	-0.568***	-0.556***	0.011	-0.180	-0.087	-0.226*	-0.582***	-0.518***
0.359*	0.186	0.003	0.599***	0.519***	0.437**	—	—	—
0.042	0.077	0.219*	-0.505***	-0.596***	-0.444***	0.227*	0.233*	0.280*
-0.034	-0.293**	-0.506***	0.232*	0.187	-0.150	0.071	-0.318**	-0.519***
0.520***	0.300	0.001	0.620***	0.486**	0.305	—	—	—
0.158	0.135	0.124	-0.331***	-0.405***	-0.408***	0.330***	0.239*	0.154
-0.530***	-0.805***	-0.659***	0.172	-0.243*	0.197	-0.548***	-0.826***	-0.661***
0.478**	0.303	0.033	0.586***	0.465**	0.299	—	—	—
-0.170	-0.165	0.044	-0.352***	-0.500***	-0.254**	-0.042	-0.013	0.088
-0.203	-0.505***	-0.715***	0.190	-0.072	-0.286*	-0.168	-0.544***	-0.673***
0.327*	0.168	0.070	0.573***	0.473**	0.473**	—	—	—
-0.107	-0.291**	-0.367***	0.129	0.015	-0.092	-0.073	-0.258**	-0.324***
	0.691***	0.552***	-0.194	-0.049	-0.507***	0.901***	0.620***	0.542***
	0.777***	0.477**	0.888***	0.689***	0.460**	—	—	—
	0.710***	0.578***	-0.182	-0.176	-0.437***	0.876***	0.602***	0.550***
		0.828***	-0.399***	0.045	-0.418***	0.711***	0.918***	0.782***
		0.751***	0.714***	0.802***	0.580***	—	—	—
		0.870***	-0.461***	-0.369***	-0.521***	0.773***	0.916***	0.830***
			-0.371***	-0.084	-0.166	0.482***	0.719***	0.901***
			0.439**	0.584***	0.688***	—	—	—
			-0.562***	-0.527***	-0.452***	0.622***	0.826***	0.936***
				0.676***	0.518***	-0.216	-0.363***	-0.435***
				0.853***	0.675***	—	—	—
				0.935***	0.762***	-0.431***	-0.624***	-0.645***
					0.390***	0.017	0.111	-0.123
					0.840***	—	—	—
					0.766***	-0.388***	-0.538***	-0.604***
						-0.507***	-0.274*	-0.257*
						—	—	—
						-0.587***	-0.556***	-0.551***
							0.688***	0.493***
							—	—
							0.763***	0.637***
								0.732***
								—
								0.851***

TABLE VI. Partial Correlation Coefficients *r* of Call Variables, Standardized for Mouth Temperature, with Morphological, Ecological, and Geographic Variables of *Acris crepitans*, *Acris gryllus*, and Both Species Combined[a]

	Morphological		Ecological		Geographic		
Call variable	Body length	Hindfoot length	Rain	Temperature	Altitude	Longitude	Latitude
SP	−0.780***	−0.690***	0.769***	0.580***	−0.552***	−0.428***	−0.395***
	−0.523***	−0.491**	—	0.478**	−0.321*	—	−0.658***
	−0.703***	−0.660***	0.490***	0.379***	−0.420***	−0.374***	−0.280**
T	0.268*	—	−0.244*	—	—	0.329**	—
	—	—	−0.537***	−0.419**	0.453**	−0.333*	0.415*
	0.186*	—	—	—	—	0.189*	—
NCT	0.245*	—	—	—	—	0.340**	—
	0.436**	0.497**	−0.326*	−0.418**	0.418**	—	0.554***
	0.274**	—	−0.428***	−0.351***	0.283**	0.366***	0.320***
NCT/T	—	—	—	—	—	—	—
	—	0.370*	0.404*	—	—	0.782***	—
	—	—	−0.437***	−0.431***	0.293**	0.270**	0.431***
Rcb	—	—	—	—	—	—	—
	—	—	0.390*	—	—	0.619***	—
	—	—	−0.357***	−0.340***	0.294**	0.195*	0.320***
Rcm	—	0.233*	−0.406***	—	0.249*	0.396***	—
	—	—	0.395*	—	—	0.642***	—
	0.215*	—	−0.515***	0.335***	0.370***	0.420***	0.288**
Rce	—	—	—	—	—	—	—
	0.407***	0.509***	—	—	—	0.734***	—
	—	0.213*	—	—	—	0.193*	—
Dcb	−0.557***	−0.547***	0.742***	—	−0.539***	−0.649***	—
	—	—	—	—	—	—	—

	−0.433***	−0.453***	0.437***	—	−0.329***	−0.521***	—
Dcm	−0.490***	−0.498***	0.729***	—	−0.586***	−0.675***	—
	0.368*	0.422**	—	−0.519***	—	—	0.537***
	−0.313***	−0.425***	0.250**	−0.239**	−0.239**	−0.459***	0.323***
Dce	−0.389***	−0.381***	0.632***	—	−0.485***	−0.533***	—
	—	0.352*	−0.545***	−0.666***	0.327*	—	0.682***
	—	−0.331***	—	−0.312***	—	−0.264**	0.359***
Npcb	0.352**	0.294**	−0.547***	—	0.665***	0.660***	−0.263*
	0.338*	0.443**	—	—	—	0.489**	—
	—	0.336***	0.279**	0.426***	—	—	−0.428***
Npcm	—	—	−0.326**	—	0.454***	0.389***	—
	0.420**	0.543***	—	−0.334*	—	0.376*	0.450**
	—	0.308***	0.352***	0.434***	−0.191*	—	−0.435***
Npce	0.452***	0.405***	−0.595***	—	0.472***	0.714***	—
	0.373*	0.523***	—	−0.396*	—	0.363*	0.554***
	0.260**	0.444***	—	0.291**	—	0.372***	−0.328***
Ngpcb	−0.535***	−0.513***	0.736***	—	−0.557***	−0.628***	—
	No variance	No variance	No variance	No variance	No variance	no variance	No variance
	−0.385***	−0.467***	0.290**	—	−0.252**	−0.444***	0.248**
Ngpcm	−0.370***	−0.399***	0.636***	—	−0.545***	−0.552***	—
	No variance	No variance	No variance	No variance	No variance	No variance	No variance
	—	−0.375***	—	−0.330***	—	−0.281**	0.393***
Ngpce	−0.384***	−0.378***	0.610***	—	−0.461***	−0.526***	—
	No variance	No variance	No variance	No variance	No variance	No variance	No variance
	—	−0.352***	—	−0.291**	—	−0.244**	0.331***

[a] *A. crepitans*, $N = 74$, top value; *A. gryllus*, $N = 38$, middle value; both species combined, $N = 112$, bottom value. Levels of significance, *: $p < 0.05$; **: $p < 0.01$; ***: $p < 0.001$.

The major correlates of the call variables for *A. crepitans* involved longitude, rainfall, body length, and altitude. In contrast, the major correlates of call variables for *A. gryllus* focused on hindfoot length, longitude, latitude, and body length. Finally, for *Acris* as a whole, the major environmental correlates with call variables are longitude, latitude, temperature, hindfoot length, and rainfall.

Multivariate Analysis

A test for the best predictors of the call variables (standardized to 22.79°C mouth temperature) was performed by a stepwise multiple regression analysis (Nie *et al.*, 1975) by employing the call characters as dependent variables and morphological, climatic, and geographic factors as independent variables. The results for *A. crepitans* and its two subspecies and for *A. gryllus* are given in Table VII. Notably, the geographic variation in a high proportion of the call variables is significantly accounted for by a combination of from one to three morphological, climatic, and geographic variables. For example, in *A. crepitans* body length *BL* was the single most important morphological variable, explaining significantly 70% of the variance in spectral peaks *SP* and 85% for *A. c. blanchardi*, whereas in *A. crepitans* 60% of the variance in *SP* was accounted for by foot length *FL*. A two-variable combination of body length *BL* and annual temperature *Ta* explained significantly 77% of the variance of *SP* in *A. crepitans*. Another example that demonstrates a substantial increase in explanation by additional variables was *Dcb* in *A. c. crepitans*. Note that *Rn* alone explained only 20% of the spatial variance, but remarkably the two-variable combination of rainfall and longitude *Rn* and *Ln* explained significantly 82% of the variance. Another striking example is *Dcm* in *A. c. crepitans*, where *Ta* alone explained nonsignificantly 16% of the variance, while *Ta* and *Rn* together explained significantly 74%, and the three-variable combination of *Ta, Rn*, and *BL* explained significantly 86% of the variance. Similar cases occurred in other variables, i.e., *Dcb, Dce, Npce,* and *Ngpcb* in *A. crepitans*; *T, NCT, Dcb,* and *Npce* in *A. c. crepitans*; *NCT* and *Npcm* in *A. c. blanchardi*; and *NCT, NCT/T, Rce, Dce, Npcm*, and *Npce* in *A. gryllus*.

We have deliberately analyzed the two subspecies of *A. crepitans* separately since in several cases they mask each other when analyzed together, and the variance is not significantly explained for the species as a whole (e.g., in the cases of *NCT, NCT/T, Rcb,* etc.). In other cases, while the variance within each subspecies was not explained significantly, nevertheless a significant result was obtained for both together (e.g., *Dce,*

TABLE VII. Coefficient of Multiple Regression R^2 with 16 Call Variables Standardized to Mouth Temperature of 22.79°C As Dependent Variables and Various Independent Variables[a]

Species and subspecies	Call variable	Stepwise model X_1	X_2	X_3	R_1^2	R_{12}^2	R_{123}^2
A. crepitans	*SP*	*BL*	*Ta*	*FL*	0.70***	0.77***	0.82***
A. c. blanchardi		*BL*	*Al*	*Ln*	0.85**	0.91*	0.95
A. c. crepitans		*Lt*	*FL*	*BL*	0.39	0.65*	0.78*
A. gryllus		*FL*	*Al*	*Rn*	0.60*	0.64	0.81
A. crepitans	*T*	*Ln*	*Lt*	*Rn*	0.24	0.32	0.38
A. c. blanchardi		*BL*	*Rn*	*Ta*	0.38	0.46	0.73
A. c. crepitans		*Ln*	*Lt*	*Al*	0.18	0.87**	0.91**
A. gryllus		*Rn*	*Ta*	*Lt*	0.72*	0.75	0.80
A. crepitans	*NCT*	*Ln*	*Al*	*Ta*	0.21	0.33	0.40
A. c. blanchardi		*Ta*	*Al*	*Rn*	0.72*	0.83	0.99**
A. c. crepitans		*Ln*	*Lt*	*Rn*	0.42	0.93***	0.95**
A. gryllus		*Lt*	*Ta*	*Al*	0.61*	0.89**	0.97**
A. crepitans	*NCT/T*	*Ln*	*Al*	*Rn*	0.05	0.18	0.22
A. c. blanchardi		*Ln*	*Ta*	*Lt*	0.56	0.65	0.75
A.c. crepitans		*Al*	*Ln*	*Rn*	0.83***	0.84**	0.87*
A. gryllus		*Ln*	*Bl*	*Lt*	0.69*	0.77*	0.85*
A. crepitans	*Rcb*	*BL*	*FL*	*Rn*	0.05	0.20	0.35
A. c. blanchardi		*BL*	*Ln*	*Al*	0.68*	0.74	0.93
A.c. crepitans		*Al*	*Ln*	*Lt*	0.62*	0.65*	0.74
A. gryllus		*Ln*	*Al*	*BL*	0.65*	0.67	0.73
A. crepitans	*Rcm*	*Ln*	*Al*	*Rn*	0.23	0.27	0.34
A. c. blanchardi		*BL*	*Ln*	*Ta*	0.67*	0.80	0.87
A. c. crepitans		*Ln*	*Rn*	*FL*	0.85***	0.95***	0.97***
A. gryllus		*Ln*	*BL*	*Lt*	0.42	0.47	0.66
A. crepitans	*Rce*	*Rn*	*Al*	*BL*	0.09	0.10	0.12
A. c. blanchardi		*Ln*	*Ta*	*Al*	0.39	0.64	0.68
A. c. crepitans		*Ta*	*Ln*	*Lt*	0.05	0.58	0.69
A. gryllus		*FL*	*Ta*	*Rn*	0.64*	0.81*	0.82
A. crepitans	*Dcb*	*Rn*	*Al*	*Ln*	0.71***	0.82***	0.87***
A. c. blanchardi		*Ln*	*Lt*	*Rn*	0.27	0.62	0.94
A. c. crepitans		*Rn*	*Ln*	*Lt*	0.20	0.82**	0.86*
A. gryllus		*BL*	*Al*	*FL*	0.24	0.47	0.58

(*Continued*)

TABLE VII. (*Continued*)

Species and subspecies	Call variable	Stepwise model X_1	X_2	X_3	R_1^2	R_{12}^2	R_{123}^2
A. crepitans	*Dcm*	*Rn*	*Ln*	*Al*	0.62***	0.66**	0.70*
A. c. blanchardi		*FL*	*Ln*	*Al*	0.58	0.61	0.92
A. c. crepitans		*Ta*	*Rn*	*BL*	0.16	0.74*	0.86*
A. gryllus		*BL*	*Al*	*Ta*	0.51*	0.55	0.74
A. crepitans	*Dce*	*Rn*	*Al*	*Ln*	0.54**	0.56**	0.58*
A. c. blanchardi		*FL*	*Ln*	*Al*	0.37	0.48	0.87
A. c. crepitans		*FL*	*Ta*	*Ln*	0.05	0.15	0.63
A. gryllus		*Lt*	*Al*	*Ta*	0.67*	0.77*	0.88*
A. crepitans	*Npcb*	*Al*	*BL*	*Rn*	0.79***	0.82***	0.82***
A. c. blanchardi		*Al*	*Ta*	*Rn*	0.88**	0.97**	0.98*
A. c. crepitans		*BL*	*FL*	*Rn*	0.31	0.38	0.42
A. gryllus		*FL*	*Rn*	*Ta*	0.44	0.63	0.74
A. crepitans	*Npcm*	*Al*	*BL*	*Ln*	0.70***	0.73***	0.76**
A. c. blanchardi		*Rn*	*Ln*	*Al*	0.86**	0.96**	0.99**
A. c. crepitans		*BL*	*FL*	*Ln*	0.51*	0.66*	0.69
A. gryllus		*FL*	*Al*	*Ta*	0.72**	0.81*	0.92*
A. crepitans	*Npce*	*Ln*	*Ta*	*BL*	0.59***	0.73***	0.73**
A. c. blanchardi		*Ln*	*Ta*	*BL*	0.37	0.52	0.64
A. c. crepitans		*Ln*	*BL*	*Rn*	0.36	0.70*	0.90**
A. gryllus		*FL*	*Al*	*Lt*	0.83**	0.86**	0.99***
A. crepitans	*Ngpcb*	*Rn*	*Ln*	*Al*	0.64***	0.69***	0.76**
A. c. blanchardi		*Al*	*Ln*	*Lt*	0.31	0.81	1.00**
A. c. crepitans		*BL*	*Rn*	*Ln*	0.09	0.13	0.41
A. gryllus				No variance			
A. crepitans	*Ngpcm*	*Rn*	*Ln*	*BL*	0.58***	0.61**	0.62*
A. c. blanchardi		*FL*	*Al*	*Rn*	0.38	0.49	0.66
A. c. crepitans		*FL*	*Rn*	*Ln*	0.11	0.15	0.47
A. gryllus				No variance			
A. crepitans	*Ngpce*	*Rn*	*FL*	*BL*	0.60***	0.63**	0.68**
A. c. blanchardi		*FL*	*Ta*	*Al*	0.36	0.50	0.61
A. c. crepitans		*Ta*	*Lt*	*Rn*	0.11	0.54	0.62
A. gryllus				No variance			

[a] Including morphological (*BL*, body length; *FL*, foot length), climatic (*Rn*, mean annual rainfall; *Ta*, mean annual temperature), and geographic (*Ln*, longitude; *Lt*, latitude; *Al*, altitude) factors. Levels of significance, *: $p < 0.05$; **: $p < 0.01$; ***: $p < 0.001$.

Ngpcm, Ngpce). In no case did a call variable remain unexplained in either *A. crepitans* (and/or one or both of its subspecies) or in *A. gryllus*. The percentage of the call variables that were explained, either singly or in combination (out of the 16 parameters analyzed), for *A. crepitans, A. c. blanchardi, A. c. crepitans*, and *A. gryllus* was 62.5, 44, 62.5, and 77%, respectively.

The percent of explanation varied among the independent variables employed in the analysis. The ranking of the seven independent variables in explaining singly a significant proportion of the call variance was annual rainfall (27%), body length (20%), altitude (17%), foot length (13%), latitude (7%), and annual temperature (3%).

The ranking of paired independent variables, regardless of their order, was *Rn–Ln* (21%), *BL–Al* (10%), *Ln–Lt* (7%), *Al–Ln* (7%), *Rn–Al* (7%), *FL–Al* (7%), *BL–Ln* (7%), while the other two-variable combinations were 3% each. Finally, the three-variable combination of rainfall, longitude, and latitude (*Rn–Ln–Lt*) accounted significantly for 24% of the possible three-variable combinations.

Clearly, relatively few environmental variables account for the substantial degree of geographic variation in the spectral and temporal parameters of the mating calls within *Acris*. Moreover, these critical environmental variables are those involved in the ecological divergence of grasslands and woodlands.

Partial Order Scalogram Analysis (POSA)

The POSA space diagram of call structure is shown in Fig. 10. Call structure across the ranges of both species partitioned the analysis space best on the basis of a profile, or structupule, of nine call variables (each categorized from 1 to 4 and representing low, medium, high, and very high values, respectively). Each profile involved three gross-temporal call structure variables, five fine-temporal call variables, and one spectral variable: (1) average click rate *NCT/T*; (2) rate of clicks in the beginning of a call *Rcb*; (3) rate of clicks in the middle of a call *Rcm*; (4) click duration in the beginning of a call *Dcb*; (5) click duration in the middle of a call *Dcm*; (6) number of pulses in a click in the end of a call *Npce*; (7) number of groups of pulses in the beginning of a call *Ngpcb*; (8) number of groups of pulses in the middle of a call *Ngpcm*; and (9) the spectral peak in a call *SP*.

The POSA generated a partial ordering of *Acris* call structure in a Cartesian space by means of five profiles:

Profile 1: 1, 1, 1, 1, 1, 2, 1, 1, 3 (lowest click rates and durations,

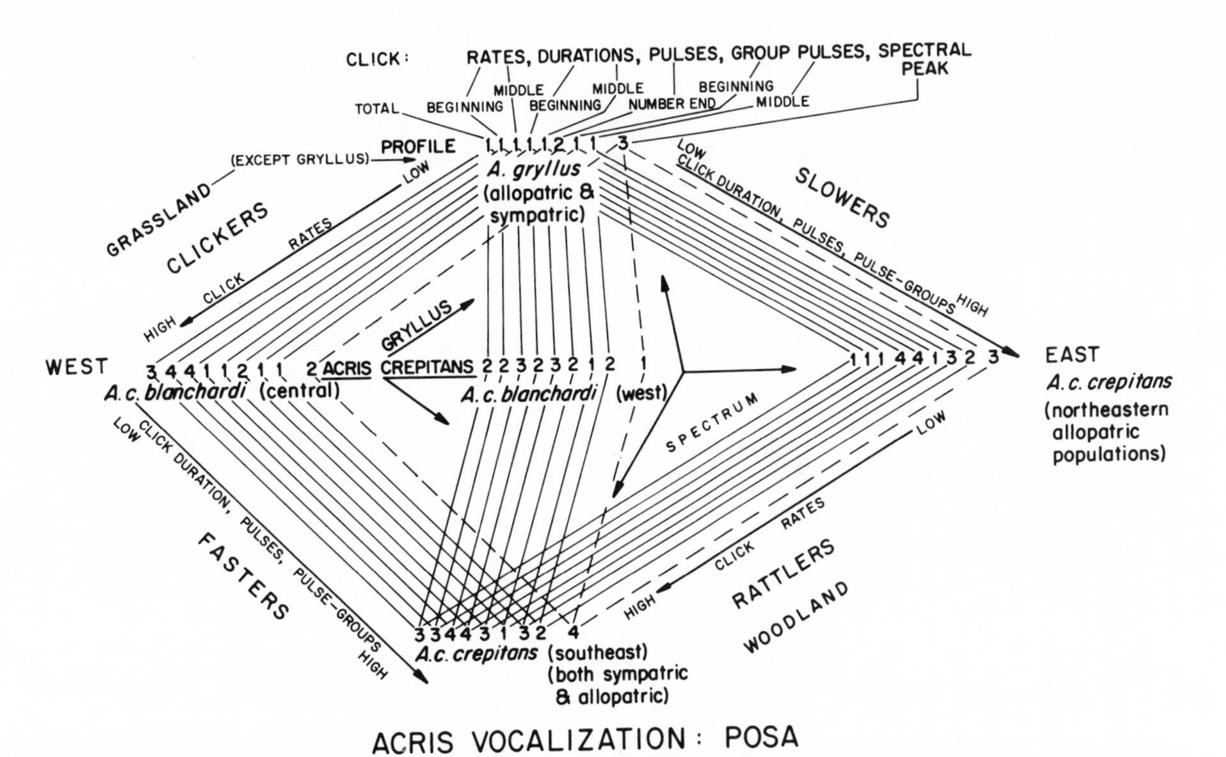

FIG. 10. Partial Order Scalogram Analysis (POSA) of nine call variables from 21 geographic populations of *Acris gryllus* and *A. crepitans*. The allopatric and sympatric populations of *A. gryllus* include localities 14–21; *A. c. blanchardi* (west) includes localities 1, 12, and 13; *A. c. blanchardi* (west central) represents locations 2, 10, and 11; *A. c. crepitans* (northeastern) includes localities 7–9; and *A. c. crepitans* (southeast) consists of localities 3–6.

high number of pulses in a click, low number of groups of pulses in a click, and high-frequency spectral peaks). This profile characterized *A. gryllus* in the southeast part of the *Acris* range (populations 14–21).

Profile 2: 2, 2, 3, 2, 3, 2, 1, 2, 1 (medium click rates and durations, high number of pulses, low groups of pulses in a click, and lowest frequency spectral peaks). This profile characterized the western allopatric populations of *A. c. blanchardi* in the xeric grasslands close to the western borders of the species range in west Texas, eastern Colorado, and South Dakota (populations 1, 12, and 13).

Profile 3: 3, 4, 4, 1, 1, 2, 1, 1, 2 (high click rates, lowest durations, high number of pulses, low number of groups of pulses, and medium-frequency spectral peaks). This profile characterized the allopatric populations of *A. c. blanchardi* ranging either in its central range (population 2) or northern borders (populations 10 and 11).

Profile 4: 1, 1, 1, 4, 4, 1, 3, 2, 3 (lowest rates, highest durations, low number of pulses, high number of groups of pulses, and high-frequency spectral peaks). This profile characterized the allopatric northeastern populations of *A. c. crepitans*, ranging near the northern border of the species range (populations 7–9).

Profile 5: 3, 3, 4, 4, 3, 1, 3, 2, 4 (high to very high rates and durations, low number of pulses and high number of groups of pulses, and very high-frequency spectral peaks). This profile characterized the sympatric southeastern populations (localities 3–6) of *A. c. crepitans* in the mesic and warm regions of the range of *A. crepitans*.

Note that each of the subspecies of *A. crepitans* is distinguished by two profiles: *A. c. blanchardi* by profiles 2 and 3, and *A. c. crepitans* by profiles 4 and 5. In contrast, the two subspecies of *A. gryllus* are largely represented by profile 1 (*A. gryllus*). Thus, call differentiation across the range of *Acris* may be grossly represented as follows: "Clicking" with only a slight audible pulsing (i.e., low number of groups of pulses in a click) characterized the grassland populations of *A. c. blanchardi* as well as those of both allopatric and sympatric *A. gryllus*; but while the former is largely a "Fast Clicker," the latter is a "Slow Clicker." In contrast, "Rattling" due to audible pulsing (i.e., high number of pulses and groups of pulses in a click) is characteristic of the woodland populations of *A. c. crepitans*; but, while the southeastern sympatric populations of *A. c. crepitans* living in the warmest part of the range are "Fast Rattlers," those allopatric populations living in the colder part of the northeastern range are "Slow Rattlers." In terms of general trends, click rates increase diagonally from northeast to southwest, whereas durations, pulses, and pulse groups increase at a right angle to rates, i.e., toward the southeast, while spectral peaks increase westward.

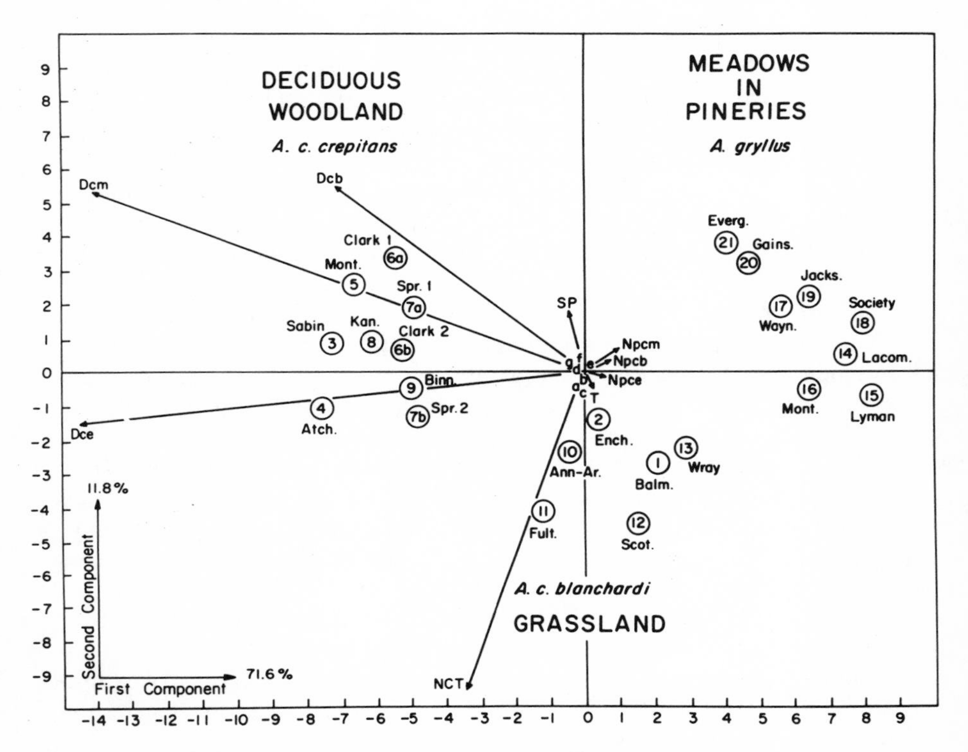

FIG. 11. Biplot based on 16 call variables and 23 populations of *A. c. crepitans, A. c. blanchardi*, and *A. gryllus*. The areas of the biplot are scaled in integral segments. The circled numbers represent the geographic localities in Table I. The vectors derive from the mean values of the call parameters listed in Table II: *SP*, spectral peak; *T*, duration; *NCT*, total number of clicks; $a = NCT/T$, overall click rate; $b = Rcb$, rate of clicks in beginning; $c = Rcm$, rate of clicks in middle; $d = Rce$, rate of clicks at end; *Dcb*, duration of clicks in beginning; *Dcm*, duration of clicks in middle; *Dce*, durations of clicks at end; *Npcb*, number of pulses in a click in beginning; *Npcm*, number of pulses in a click in middle; *Npce*, number of pulses in a click at the end; $e = Ngpcb$, number of groups of pulses in clicks in beginning; $f = Ngpcm$, number of groups of pulses in clicks in the middle; and $g = Ngpce$, number of groups of pulses in clicks at end. The radii of the "significant comparison circles" for $p < 0.05$ (see text for explanation) are 2.7 on the axes indicated (except for number 8, Kanawaukee Lake, New York, where it is 3.0, and for number 17, Waynesboro ponds, Georgia, where it is 3.5, where sample sizes were four and three, respectively).

Canonical Decomposition (Candec) and Biplot

The biplot of call structure and differentiation, based on the 16 call variables for the sampled populations, is shown in Figure 11. The call biplot is a graphical display of a two-dimensional approximation to the data matrix. In our biplot analysis, the first component accounted for 71%

of call variance and the second component 12%, amounting to an impressive explanation of 83% of the geographic variation in call differentiation. The display consists of 23 points each representing a different population sample (see Table II), and a vector corresponding to the various call parameters. The distance between the populations indicates measures of dissimilarity or Mahalanobis-type distance, based on call variables. Thus, if populations have similar overall call structure, they appear close together on the biplot. Similarly, it shows clustering of variables characterizing specific populations, their variances, and correlations (approximated by the cosine of the angles between vectors). Tests of statistical significance between populations, represented at the 0.05 significance level, can be obtained by drawing "comparison circles" surrounding the populations (where the radii of circles appear in the caption of the figure). Note that, whereas most populations between species or subspecies (except in *A. gryllus*) differ significantly, they are largely nonsignificant within species (or in the two subspecies of *A. crepitans*).

It is obvious that the populations in the biplot fall into three distinct clusters, which correspond to a combination of the geographic location and the species (or subspecies for *A. crepitans*). The three clusters are distinguished primarily by the following call variables: *A. c. crepitans*, high values of *Sp, Dcb, Dcm*, and *Dce*; *A. c. blanchardi*, high values of *T* and *NCT* and low values of *SP*; *A. gryllus*, high values of *Npcb, Npcm,* and *Npce* and low values of *Dcb, Dcm,* and *Dce*. The geographic–taxonomic clustering was based strictly on call analysis—the geographic location itself *never entered* the biplot analysis. This is a rather remarkable result. Thus we find that all grassland populations of *A. c. blanchardi* cluster in the middle bottom of the biplot in their original geographic order. Similarly, *A. c. crepitans* appears in the upper left of the biplot. Finally, the eight populations of *A. gryllus* cluster within the upper right portion of the plot without a clear-cut separation between *A. g. dorsalis* and *A. g. gryllus*. Nevertheless, the different populations within *A. gryllus* display their rough geographic distribution on the biplot.

The biplot therefore essentially reflects the three geographic regions within the genus *Acris*: grasslands for *A. c. blanchardi*, woodlands for *A. c. crepitans*, and open spaces or meadows in pineries for *A. gryllus*. The biplot indicates that, on the basis of overall call differentiation, *A. c. crepitans* as a whole is quite distinct from *A. c. blanchardi*, more so for some populations than even from populations within *A. gryllus*. It should be appreciated that the biplot analysis is an exact representation of the best two-dimensional fit, i.e., planar fit to that matrix data. In the present case more than 83% of the variation is accounted for by this plane, so that the representation is very good. Thus, the biplot provides graphical

TABLE VIII. Field-Recorded Natural Calls Tested in Female Discrimination Tests

Locality of recording	Mean spectral peak, Hz	Duration, sec	Silent interval, sec	Recording temperature, °C	
				Air	Water
Acris crepitans					
Balmorhea, West TX	2700	29	14.5	21.5	21.5
Enchanted Rock, central TX	3350	20	10.0	22.5	18.5
Sabine River, east TX	4700	15	7.5	25.0	27.0
Atchafalaya River, LA	4100	20	10.0	16.0	24.0
Montgomery ponds, AL	4300	16	8.0	22.0	22.5
Springdale ponds, NJ	3350	12	6.0	18.0	16.5
Clark Hill Reserve, GA	4200	10	5.0	22.0	25.0
Acris gryllus					
Wallace, GA	3600	10	5.0	21.0	24.0
Waynesboro, GA	3500	22	11.0	21.0	23.0
Montgomery ponds, AL	3300	11	6.5	22.0	22.5

representation of our rather complex data as well as simultaneous significance tests. It also provides a clear picture of the multivariate structure of the data, namely insight into the complicated relationships within samples and variables. Such an overall structural call differentiation is hardly attainable on the basis of univariate and/or purely analytical methods.

Female Mating Call Discrimination

Natural Calls

Representative calls for female call discrimination tests were selected from seven localities of *A. crepitans* and three localities of *A. gryllus* (Table VIII). The results with tests of *A. crepitans* females in three widely separated populations, one sympatric (*A. c. crepitans* at Clark Hill, Georgia) and two allopatric (*A. c. crepitans* at Springdale, New Jersey and *A. c. blanchardi* at Palmetto Fish Hatcheries, Texas), are given in Table IX. The entries indicate the number of positive phonotactic scores for each female to various paired calls (positive trials to the local call are recorded in the lower row as compared to positive trials for the alternative call in the upper row for each stimulus pair). For example, in the experiments in Clark Hill, Georgia involving a choice between the mating call of *A. gryllus* from South Carolina and *A. crepitans* from the local Georgia pop-

TABLE IX. Results of Female Discrimination Tests of *Acris crepitans*, Conducted in Three Localities[a]

Pairs of calls tested (species and population)	Results	Total	Discrimination index	Wilcoxson matched pairs signed-ranked test
Clark Hill Reserve, Georgia				
A *gryllus* (SC)	0 1 0 0 1 0	2	0.90	$p < 0.02$
A. *crepitans* (GA)	1 4 5 2 3 3	18		
A. *crepitans* (W. TX)	0 0 0 0 0 1	1	0.93	$p < 0.02$
A. *crepitans* (GA)	2 4 3 1 1 4	15		
A. *crepitans* (C. TX)	0 0 0 0 0 0	0	1.00	$p < 0.02$
A. *crepitans* (GA)	1 9 2 3 2 3	20		
A. *crepitans* (E. TX)	0 0 1 0 1 0 0 0	2	0.91	$p < 0.05$
A. *crepitans* (GA)	1 9 0 3 0 5 1 2	21		
A. *crepitans* (LA)	1 2 0 0 0 0 2 0 0	5	0.85	$p < 0.005$
A. *crepitans* (GA)	8 3 3 2 7 1 2 3 1	30		
A. *crepitans* (AL)	0 1 2 0 1 1	5	0.61	NS
A. *crepitans* (GA)	1 2 1 3 1 0	8		$p = 0.291^b$
A. *crepitans* (NJ)	0 1 1 1 1 6 0 0 0	10	0.71	$p < 0.05$
A. *crepitans* (GA)	3 1 0 3 8 4 3 2 1	25		
Springdale, New Jersey				
A. *gryllus* (AL)	0 0 0 0 0	0	1.00	$p < 0.05$
A. *crepitans* (NJ)	1 1 3 5 4	14		
A. *crepitans* (W. TX)	3 0 0	3	0.85	$p < 0.001^b$

(Continued)

TABLE IX. (*Continued*)

Pairs of calls tested (species and population)	Results	Total	Discrimination index	Wilcoxson matched pairs signed-ranked test
A. crepitans (NJ)	15 2 1	18		
A. crepitans (C. TX)	0 0 0 0	0	1.00	$p < 0.001$[b]
A. crepitans (NJ)	2 6 2 5	15		
A. crepitans (E. TX)	1 5 0 0	6	0.66	NS
A. crepitans (NJ)	0 2 7 3	12		$p = 0.119$[b]
A. crepitans (LA)	1 1 0 0 0	2	0.87	$p < 0.01$
A. crepitans (NJ)	1 3 2 4 4	14		
A. crepitans (GA)	3 11 2 6 0 0	22	0.45	NS
A. crepitans (NJ)	4 0 8 0 3 3	18		$p = 0.50$[b]
Palmetto Fish Hatcheries, Central Texas				
A. gryllus (SC)	3 3 0 1 0	7	0.63	$p < 0.10$
A. crepitans (C. TX)	6 3 2 0 1	12		
A. crepitans (GA)	0 1 1 0 0 0 0	2	0.91	$p < 0.01$
A. crepitans (C. TX)	1 5 3 6 3 2 2	22		
A. crepitans (NJ)	2 2 0	4	0.80	$p = 0.006$[b]
A. crepitans (C. TX)	11 4 1	16		

A. crepitans (LA)	1	1	2	0							4	0.82	$p = 0.002$
A. crepitans (C. TX)	6	5	3	4							18		
A. gryllus (GA)	0	0	1	0	0						1	0.93	$p < 0.05$
A. crepitans (C. TX)	2	3	3	3	2						13		
A. crepitans (C. TX) Played backward	0	0	0	0	0						0	1.00	$p < 0.05$
A. crepitans (C. TX) Played forward	5	7	3	3	2						20		
A. crepitans (C. TX) Synthetic call with spectral peak at 2966 Hz	2	1	0	1	0	0					4	0.70	$p < 0.05$
A. crepitans (C. TX) Synthetic call with spectral peak at 3466 Hz (normal)	2	2	1	3	1	1					10		
A. crepitans (C. TX) Synthetic call with spectral peak at 3966 Hz	1	1	0	0	0	1	4	2	3	2	14	0.60	$p < 0.10$
A. crepitans (C. TX) Synthetic call with spectral peak at 3466 Hz (normal)	2	2	2	2	1	0	2	3	3	4	21		

[a] For characterization of populations see Table I.
[b] According to binomial test, since sample size is too small for Wilcoxon's test.

ulation, the first female performed only one trial and then stopped responding (i.e., hopped away from the testing arena during all subsequent trials). In her only response, she selected the call of her local male, so that the first column entry under Results in Table IX is entered as 0 (above) vs. 1 (below) for *A. crepitans* (GA). The second female in tests with the same paired calls completed five trials: one positive score to the foreign call and four positive trials to the call of her local male. The column listed "Total" to the right side of the two-paired row entries shows the final tabulation for all of the females tested with that particular paired-call selection [in this case a total of six females were tested with the resultant outcome of two positive responses to *A. gryllus* (SC) vs. 18 positive trials to the call of *A. crepitans* (GA)]. The next column presents the discrimination index, on a scale of +1 to −1, for the various pairs of calls. A value of +1.00 reflects complete discrimination for the local call (e.g., the third set of paired calls in Table IX involving *A. crepitans* of central Texas vs. Georgia); −1.00 would correspond to a complete preference of the alternative call (note there are no negative values in this column of Table IX, so that the alternative call was never reliably preferred), and an index of zero would indicate no preference for either call. Low values, e.g., 0.61 in the sixth set of pair calls tested (*A. crepitans* AL vs. *A. crepitans* GA), indicate a nonsignificant outcome. The last column therefore gives the probability, based on a nonparametric statistical test, that the resultant trials could have been obtained by chance alone.

Acris c. crepitans females in sympatric Clark Hill, Georgia and in allopatric Springdale, New Jersey discriminated significantly against the calls of *A. gryllus*, but in the allopatric Palmetto Fish Hatcheries in central Texas the level of discrimination against *A. gryllus* was only $p < 0.10$, i.e., nonsignificant on the basis of our sample size. Remarkably, within *A. crepitans* itself both subspecies discriminated one against the other, and the discrimination index generally increased with geographic distance. Thus, the discrimination index for the Clark Hill, Georgia tests increased in response to calls from Alabama to Louisiana, east Texas, central Texas, and west Texas (0.61, 0.85, 0.91, 1.00, and 0.93, respectively), namely with geographic separation. A similar trend in the discrimination index occurred also in our second allopatric *A. c. crepitans* tests in Springdale, New Jersey. The discrimination index for Springdale females increased, though inconsistently, to calls from Georgia to Louisiana, east Texas, central Texas, and west Texas (0.45, 0.87, 0.66, 1.00, and 0.85, respectively). Finally, females in Palmetto, central Texas showed a significant preference for calls for their own local males compared to calls of *A. c. crepitans* from other parts of its range (e.g., Louisiana, Georgia, New Jersey). Therefore, while within *A. c. crepitans* discrimination against other population calls increased with distance, the discrimination of *A. c. crepitans* against *A. c. blanchardi* and *vice versa* was complete and unrelated to distance. Furthermore, discrimination was sharper, at least in *A. c. blanchardi*, between the two subspecies of *A. crepitans* than between the two species *A. crepitans* and *A. gryllus*. Thus there is evidence for the development of positive assortative mating between the two species *A. crepitans* and *A. gryllus* as well as between the two subspecies of *A. crepitans* (which live in two different biomes, grassland vs. woodland).

Synthetic Calls

Electronically generated synthetic calls were also tested in order to pinpoint the parameters by which females discriminate male calls. As we have mentioned, these studies are the subject of a separate report dealing with the communication behavior of *Acris*. Here we will only briefly summarize some of the results of tests conducted in central Texas as an illustration of our conclusions. One of the pairs of tests listed in Table IX compared the local call played forward (normal) with the *same* call played backward (reverse). The discrimination index, based on 20 trials, was 1.00 for this comparison and is obviously significant. In no case did a female respond positively to the reversed call. Now, in playing the call backward, clearly the spectral (dominant frequency) peak remained the same. Thus, a female's preference must reside in the (normal) *temporal pattern* of clicks in the calls of her local males. This conclusion was verified by generating calls having the temporal patterns of other more distant populations compared to the local pattern, even though the spectral energy in each of the clicks in the two calls was *identical*. This is one of the advantages of using synthetic calls, namely various parameters can be varied independently.

A number of synthetic calls having the appropriate local temporal pattern of clicks but with different spectral peaks were compared. Table IX shows the results for two such comparisons conducted in Palmetto, central Texas: a synthetic call with a peak at 2966 Hz vs. a synthetic call with a peak at 3466 Hz, and then a synthetic call with a peak at 3966 Hz. vs. a call with a peak at 3466 HZ. The females preferred the call with a peak at 3466 Hz, which is close to the value in a typical natural call for the Palmetto population. Thus, the basis for a female's selectivity to the calls of local males resides in both the temporal pattern of clicks and the location of the spectral energy peak in each click. Since these patterns change geographically, the behavioral selectivity nicely confirms our call analyses.

DISCUSSION

The discussion of our studies is subdivided into the following sections: evolution of anuran mating calls, evolution of *Acris* mating calls, and evolutionary history of *Acris*.

Evolution of Anuran Mating Calls and *Acris* Call Patterns

Bioacoustic Aspects

Mating calls of anurans serve primarily in species identification (Blair, 1964*a,b*). The male's call has been shown to be a homospecific mate attractant in five major anuran families (Blair, 1974). The signal characteristics of a mating call can be defined by several descriptive variables, which are basically related to intensity patterns in the two dimensions of frequency and time (Straughan, 1973). Generally, the principal variables involved in the interspecific differentiation of a call are *dominant frequency, duration, pulse* or *click rate*, and *degree of modulation* (Blair, 1964*a*). The dominant frequencies, i.e., spectral peaks, are one of the outstanding components of mating calls. Recent studies of auditory function in anurans have shown that their ear behaves as a peripheral filter which is tuned to particular frequency bandwidths (Capranica, 1965, 1976, 1977; Straughan, 1973). In other words, an anuran's ear is most sensitive to sounds of rather restricted frequency range and in most cases [see Loftus-Hills and Johnstone (1970) for review] this "best frequency" matches the dominant frequency in its mating call (Straughan, 1973). In the present study we found that female *A. crepitans* would no longer respond to a synthetic mating call, with an intensity of 92 dB SPL at a distance of 1 m, if the spectral peak was shifted by as little as 500 Hz either above or below the mean for the local population, even though the temporal pattern was left unchanged. Furthermore, the frequency sensitivity of the auditory system of female cricket frogs varies geographically in accord with the geographic variation of the spectral peak in the male's mating calls, thus enabling females to respond preferentially to the calls of their local dialect (Capranica *et al.*, 1973).

While the spectral peak in the call may define the frequency channel through which species-specific messages are transmitted, discrete temporal events (such as clicks and pulses) with a specific pattern in time can also convey species-specific information. For example, pulse repetition rate and note or click duration are important discriminatory cues in the *Hyla ewingi* complex and closely related hybrids (Loftus-Hills and Littlejohn, 1971; Straughan, 1973). Our field discrimination tests with synthetic calls and reversed natural calls (Table VIII) indicate that the click pattern is a very important feature for female call recognition in *Acris*. Interspecific differences in mating calls are generally attributed more to neural and muscular control of the vocal structures than to morphological differences (Blair, 1964*a*), but this may be only partly true. Blair (1964*b*, p. 334) emphasized that "Some important parameters of the call (viz. frequency and pulse rate) are usually functions of body size, and hence differentiation in these attributes may be a secondary effect of adaptive differentiation in body size. Other parameters of the call appear to have been the subject of selection per se." The dominant frequencies in anuran mating calls are in general inversely proportional to body size (Blair, 1964*a*). Similarly, a variation in the temporal call pattern may result from anatomic constraints in addition to variation in the central nervous system's motor control of the vocal apparatus (Martin, 1972).

Our results in *Acris* showed frequent correlations between spectral and temporal parameters (Table V), as well as significant correlations of spectral and temporal call variables with body size (Table VI). Furthermore, significant correlations exist among all call parameters and the environment (Table VII). The major environmental factors include rainfall, longitude, and latitude. Similarly, body size varies geographically with the environment. Fifty-nine percent of the variance in body length of *A. crepitans* is explained by rainfall alone, and 61% is explained by a two-variable combination of rainfall and longitude. Even more striking is the change in body length in *A. gryllus*. Seventy-four percent of it size variance is explained by latitude alone and 89% by the two-variable combination of latitude and rainfall.

Ambient temperature is the one environmental factor that has been clearly shown to affect the call (Blair, 1964*b*; Schneider and Nevo, 1972; Nevo and Schneider, 1976). It may be instructive in this connection to consider the regression coefficients (the values for both species of *Acris* appear in parentheses) between each of the 16 call parameters and mouth temperature: *SP* (71.9), *T* (0.15), *NCT* (3.55), *NCT*/*T* (0.19), *Rcb* (0.09), *Rcm* (0.22), *Rce* (0.21), *Dcb* (−5.45), *Dcm* (−8.72), *Dce* (−10.71), *Npcb* (0.06), *Npcm* (0.16), *Npce* (0.17), *Ngpcb* (−0.04), *Ngpcm* (−0.03), and *Ngpce* (−0.10). The regression coefficients of *Acris* are in line with the evidence from other anurans. Dominant frequency, click rates, and number of pulses per click increase and durations decrease with temperature, though these coefficients vary between species and subspecies. Accordingly, in our study we have presented the results of our call analyses in the following forms: (1) raw data, (2) analysis in which populations that were recorded in either very low or very high temperature above the average (22.79°C) were excluded (we included only populations recorded between 19.7 and 24.8°C), and (3) regressed estimates standardized to a mean mouth temperature of 22.79°C for both species (only "durations" were standardized for each subspecies separately, since their regression values differed drastically).

Ecological Aspects

Call structure and evolution are affected by both the biotic (acoustic) and the abiotic components of the environment (Littlejohn, 1965). Since the character of the mating call is presumably determined genetically, as deduced from hybrid calls (Blair, 1955), natural selection may affect it either directly or indirectly. Furthermore, provided that the call variables are standardized to the same ambient temperature, geographic variation in call structure within a species reflects local genetic influence.

The notion that the type of call should vary with the habitat and generally with the environment exclusive of other sound sources in terms of efficient communication has been discussed by Schiøtz (1973). Consequently, divergence and convergence of call structures in different and similar habitats, respectively, should be expected. Obviously, for a mating call to be functional, it must have the design features appropriate for transmission in the particular physical environment, namely avoidance of masking by local noise, contain species-specific features, and be locatable (Straughan, 1973). Thus, anurans that live in dense forests generally have a different call structure than species from more open situations such as a savanna region (Schiøtz, 1973). While such correlates of call structure and habitat are largely unknown, Schiøtz has advanced several possible speculations for the differences in anuran calls from open as opposed to closed regions, including the large distances between breeding sites in open compared to closed areas, and the large vs. small breeding sites in savanna compared to forest ranges. Savanna species, with greater distance between neighboring breeding sites and greater body size, may have a louder voice and lower dominant frequencies. Similarly, some species of forest anurans that do not form breeding aggregations have a higher pitched voice with a distinct insectlike quality. Localization of these calls by more advanced vertebrates may be more difficult, which in turn may help the anurans elude larger predators (Schiøtz, 1964).

Reproductive Character Displacement

Reproductive character displacement (Brown and Wilson, 1956; but see also Wallace, 1889; Gulick, 1890; Fisher, 1930; Dobzhansky, 1940;

Muller, 1942; P. R. Grant, 1972) or reinforcement (Blair, 1955, 1974) may be defined as a process whereby attributes of one biparental species are changed through natural selection due to an interaction with individuals of one or more other biparental species that have similar or overlapping requirements of ecology, reproduction, or both (Littlejohn, 1980, p. 36). The available records of reinforcement in closely related anuran species include the *Hyla ewingi* group (Littlejohn, 1965, 1969; Littlejohn and Loftus-Hills, 1968), *Microhyla, Acris*, and *Bufo americanus* (Blair, 1955, 1958*a,b*, 1962, 1974). *Pseudacris* (Fouquette, 1975), and the diploid–tetraploid species complex of North American *Hyla* (Ralin, 1977). In all of these cases the differences in call structure between two closely related species were reported to be greater in sympatry than in allopatry.

Our study of cricket frogs raises the fundamental question of the origin of ethological reproductive isolating mechanisms. In other words, the present problem may be restated as follows: What are the major causes for the distinct calls of *A. c. crepitans* and *A. g. gryllus*? Did they arise primarily due to ecological adaptations or alternatively to reproductive character displacement and reinforcement? Since none of the proponents of reproductive character displacement have ever required that it should be a universal process (Blair, 1974), each case must be assessed separately, or, as stated by Dobzhansky (1940, p. 320), "The basic problem which remains to be settled is how frequently and to what extent can the isolating mechanisms be regarded adaptational by-products arising without the intervention of the special selective processes postulated above. Only experimental data could elucidate the situation further." In particular, it should be rewarding to analyze comprehensively cases, such as that of cricket frogs, in which reproductive character displacment has been suggested as the major explanatory model for call differentiation (Blair, 1958*a,b*, 1964*a,b*, 1974).

Evolution of *Acris* Mating Calls

Past and Present Studies of Acris Call Structure

The calls of *Acris* were first described by Blair (1958*a,b*). His analysis was based on a mixed population in southwestern Mississippi, a series of recordings of *A. crepitans* from Texas, Oklahoma, and Louisiana, and a few recordings of *A. gryllus* from peninsular Florida. The results then available indicated that the calls of Florida *A. gryllus* and western *A. crepitans* showed only slight differences. In contrast, their calls in sympatric Mississippi differed in several characteristics (Blair, 1958*a*, Figs.

5 and 6). These included both quantitative (i.e., interval between clicks and dominant frequency) and qualitative (i.e., pulsed call of *A. crepitans* and unpulsed call of *A. gryllus*) characteristics (Blair, 1958*a*, Fig. 7). This evidence was interpreted as a demonstration of strong call differentiation due primarily to reinforcement in the sympatric zone, similar to the case of two species of *Microhyla* (Blair, 1955). The differences in call structure between *A. crepitans* and *A. gryllus* were "readily explainable only by the theory that selection against hybridization in the overlap zone has served to sharpen the call differences and thus increase the effectiveness of the isolation mechanism. The only alternative explanation for the increased differences in call in the overlap zone would be that the differences are purely fortuitous, but this explanation seems hardly defensible" (Blair, 1958*a*, pp. 44–45).

The present study of *Acris* call structure and differentiation was conducted with the conviction that only comprehensive multidisciplinary studies can evaluate the forces that mold call structure and thereby indicate the best explanatory model(s) out of the several possible alternatives. The latter include: (1) direct selection in a narrow sympatric zone, namely reinforcement; (2) direct selection in allopatry in accord with the biotic and/or abiotic environment; (3) incidental to adaptive differentiation of other characters such as body size; and (4) stochastic differentiation.

Our comprehensive approach has involved:

1. A detailed analysis of gross and fine call structure across the entire range of the two species involved, *A. crepitans* and *A. gryllus*, including their four subspecies in sympatry and allopatry.

2. An attempt to analyze their calls not only as a structure comprising independent spectral and temporal variables, but also as a multivariate structure in which correlations exist among the individual components. We employed two multivariate approaches: one qualitative in trying to extract geographic ordering of call structure (POSA), and the other quantitative in trying to evaluate statistical similarities and multidimensional distances between the various populations on the basis of specific call attributes and their clustering (Candec).

3. An attempt to relate (by univariate Pearsonian correlations) and predict (by multivariate stepwise multiple regression) the variation-independent call variables by morphological and environmental factors.

4. Mating call discrimination field tests and phonotaxis by females both in allopatry and sympatry, including natural and synthetic calls.

5. Electrophysiological recordings from the auditory nervous system in order to determine the frequency and threshold sensitivity of cricket frogs derived from distant allopatric populations compared to extreme

borders of the range (Capranica and Frishkopf, 1966; Capranica *et al.*, 1973).

6. The study of genetic structure and differentiation of both allozyme (Dessauer and Nevo, 1969; Salthe and Nevo, 1969) and visual (Nevo, 1973*a,b*) polymorphisms to assess the genetic structure of populations and species.

We will first discuss the analysis of size variation, and then summarize the findings concerning call structure and differentiation.

Adaptive Variation in Size of Cricket Frogs

Since both spectral and temporal call parameters in *Acris* were found to be correlated with body size (Table VI), it is of paramount interest to follow size differentiation across the entire range and assess its ecological correlates. If size in *Acris* varies in accord with different environments, then call structure must have evolved either directly and/or incidentally with the environment.

Geographic variation in body size *BL* and relative foot length *FL/BL* were studied across the ranges of the two *Acris* species. Measurements of *BL* and *FL/BL* of 5286 frogs from 108 localities were divided into 11 biogeographic regions and subjected to discriminant and multiple regression analyses. Mean *BL* of both sexes increased progressively westward in both the southern, allopatric–sympatric transect and in the northern, solely allopatric transect. Overall, mean *BL* of *A. crepitans* males increased from 22.1 to 25.8 mm (3.7 mm or 16.7%), whereas that of females increased from 22.5 to 28.7 mm (6.2 mm or 27.5%) (Nevo, 1973*a*, Figs. 1 and 2 and Table 1). Mean *BL* of *A. gryllus* males was 22.5 mm, SD 0.9, and it increased from 22.1 mm in *A. g. dorsalis* in southern Florida to 22.8 mm in the West Gulf Plain (Louisiana and Mississippi). Size increased in south and north alike: mean *BL* of *A. c. crepitans* increased from 22.6 mm in the south Atlantic East Gulf Plain (Mississippi, Alabama, Georgia, South Carolina, North Carolina) to 25.5 mm in *A. c. blanchardi* in west Texas. Similarly, mean *BL* increased in the northern transect from 22.5 mm in the northeastern range of *A. c. crepitans* in New Jersey and New York to 25.8 mm in *A. c. blanchardi* in the northern Great Plains (Iowa, South Dakota, Nebraska, and Colorado). Mean relative foot length *FL/BL* increased clinally from a minimum of 75.9 [$(FL/BL) \times 100$] for northeastern *A. crepitans* and 81.04 in southeastern *A. gryllus* to 83.3 and 86.9, respectively, in the Mississippi River embayment. The *FL/BL* value was significantly higher in *A. gryllus* than in *A. crepitans* (83.95 and 79.55 mean values, respectively; $p < 0.01$) (Nevo, 1973*a*, Fig. 2 and Table 1).

Body length *BL* varied in *A. crepitans*, negatively correlated ($r =$

−0.72 and −0.70 for male and female, respectively) with a sharp regression with annual rainfall (Nevo, 1973*a*, Fig. 4). Rainfall was the best single predictor, accounting for 0.47 ($p < 0.001$) of size variation of *A. crepitans* males across the range. A combination of three humidity variables explained 0.59 ($p < 0.001$) of male *BL* variance and the two-variable combination of rainfall and longitude (*Rn–Ln*) accounted for 0.61 ($p < 0.001$) of size variance. Size of *A. crepitans* increased from *A. c. crepitans* to *A. c. blanchardi* mainly with longitudinal variables related to available humidity rather than to latitudinal variables related to temperature. In *A. gryllus, BL* variation in the present study was primarily accounted for by latitude ($r^2 = 0.74, p < 0.01$) and by a two-variable combination of latitude and rainfall ($r^2 = 0.89$, $p < 0.01$).

Cricket frogs are smallest in the marginal northeast as well as in the warm southeastern parts of their range. Size in *Acris* seems to increase primarily in response to increasingly arid environments. For *A. crepitans* this trend culminates in *A. c. blanchardi* in the semiarid western Great Plains along the western species' borders, both in the dry, warm south and in the dry, cold north (Nevo, 1973*a*, Figs. 1 and 2). This remarkable increase in *A. crepitans* size from east to west is correlated with the ecological gradient of increasing aridity from the mesic woodlands, where *A. c. crepitans* lives, to the xeric grasslands, where *A. c. blanchardi* ranges. Although the increase in size of *A. gryllus* is much smaller than in *A. crepitans* (Nevo, 1973*a*, Table 2), presumably because of the smaller variation in humidity over its range, its size does increase from *A. g. dorsalis* to *A. g. gryllus*. Notably, *A. gryllus* is similar in size to eastern *A. c. crepitans* populations in their overlap zone, as predicted by the regression line of size on rainfall with no apparent character displacement in size (Nevo, 1973*a*, Fig. 4). However, the two species differ significantly in their overlap zone with regard to relative hindfoot length.

The significance of large size of *Acris* as an adaptation to arid climates is implied also by the results of desiccation experiments (Nevo, 1973*a*, Figs. 5 and 6). Since evaporative water loss is a function of surface-to-volume ratio (Farrell and MacMahon, 1969, and references therein), large frogs lose relatively less water and their vital time limit is considerably greater than that of small ones. Large frogs are therefore superior in dry habitats, hot or cold, because of their relatively low surface-to-volume ratio and correspondingly low rates of water loss. This is particularly true in xeric environments, where successful migration between temporary bodies of water depends on efficient moisture conservation.

Relative foot length *FL*/*BL* in both species is positively correlated and best predicted by a combination of temperature, rainfall, and longitude. Therefore, while *BL* is causally related to humidity, large size being an adaptation to arid climates, relative foot length may be causally related to predation pressure and/or competition for food, since large *FL/BL* is a distinctive advantage for increased jumping efficiency (Nevo, 1973*a*). In any event, the difference in relative hindfoot length between *A. crepitans* and *A. gryllus* may reflect their differential habitats and microhabitats, not only in the overlap zone, but also in allopatry.

Comparative Ecology of Cricket Frogs in Allopatry and Sympatry

The differential ecologies of cricket frogs in allopatry are reflected by their lives in different biomes: *A. gryllus* in southeastern pineries, *A. c. blanchardi* in open grasslands, and *A. c. crepitans* in the eastern deciduous forests. *Acris gryllus*, the pineries species, is a terrestrial, shade-loving frog that lives in meadows, along creeks and pond margins in open vegetation, and mats or wooded edges (Wright and Wright, 1949). Various degrees of ecological separation between *A. crepitans* and *A. gryllus* have been described in sympatry. Fairly sharp ecological separation has been reported in Mississippi (Blair, 1958*a,b*; Boyd, 1964; Ferguson *et al.*, 1967) and in some areas of Alabama (Viosca, 1944; Mecham, 1964), although elsewhere, in central and southern Alabama, mixed choruses of both species have been recorded (Mecham, 1964). A detailed comparative ecological study of both species was conducted in Louisiana by Bayless (1966, 1969). In southeastern Louisiana, *A. crepitans* inhabits the shrub-dominated pond margins, while *A. gryllus* tends to avoid areas overhung with shrub growth and instead breeds successfully in a nearby meadow with temporary pools devoid of *A. crepitans*. In addition, *A. gryllus* occupies many more drainage ditches as nonbreeding habitats than *A. crepitans*. Similarly, statistically significant differential food intake has been demonstrated for the two species in mixed groups, although the same food types are taken. It was suggested that *A. c. crepitans* may be adaptively superior to *A. g. gryllus* in extensive permanent or semipermanent wetlands and in marshes, as is also reflected by its greater amount of toe webbing, whereas *A. g. gryllus* may be adaptively superior in dry uplands consisting of well-drained areas where temporary pools with variable moisture conditions form the only available breeding sites.

Reported cases of local sympatry may reflect various local habitat and microhabitat types. Thus, partial Gaussian competitive exclusion or microhabitat isolation, possibly through habitat selection, appears to reduce interspecific competition through ecological divergence in habitat and microhabitat utilization and in food intake (Bayless, 1966, 1969). In the present study, a sympatric zone of mixed choruses of both species was found in Montgomery, Alabama, but the other four *A. gryllus* sites and that of *A. crepitans* in Clark Hill, Georgia consisted of separated breeding aggregations (Table I), thus reinforcing the idea that these species occupy different habitats.

Acris Call Structure and Differentiation

The major findings concerning call structure and differentiation in *Acris* are summarized as follows:

1. A typical call (Fig. 2) consists of a series of audible pulsed clicks separated by silent intervals.
2. The dominant frequency (i.e., spectral peak) and the duration of clicks decrease with temperature (Table II).
3. Geographic variation was found in all call variables within and between species and subspecies except in the number of groups of pulses in a click, which were invariant in *A. gryllus* (Tables II and III and Figs. 4–9). Thus, the spectral peak *SP* decreased progressively northward and westward in both species, regardless of allopatric–sympatric situations. With regard to temporal features, the total call duration *T*, total number of clicks in a call *NCT*, rate of clicks in the entire call *NCT/T*, rates of clicks in the beginning, middle, and end of a call *Rcb, Rcm,*and *Rce*, duration of a click in the beginning, middle, and end of a call *Dcb, Dcm*, and *Dce* (but only in *A. gryllus*), and number of pulses in a click at the beginning, middle, and end of a call *Npcb, Npcm*, and *Npce* all increased westward. In contrast, *Dcb, Dcm*, and *Dce* as well as the number of groups of pulses in a click at the beginning, middle, and end of a call *Ngpcb, Ngpcm*, and *Ngpce* in *A. crepitans* (invariant in *A. gryllus*) decreased westward.
4. In 12 call variables (*NCT, NCT/T, Rcb, Rcm, Dcm, Dce, Npcb, Npcm, Npce, Ngpcb, Ngpcm, Ngpce*) the mean values for *A. crepitans* and *A. gryllus* were nonoverlapping, whether in allopatry or sympatry. In the remaining four call variables (*SP, T, Rce, Dcb*) their mean values overlapped and, in fact, *Dcb* tended to have a parallel pattern in the two species (Table III).
5. In five call variables (*NCT, NCT/T, Rcb, Rcm, Rce*) the difference in mean values between *A. gryllus* and *A. crepitans* was smaller in their zone of sympatry than in allopatry. For ten of the call variables the pattern was opposite, namely a larger gap in sympatry than in allopatry. Of the four call variables that displayed overlap in values, only *T* and *Dcb* showed a larger difference in sympatry (Table III).
6. The spectral peak is significantly correlated with certain temporal variables (rates, durations, numbers of pulses, and groups of pulses; see Table V).

7. Each of the 16 call variables displayed an impressive degree of significant correlation with, and prediction by, morphological (body size and hindfoot length), climatic (mean annual rainfall and temperature), and geographic (altitude, longitude, latitude) variables (Tables VI and VII).

8. In our nonmetric POSA analysis (Fig. 10), we conclude that call structure across the range of both species is best partitioned into multidimensional space by five profiles of nine call variables each (*SP, NCT/T, Rcb, Rcm, Dcb, Dcm, Npce, Ngpcb*, and *Ngpcm*). One profile characterizes *A. gryllus* by lowest click rates, durations, high number of pulses per click, low number of groups of pulses, and high-frequency spectral peaks. Two profiles characterize *A. c. blanchardi*: medium click rates and durations, high number of pulses and low number of groups of pulses, lowest frequency spectral peaks for western allopatric *A. c. blanchardi*; and high click rates, lowest durations, high number of pulses and low number of groups of pulses, and medium-frequency spectral peaks for central allopatric *A. c. blanchardi*. Two profiles characterize *A. c. crepitans:* lowest rates, highest durations, low number of pulses, high number of groups of pulses, and high-frequency spectral peaks for the allopatric northeastern populations; and high to very high rates and durations, low number of pulses, high number of groups of pulses, and very high-frequency spectral peaks for southwestern allopatric and southeastern sympatric populations. Call differentiation may therefore be represented by: Medium to Fast Clicking (western and central *A. c. blanchardi*); Slow Clicking (*A. gryllus*, both sympatric and allopatric); and Fast Rattling (southeastern allopatric and sympatric *A. c. crepitans*). In terms of general trends across the *A. crepitans* range, click rates increase southwestward, whereas durations and groups of pulses increase southeastward.

9. The metric multivariate analysis (Candec, Fig. 11), in the form of a two-dimensional biplot graphic representation, accounts for more than 84% of the call variation in *Acris*. Remarkably, the biplot resulted in three call clusters corresponding to the three biomes occupied by *Acris*: *A. c. blanchardi* in grasslands, *A. c. crepitans* in deciduous forests, and *A. gryllus* in subtropical pineries. For overall call differentiation *A. c. crepitans* as a whole is quite distant from *A. c. blanchardi*, more so for some populations than even for *A. gryllus*.

10. With regard to mating call discrimination by females, (a) *A. c. crepitans* females, either sympatric or allopatric, discriminated significantly against mating calls of *A. gryllus*, whereas allopatric *A. c. blanchardi* discriminated only $p < 0.10$ against *A. gryllus*; (b) within *A. crepitans*, discrimination against other populations increased with geographic separation; and (c) electronically generated calls, tested in the

field, indicated that both spectral and temporal parameters are crucial for female discrimination.

Acris Call Structure and the Allopatric and Sympatric Theories of the Origin and Evolution of Premating Reproductive Isolation

These condensed results support the allopatric and reject the sympatric theory of the origin of premating reproductive isolation as a major explanatory model for call differentiation in *Acris*. The major reasons for our conclusion are: (1) Both call structure and female preference in allopatric *A. c. crepitans*, whether southwest or northeast, are equally distinct from sympatric *A. c. crepitans* as from *A. gryllus*. The characteristic Rattling call structure of *A. c. crepitans* separates it drastically from either the Medium to Fast Clicking of *A. c. blanchardi* or the Slow Clicking of *A. Gryllus*. The latter remains a "Slow Clicker," either in sympatry or allopatry. (2) Geographic variations in call variables are often clinal in both allopatry and sympatry, and are largely correlated with and predicted by morphological, ecological, and geographic factors, independent of sympatry. (3) Most call variables of *A. gryllus* and *A. crepitans* in both allopatry and sympatry are nonoverlapping and represent distinct universes, namely woodland *A. c. crepitans* and grassland *A. c. blanchardi*. (4) The multivariate analyses suggest that there are three largely separated call clusters: *A. c. blanchardi, A. c. crepitans*, and *A. gryllus*. Most importantly, these three call clusters are correlated with the three different biomes: open grasslands, covered deciduous forests, and semiopen meadows in pineries, respectively.

It therefore appears quite plausible that the origin and evolution of these three call universes are primarily ecological and relate to the three different environments rather than to reinforcement in sympatry. In fact, the very broad sympatric zone between *A. c. crepitans* and *A. g. gryllus*, with apparently no substantial natural hybridization, does not lend support to the reinforcement model as a major selective force for sharpening their call differences. It suggests that most if not all call differentiation evolved in allopatry, either directly or incidentally, and preceded the sympatric contact between *A. gryllus* and *A. crepitans*. However, some reinforcement may be operating along the grassland forest border between *A. c. blanchardi* and *A. c. crepitans*, and this is precisely the zone where it should be investigated. However, even the major call differentiation seems to have developed due to their divergent grassland and woodland ecologies, rather than to reinforcement as a prime differentiator.

If call differentiation reflects adaptive radiation into a new ecological zone, such as the expansion of grassland *A. c. blanchardi* into the deciduous eastern forest where *A. c. crepitans* currently ranges, then selection of the call structure components that are optimal for the new environment and/or incidental divergence related to size could alone explain call differentiation. The two divergent call sets assumed parapatric distributions with little or no interaction, because each presumably is better adapted to its ecological regime. The latter certainly involves physical as well as biotic factors that affect call differentiation. The existence of the typical "Rattling" call type of *A. c. crepitans* in the warm southeast part of its range from east Texas to the Mississippi River and at the colder northeastern part of the range, some 400 miles from sympatry, supports the hypothesis that call differentiation in *Acris* is primarily ecological. This type of call has either been directly selected for efficient communication in a predominant woodland and shrub environment, and/or it is incidental to adaptive differentiation in body size, being smaller in forests and larger in grasslands. Thus the two systems of *A. c. blanchardi* and *A. c. crepitans*, which are parapatrically distributed, presumably maintain their call distinctness because of the continuous action of the two alternative selective regimes, grasslands vs. woodlands. Each occupies a distinct adaptive peak without as yet the development of intrinsic genetic incompatibility. Furthermore, the reproductive efficiency of individuals is maintained, as shown in our female discrimination tests, by effective homogamic mechanisms based on set-specific communication. In other words, both *A. c. blanchardi* and *A. c. crepitans* may be regarded as incipient ecospecies rather than subspecies (to be discussed in more detail in the subsequent section).

The explanation put forward for the two "subspecies" of *A. crepitans* may be at least partly relevant to the interaction between *A. c. crepitans* and *A. g. gryllus*. The partial habitat and microhabitat separation of these two species is similar to that between the two subspecies of *A. crepitans*. That is, while *A. c. crepitans* primarily occupies shrub-dominated wetlands, *A. g. gryllus* primarily occupies open meadows. Even in a mixed breeding population, these microhabitat differences were found between the two species. If hybrids are created (and apparently such a process is negligible), they will be inferior in fitness compared to either parental type in their respective habitats. Thus, natural selection may operate in mixed populations that are only partly separated by habitats or microhabitats in order to prevent errors of mate selection between the previously preadapted two divergent sets, and to enable sympatry with the lowest genetic interaction possible. Similarly, elimination of hybrids with lower fitness, due to positive assortative mating, may lead to greater reproductive success for individuals of each parental species because it results in the most efficient production of the fittest progeny (viewed retrospectively) (Littlejohn, 1980, p. 33).

The distinct phenotypic calls of *A. c. crepitans* and *A. g. gryllus* are apparently maintained, therefore, by the sustained influence of ecological factors on the maintenance of adaptive peaks in discrete niches, rather than by genetic incompatibility [since hybrids are easily formed artificially (Mecham, 1964; E. Nevo, unpublished results)]. The positive assortative mating demonstrated in both allopatric and sympatric *A. c. crepitans* when tested with *A. gryllus* provides strong evidence for the ecological allopatric theory of call differentiation as opposed to the genetical sympatric theory. This interpretation does not rule out the operation of continued selection against hybridization in locally sympatric mixed populations and/or along the grassland/forest border. Rather, it suggests that substantial gaps in a potential continuum of phenotypic call variability preceded sympatry, and enabled the successful coexistence of the two species whose call differentiation originated and was substantially perfected in allopatry before assuming sympatry. Therefore, synchronic and syntopic reproductively mixed populations each safeguard their own species' integrity by premating reproductive isolation. These presumably operate to maximize the fitness of individuals characterized primarily by similarity of ecological adaptation (i.e., ecospecies) (Littlejohn, 1980).

Evolutionary divergence across the grassland/forest border in eastern Texas has been amply documented in a variety of vertebrates, including anurans (Blair, 1958*b*). For example, the chorus frogs *Psuedacris triseriata* and *P. clarki* show limited sympatry along the grassland/forest border in eastern Texas. They "differ in color pattern and size, and strikingly in pulse rate of call which provides the basis for the discrimination by females" (Michaud, 1962), and in various attributes that serve as ecological isolating mechanisms, yet fertility of their interspecific hybrids has been demonstrated (Lindsay, 1958; Blair, 1964*b*) [see also distributional maps in Conant (1975), pp. 403–412]. Grassland/forest call differentiation and east–west species-pair call differentiation, similar to that between *A. crepitans* and *A. gryllus*, also occurs in other hylids (*Hyla cinerea* and *H. gratiosa, H. chrysoscelis* and *H. avivoca*, and *H. versicolor* and *H. chrysoscelis*) (Blair, 1964*a,b*, and references therein). However, unless call structure is analyzed across the *entire* range in as many populations as possible in allopatry as well as in sympatry, it is difficult to resolve the question of whether the origin and differentiation of the call is due to incidental factors or to reinforcement in sympatry.

EVOLUTIONARY HISTORY OF *ACRIS*

Systematics and the Fossil Record

Treefrogs (Hylidae) originated and radiated in Cretaceous South America (Estes and Reig, 1973) and invaded North America in the Paleocene, where they developed into several distinctive extratropical species groups (Savage, 1973). Hylidae currently includes 30 genera and 395 species (Duellman, 1979), ranging from cold to tropical regions in the New and Old Worlds. The genus *Acris*, or its immediate ancestors, is known in North America from the lower Miocene, and it originated as a semiaquatic subtropical hylid probably in Oligocene times (Chantell, 1964).

The long evolutionary history of *Acris* in the U. S. and its continuity from lower Miocene to Recent times is documented by the fossil record in both Florida (Holman, 1961) and the Great Plains, including Colorado, Nebraska, and Kansas (Chantell, 1964, 1966, 1971, and references therein). Osteologically, *Acris* is well differentiated from the other genera of temperate North American Hylidae, but it shows only little intrageneric differentiation (Chantell, 1968). Furthermore, while some of the fossil material, even as old as lower Miocene, is modern in morphology (Chantell, 1965), other material (i.e., lower Miocene *Proacris* from Florida and the Mio-Pliocene Valentine *Acris* from Nebraska) suggests that the present osteological homogeneity may be a relatively recent phenomenon (Chantell, 1968). In any case, speciation in hylids has often been associated with ecological, behavioral, or cytological, rather than morphological, factors (Johnson, 1961, 1963; Schneider and Nevo, 1972). *Acris crepitans* and *A. gryllus* are closely related and were assigned to a common species by Schmidt (1953). More recently, studies of differences in reproduction, call behavior, ecology, and distribution have favored recognition as separate species (Boyd, 1964; Neill, 1954; Mecham, 1964).

Speciation of *Acris* during Tertiary times must have preceded that of the Pleistocene. The present distributional pattern suggests that the range of *A. gryllus* is of a presumed Pleistocene isolate in Florida (Neill, 1950; Blair, 1958*a,b*). However, the distribution of *A. crepitans* suggests that the evolution of *A. c. blanchardi* and *A. c. crepitans* may have been associated with the development of grasslands in the Pliocene and to the establishment of the grassland/forest ecotone. *Acris c. blanchardi* is a grassland-adapted form, whereas *A. c. crepitans* is a forest-adapted form. These two subspecies may well be incipient parapatric ecospecies, as suggested by their differential call structure, homogamic mate selection, and genetic differentiation.

Genetic Differentiation and Speciation in *Acris*

Population genetic structure of the cricket frogs *A. crepitans* and *A. gryllus* has been examined in terms of protein variation throughout their geographic ranges (Dessauer and Nevo, 1969; Salthe and Nevo, 1969). A total of 850 animals, collected from 32 populations in 20 states, was processed to obtain electrophoretic evidence from 21 blood and liver proteins. Molecular data confirm the reproductive isolation of the two species *A. gryllus* and *A. crepitans*. They have different albumins, transferrins, and hemoglobins. Frogs with hybrid protein patterns were not found in regions where the two species are sympatric. Within *A. crepitans* nine proteins are invariant across the species range and 13 exhibit one or more variant phenotypes. However, the most remarkable phenomenon in the genetic structure of *A. crepitans* is its geographic divergence into Plains, Delta, and Appalachian groups as suggested by the presence of specific hemoglobin, transferrin, and liver esterase variants, which characterize animals from each region (Dessauer and Nevo, 1969, Tables 1 and 2 and Figs. 2–8). Even *Ldh-1*, which varies geographically across the range of *A. crepitans* and thus suggests some genetic continuity, shows distinct genic differentiation. One allele predominates in eastern populations of *A. c. crepitans* and becomes less common progressively westward through the Middle West (Salthe and Nevo, 1969).

Although the color polymorphism in *Acris* (Nevo, 1973*b*) does not distinguish the two subspecies of *A. crepitans*, it does provide additional supportive evidence for the ecological east–west genetic differentiation. All *A. gryllus* and *A. c. crepitans* populations and the central population of *A. c. blanchardi* are trimorphic and highly heterozygous. The red and green alleles disappear toward the western and northern range of *A. c. blanchardi*, where gray monomorphism prevails. Regional morph clines are detectable and strongly correlate with rainfall and to a lesser degree with temperature. The frequency of the gray morph increases along increasingly arid habitats until it culminates and characterizes *A. c. blanchardi* in the west and in the north.

Thus, *A. c. crepitans* and *A. c. blanchardi* demonstrate substantial genetic divergence, which is primarily related to forest vs. grassland biomes. Though gene exchange is apparently continuing to some extent between the forest and grassland populations, it may be partly restricted by both premating isolating mechanisms (call difference) and hybrid inferiority along the grassland/forest borders. We have previously shown (Capranica *et al.*, 1973) that the geographic variation in frequency sensitivity of the auditory nervous system of females of the two "subspecies" closely matches the spectral energy in their mating calls, thus enabling them to respond preferentially to the calls of their local dialect (as demonstrated by our female discrimination tests). In other words, we suggest that *A. c. crepitans* and *A. c. blanchardi* are in some degree sexually isolated, and may be legitimately considered incipient ecospecies.

If this interpretation is correct, then the evolutionary divergence of *A. c. crepitans* and *A. c. blanchardi* may relate to development of the arid Great Plains some time in the Pliocene, whereas the evolutionary divergence of *A. crepitans* and *A. gryllus* must have preceded it and may have taken place in late Miocene or early Pliocene times. A thorough study of genetic, ecological, and call structure differentiation along the grassland/forest boundary where *A. c. blanchardi* and *A. c. crepitans* are contiguous might be very profitable. It may provide an active case of the final stages of ecological speciation. Furthermore, this is precisely the zone in which some reinforcement might operate if indeed hybridization is ongoing between the two ecospecies along a relatively narrow contact zone. *Acris c. crepitans* and *A. c. blanchardi* may well be incipient parapatric ecospecies that have undergone genetic divergence to their forest and grassland biomes and thus developed substantially different call structures and homogamic mate selection, which presumably safeguard to a large extent their species identification.

SUMMARY

The evolutionary origin of ethological reproductive isolation in cricket frogs, genus *Acris*, has been studied by call analysis across their range in central and eastern U. S. and by field discrimination tests with females in three widely separated populations. Mating calls of 112 males (out of 393 recorded in 46 localities) were analyzed in terms of their spectral and temporal signal characteristics. These representative males consisted of 74 *A. crepitans* in two subspecies, *A. c. blanchardi* and *A. c. crepitans* from 13 localities (11 allopatric and two sympatric), and 38 *A. gryllus* (in two subspecies, *A. g. dorsalis* and *A. g. gryllus*) from eight localities (three allopatric and five sympatric; Fig. 1 and Table I). Twenty-six variables—ten background variables and 16 call variables—entered in the analysis. The relationships between these variables were subjected to univariate and multivariate statistical analyses.

The mating call of a male cricket frog consists of a patterned series of audible clicks, each comprised of several pulses. A comparison of 16 spectral and temporal call parameters, standardized to 22.79°C mouth temperature, revealed the following results:

1. Geographic variation was found in all 16 spectral and temporal call variables within and between species (except for the number of pulses within each click in *A. gryllus*): spectral peak *SP*, total call duration *T*, total number of clicks in a call *NCT*, click rate at beginning, middle, and end of a call *Rcb, Rcm,* and *Rce*, duration of a click in the beginning, middle, and end of a call *Dcb, Dcm,* and *Dce*, number of pulses in a click in the beginning, middle, and end of a call *Npcb, Npcm,* and *Npce*, and number of groups of pulses in a click at the beginning, middle, and end of call *Ngpcb, Ngpcm,* and *Ngpce*.
2. In 12 call variables the values of *A. crepitans* and *A. gryllus* were nonoverlapping, whether in sympatry or allopatry. Only in the remaining four variables (*SP, T, Rce, Dcb*) did their values overlap, and of these, only *T* and *Dcb* showed a larger difference in sympatry than in allopatry.
3. The frequency of the spectral peak was significantly correlated with the temporal features.
4. All 16 variables displayed significant correlations with morphological, ecological, and geographic parameters.

Nonmetric multivariate analysis (POSA) partitioned the call structure across the range into five different profiles based on nine differentiating call variables. This analysis led to four major categories of calls: Medium to Fast Clicking (*A. c. blanchardi*), Slow Clicking (*A. gryllus*, both sympatric and allopatric); Fast Rattling (sympatric and southwestern allopatric *A. c. crepitans*), and Slow Rattling (allopatric northeastern *A. c. crepitans*). The metric multivariate analysis (Candec) explained in a two-dimensional graphic biplot more than 83% of the call variation. Remarkably, the biplot resulted in three call clusters corresponding to the three biomes occupied by *Acris*: *A. c. blanchardi* in grasslands, *A. c. crepitans* in deciduous woodlands, and *A. gryllus* in subtropical pineries. In overall call differentiation *A. c. crepitans* as a whole is quite distant from *A. c. blanchardi*, even more so for some populations than from *A. gryllus*.

Field studies of phonotaxis to playback of recorded natural and synthetic mating calls were conducted with 123 gravid females in Texas, Georgia, and New Jersey. Based on 486 discrimination tests, *A. c. crepitans* females (either sympatric or allopatric) showed a highly significant preference for calls of their own males compared to the calls of *A. gryllus*, whereas the degree of discrimination by allopatric *A. c. blanchardi* against calls of *A. gryllus* was much weaker (only $p < 0.10$). For *A. crepitans* as a whole, each of the two subspecies discriminated completely against the other. Furthermore, the level of discrimination by *A. c. crepitans* females against the calls of males within this same subspecies increased with their geographic separation. Tests with electronically synthesized calls in these field studies verified that both the spectral and the temporal characteristics are crucial cues for a female's discriminatory abilities.

The results of our study support the allopatric and reject the sympatric theory of the origin of premating reproduction isolation as a major explanatory model of call differentiation in *Acris*, chiefly because: (1) allopatric *A. c. crepitans*, whether in the southwestern or northeastern U. S., is just as distinct from *A. gryllus* as sympatric *A. c. crepitans*, both in call structure and in female preference; and (2) most call variables of *A. gryllus* and *A. crepitans* do not overlap and are largely explained in each species by environmental factors independent of sympatry. Our analyses indicate that call structure in the genus *Acris* is correlated primarily with the original environment from which the allopatric species originated before encountering sympatry. Our results also suggest that the two subspecies *A. c. blanchardi*, which is a grassland-adapted form, and *A. c. crepitans*, which is a forest-adapted form, may be incipient ecospecies. Their call differences, presumably related to grassland/forest biomes, may provide sufficient basis for positive assortative mating to safeguard their advanced genetic differentiation and identification.

Acknowledgments

E. N. gratefully acknowledges the hospitality of Prof. W. F. Blair, University of Texas, where this work started in 1965 during his tenure as a visiting professor. His subsequent appointment as Fellow in Biology at Harvard University enabled the continuation of this study; the encouragement and assistance of Prof. Ernst Mayr is greatly appreciated. R. Oldham first introduced E. N. to *Acris*. R. R. C. thankfully acknowledges the support and exceptional facilities of the Bell Telephone Laboratories during a major part of his study while he was a member of their technical staff.

We are deeply indebted to Prof. L. Guttman and K. R. Gabriel for assistance in statistical analysis as well as for fruitful discussions. Our special thanks go to A. Beiles for his statistical assistance and stimulating conversations as well as his valuable advice for improving the manuscript, and to D. Adler for commenting on the manuscript.

This research was supported in part by grant GB-3167 from the National Science Foundation to the Committee of Evolutionary Biology, Harvard University; Grants-in-Aid from Sigma Xi; by a grant from the

United States–Israel Binational Science Foundation (BSF), Jerusalem, Israel; by funds from the Bell Telephone Laboratories; and by National Science Foundation grant GB-18836 and National Institutes of Health grant NS-09244 to R. R. C.

E. N's deceased son, Tal, greatly helped us in many collecting and recording trips across the *Acris* range. His field assistance and companionship were indispensable for the completion of this study and we dedicate this chapter to his memory.

While this manuscript was in press we received the sad news of Frank Blair's death. We therefore also wish to dedicate this paper to his memory. He was a remarkable man, scientist, and teacher. His innumerable contributions to evolutionary theory, including, among many other aspects, vocal communication in frogs and toads, are shining examples of the originality and deep insight he brought to the common human endeavor to understand nature and the world in which we live.

REFERENCES

Awbrey, F. T., 1965, An experimental investigation of the effectiveness of anuran mating calls as isolating mechanisms, Ph.D. Thesis, University of Texas at Austin.

Bayless, L. E., 1966, Comparative ecology of two sympatric species of *Acris* (Anura: Hylidae) with emphasis on interspecific competition, Ph.D. Thesis, Tulane University, Louisiana.

Bayless, L. E., 1969, Ecological divergence and distribution of sympatric *Acris* populations (Anura; Hylidae), *Herpetologica* **25:**181–187.

Blair, W. F., 1955, Mating call and stage of speciation in the *Microhyla olivacea–M. carolinensis* complex, *Evolution* **9:**469–80.

Blair, W. F., 1958*a*, Mating call in the speciation of anuran amphibians, *Am. Nat.* **92:**27–51.

Blair, W. F., 1958*b*, Distributional patterns of vertebrates in the southern United States in relation to past and present environments, in: *Zoogeography* (C. L. Hubbs, ed.), pp. 433–468, American Association for the Advancement of Science, Washington, D.C.

Blair, W. F., 1962, Non-morphological data in anuran classification, *Syst. Zool.* **11:**72–84.

Blair, W. F., 1964*a*, Acoustic behavior of Amphibia, in: *Acoustic Behavior of Animals.* (R. G. Busnel, ed.), pp. 694–708, Elsevier, Amsterdam.

Blair, W. F., 1964*b*, Isolating mechanisms and interspecies interactions in anuran amphibians, *Q. Rev. Biol.* **39:**334–44.

Blair, W. F., 1974, Character displacement in frogs, *Am. Zool.* **14:**1119–1125.

Boyd, C. E., 1964, The distribution of cricket frogs in Mississippi, *Herpetologica* **20:**201–202.

Brown, W. L., and Wilson, E. O., 1956, Character displacement, *Syst. Zool.* **5:**49–64.

Capranica, R. R., 1965, *The Evoked Vocal Response of the Bullfrog*, MIT Press, Cambridge, Massachusetts.

Capranica, R. R., 1976, Morphology and physiology of the auditory system, in: *Frog Neurobiology* (R. Llinás and W. Precht, eds.), pp. 551–575, Springer, Berlin.

Capranica, R. R., 1977, Auditory processing of vocal signals in anurans, in: *The Reproductive Biology of Amphibians* (D. H. Taylor and S. I. Guttman, eds.), pp. 337–355, Plenum Press, New York.

Capranica, R. R., and Frishkopf, L. S., 1966, Responses of auditory units in the medulla of the cricket frog, *J. Acoust. Soc. Am.* **40:**1263.

Capranica, R. R., and Moffat, A. J. M., 1975, Selectivity of the peripheral auditory system of spadefoot toads (*Scaphiopus couchi*) for sounds of biological significance, *J. Comp. Physiol.* **100:**231–249.

Capranica, R. R., Frishkopf, L. S., and Nevo, E., 1973, Encoding of geographic dialects in the auditory system of the cricket frog, *Science* **182:**1272–1275.

Chantell, C. J., 1964, Some Mio-Pliocene hylids from the Valentine Formation of Nebraska, *Am. Midl. Nat.* **72:**211–225.

Chantell, C. J., 1965, A lower Miocene *Acris* (Amphibia: Hylidae) from Colorado, *J. Paleontol.* **39:**507–508.

Chantell, C. J., 1966, Late Cenozoic hylids from the Great Plains, *Herpetologica* **22:**259–264.

Chantell, C. J., 1968, The osteology of *Acris* and *Limnaoedus* (Amphibia: Hylidae), *Am. Midl. Nat.* **79:**169–182.

Chantell, C. J., 1971, Fossil Amphibians from the Egelhoff local fauna in north-central Nebraska, *Contrib. Mus. Paleontol. Univ. Michigan* **23:**239–246.

Conant, R., 1975, *A Field Guide to Reptiles and Amphibians*, Houghton Mifflin, Boston.

Darwin, C., 1859, *The Origin of Species by Means of Natural Selection*, John Murray, London.

Dessauer, H. C., and Nevo, E., 1969, Geographic variation of blood and liver proteins in cricket frogs, *Biochem. Genet.* **3:**171–188.

Dobzhansky, T., 1940, Speciation as a stage in evolutionary divergence, *Am. Nat.* **74:**312–321.

Dobzhansky, T., 1951, *Genetics and the Origin of Species*, 3rd. ed., Columbia University Press, New York.

Dobzhansky, T., Ayla, F. J., Stebbins, G. L., and Valentine, J. W., 1977, *Evolution*, Freeman, San Francisco.

Draper, N. R., and Smith, H., 1966, *Applied Regression Analysis*, Chapters 3, 5, and 6, Wiley, New York.

Duellman, W. E., 1979, The number of amphibians and reptiles, *Herpetol. Rev.* **10:**83–84.

Estes, R., and Reig, O., 1973, The early fossil record of frogs: A review of the evidence, in: *Evolutionary Biology of the Anurans* (J. L. Vial, ed.), pp. 11–63, University of Missouri Press, Columbia, Missouri.

Farrell, M. P., and MacMahon, J. A., 1969, An ecophysiological study of water economy in eight species of tree frogs (Hylidae), *Hereptologica* **25:**279–294.

Ferguson, D. E., Landreth, H. F., and McKeown, J. P., 1967, Sun compass orientation of the northern cricket frog, *Acris crepitans, Anim. Behav.* **15:**45–53.

Fisher, R. A., 1930, *The Genetical Theory of Natural Selection*, Clarendon Press, Oxford.

Fouquette, M. J., 1975, Speciation in chorus frogs. I. Reproductive character displacement in the *Pseudacris nigrita* complex, *Syst. Zool.* **24:**16–23.

Frishkopf, L. S., and Goldstein, M. H., Jr., 1963, Responses to acoustic stimuli from single units in the eighth nerve of the bullfrog, *J. Acoust. Soc. Am.* **35:**1219–1228.

Frishkopf, L. S., Capranica, R. R., and Goldstein, M. H., Jr., 1968, Neural coding in the bullfrog's auditory system—A teleological approach, *Proc. IEEE* **56:**969–980.

Gabriel, K. R., 1971, The biplot–Graphic display of matrices with applications to principal component analysis, *Biometrika* **58:**453–467.

Gabriel, K. R., 1973, Canonical Decomposition and Biplots: Notes, Examples, and Computer Program CANDEC, Department of Statistics, Hebrew University, Jerusalem, Israel.

Gerhardt, H. C., 1974, The significance of some spectral features in mating call recognition in the green treefrog (*Hyla cinerea*), *J. Exp. Biol.* **61**:229–241.

Grant, P. R., 1972, Convergent and divergent character displacement, *Biol. J. Linn. Soc.* **4**:39–68.

Grant, V., 1971, *Plant Speciation*, Columbia University Press, New York.

Gulick, J. T., 1890, Divergent evolution through cumulative segregation, *J. Linn. Soc. Zool.* **20**:189–274.

Guttman, L., 1966, Order analysis of correlation matrices, in: *Handbook of Multivariate Experimental Psychology* (R. B. Cattell, ed.), pp. 444–458, Rand McNally, Chicago.

Holman, J. A., 1961, A new hylid genus from the lower Miocene of Florida, *Copeia,* **1961**(3):354–355.

Johnson, F. C., 1961, Cryptic speciation in the *Hyla versicolor* complex, Ph.D. Thesis, University of Texas at Austin.

Johnson, F. C., 1963, Additional evidence of sterility between call-types in the *Hyla versicolor* complex, *Copeia* **1963**:139–143.

Levin, D. A., 1971, The origin of reproductive isolating mechanisms in flowering plants, *Taxon* **20**:91–113.

Levin, D. A., 1978, The origin of isolating mechanisms of flowering plants, *Evol. Biol.* **11**:185–317.

Lindsay, H. L., Jr., 1958, Analysis of variation and factors affecting gene exchange in *Pseudacris clarki* and *Pseudacris nigrita* in Texas, Ph.D. Thesis, University of Texas at Austin.

Lingoes, J. C., 1973, *The Guttman–Lingoes Nonmetric Program Series*, Mathesis Press, Ann Arbor, Michigan.

Littlejohn, M. J., 1965, Premating isolation in the *Hyla ewingi* complex (Anura, Hylidae), *Evolution* **19**(2):234–243.

Littlejohn, M. J., 1969, The systematic significance of isolating mechanisms, in: *Systematic Biology Proc. Inter. Conf.,* pp. 459–482, National Academy of Science, Washington, D.C.

Littlejohn, M. J., 1980, Reproductive isolation—A critical rview, in: *Evolution and Speciation: Essays in Honor of M. J. D. White* (W. Atchley and D. S. Woodruff, eds.), pp. 298–334, Cambridge University Press, Cambridge.

Littlejohn, M. J., and Loftus-Hills, J. J., 1968, An experimental evaluation of premating isolation in the *Hyla ewingi* complex (Anura: Hylidae), *Evolution* **22**:659–663.

Littlejohn, M. J., and Martin, A. A., 1969, Acoustic interaction between two species of leptodactylid frogs, *Anim. Behav.* **17**:785–791.

Littlejohn, M. J., and Michaud, T. C., 1959, Mating call discrimination by females of Strecker's chorus frog (*Pseudacris streckeri*), *Tex. J. Sci.* **11**:86–92.

Loftus-Hills, J. J., and Johnstone, B. M., 1970, Auditory function, communication, and the brain-evoked response in anuran amphibians. *J. Acoust. Soc. Am.* **47**:1131–1138.

Loftus-Hills, J. J., and Littlejohn, M. J., 1971, Pulse repetition rate as the basis for mating call discrimination by two sympatric species of *Hyla, Copeia* **1971**(1):154–156.

Martin, W. F., 1972, Evolution of vocalization in the genus *Bufo,* in: *Evolution in the Genus Bufo* (W. F. Blair, ed.), pp. 279–309, University of Texas Press, Austin.

Martof, B. S., and Thompson, E. F., 1958, Reproductive behavior of the chorus frog, *Pseudacris nigrita, Behavior* **13**:243–258.

Mayr, E., 1942, *Systematics and the Origin of Species*, Columbia University Press, New York.

Mayr, E., 1970, *Populations, Species, and Evolution*, Belknap Press of Harvard University Press, Cambridge.
Mecham, J. S., 1964, Ecological and genetic relationships of the two cricket frogs, genus *Acris* in Alabama, *Herpetologica* **20:**84–91.
Michaud, T. C., 1962, Call discrimination by females of the chorus frogs, *Pseudacris clarki* and *Pseudacris nigrita, Copeia* **1962:**213–215.
Mudry, K. M., Constantine-Paton, M., and Capranica, R. R., 1977, Auditory sensitivity of the diencephalon of the leopard frog *Rana p. pipiens, J. Comp. Physiol.* **114:**1–13.
Muller, H. J., 1942, Isolating mechanisms, evolution and temperature, *Biol. Symp.* **6:**71–125.
Neill, W. T., 1950, Taxonomy, nomenclature, and distribution of southeastern cricket frogs, genus *Acris, Am. Midl. Nat.* **43:**152–156.
Neill, W. T., 1954, Ranges and taxonomic allocations of amphibians and reptiles in the southeastern United States, *Publ. Res. Div. Ross Allen's Reptile Inst.* **1:**755–796.
Nevo, E., 1969, Discussion of the systematic significance of isolating mechanisms, in: *Systematic Biology*, pp. 485–489, National Academy of Science, Washington, D.C.
Nevo, E., 1973*a*, Adaptive variation of size in cricket frogs, *Ecology* **54:**1271–1281.
Nevo, E., 1973*b*, Adaptive color polymorphism in cricket frogs, *Evolution* **27:**353–367.
Nevo, E., and Schneider, H., 1976, Mating call pattern of green toads in Israel and its ecological correlate, *J. Zool. Lond.* **178:**133–145.
Nie, N. H., Hull, C. H., Jenkins, J. G., Steinberger, K., and Bent, D. H., 1975, *SPSS Statistical Package for the Social Sciences*, 2nd ed., McGraw-Hill, New York.
Oldham, R. S., and Gerhardt, H. C., 1975, Behavioral isolating mechanisms of the treefrogs *Hyla cinerea* and *H. gratiosa, Copeia* **1975:**223–230.
Otte, D., 1974, Effects and functions in the evolution of signalling systems, *Ann. Rev. Ecol. Syst.* **5:**385–417.
Patterson, J. T., and Stone, W. S., 1952, *Evolution in the Genus Drosophila*. Macmillan, New York.
Ralin, D. B., 1977, Evolutionary aspects of mating call variation in a diploid–tetraploid species complex of treefrogs (Anura), *Evolution* **31:**721–736.
Salthe, S. N., and Nevo, E., 1969, Geographic variation of lactate dehydrogenase in the cricket frog, *Acris crepitans, Biochem. Genet* **3:**335–341.
Savage, J. M., 1973, The geographic distribution of frogs: Patterns and predictions, in: *Evolutionary Biology of the Anurans* (J. L. Vial, ed.), pp. 351–454, University of Missouri Press, Columbia, Missouri.
Sawyer, S., and Hartl, D., 1981, On the evolution of behavioral reproductive isolation: The Wallace effect, *Theor. Popul. Biol.* **19:**261–273.
Schiøtz, A., 1964, The voices of some West African amphibians, *Vidensk. Medd. Naturalist. Foren.* **127:**3–83.
Schiøtz, A., 1973, Evolution of anuran mating calls: Ecological aspects, in *Evolutionary Biology of the Anurans* (J. L. Vial, ed.), pp. 311–319. University of Missouri Press, Columbia, Missouri.
Schmidt, K. P., 1953, *A Check List of North American Amphibians and Reptiles*, American Society of Ichthyologists and Herpetologists, Chicago.
Schneider, H., and Nevo, E., 1972, Bio-acoustic study of yellow-lemon tree frog, *Hyla arborea savignyi, Audouin. Zool. Jahrb.* (*Zool. Physiol.*) **76:**497–506.
Straughan, I. R., 1966, An analysis of the mechanism of sex and species recognition and species isolation in certain Queensland frogs, Ph.D. Dissertation, University of Overland, Brisbane, Australia.
Straughan, I. R., 1973, Evolution of anuran mating calls: Bioacoustical aspects, in: *Evo-*

lutionary Biology of the Anurans (J. L. Vial, ed.), pp. 321–327, University of Missouri Press, Columbia, Missouri.

Viosca, P. A., Jr., 1944, Distribution of certain cold-blooded animals in Louisiana in relation to the geology and physiography of the state, *Proc. Louisiana Acad. Sci.* **8**:47–62.

Wallace, A. R., 1889, *Darwinism, An Exposition of the Theory of Natural Selection with Some of its Applications*, Macmillan, London.

Wright, A. H., and Wright, A. A., 1949, *Handbook of Frogs and Toads*, Comstock, Ithaca, New York.

5

The Development of Behavior from Evolutionary and Ecological Perspectives in Mammals and Birds

MARC BEKOFF

and

JOHN A. BYERS

> Modification of developmental patterns through natural selection may thus be regarded as the major source of evolutionary novelty and the primary phenotypic expression of evolutionary change (Mason, 1979, p. 8).

INTRODUCTION

Comprehensive studies of behavior must consider at least four major topics, *evolution, adaptation, causation,* and *development* (Tinbergen, 1951, 1963). In this chapter, we argue that separation of developmental questions from consideration of (1) the historical routes by which behav-

MARC BEKOFF • Department of Environmental, Population, and Organismic Biology, University of Colorado, Boulder, Colorado 80309. JOHN A. BYERS • Department of Biological Sciences, University of Idaho, Moscow, Idaho 83843.

ioral traits evolve, (2) their adaptive significance, somehow, but *rarely,* measured in terms of inclusive fitness, and (3) the factors (internal and external) responsible for causing the trait to be expressed, though possible, produces results of limited applicability. Because all populations are age-structured (Charlesworth, 1980), developmental questions are important to consider in evolutionary and ecological research in behavioral biology (e.g., Alexander, 1974; Gould, 1977; Burghardt and Bekoff, 1978; Cairns, 1979; Fagen, 1981; Gubernick and Klopfer, 1981; Immelmann *et al.,* 1981; Hinde, 1982, 1983; Cheverud *et al.,* 1983; Stamps, 1983; A. Bekoff, 1981, 1985; Patenaude, 1984; M. Bekoff, 1985; Price and Grant, 1985).

In this chapter we review select, but diverse, areas of research in which relationships among development, evolution, and ecology are apparent. Where possible, ideas from population biology (demographic and life-history studies) are introduced. General topics to be discussed include the development of comfort behavior, helping [alloparental behavior (E. O. Wilson, 1975)], agonistic behavior, social play, life-history tactics, and the relationship between social development and social organization. There are numerous other areas [imprinting, individual (including kin) discrimination, predatory and antipredatory behavior, tool use, echolocation and other forms of communication such as bird song] that potentially could be considered, and our choice merely reflects those areas with which we are most familiar and in which age-related changes in behavior or developmental mechanisms can be highlighted, and evolutionary and ecological interpretations are possible.

Because, to the best of our knowledge, there are no field data that establish a *direct, causal* relationship between individual variations in development and differences in fitness [see Henderson (1981) for an elegant study on captive mice], the assumption that there is some link between ontogeny and fitness must be accepted solely on faith in modern evolutionary theory. The often quoted maxim that patterns of behavioral development evolve is predicated on the undemonstrated requirements that (1) heritable variation in ontogeny exists and (2) such variation is linked to differences in individual survival and reproductive success. Even the most solid *correlational* studies will require *detailed* analyses of ontogeny over an individual's life span. In the absence of these data, attempts to discuss the possible adaptive significance (in terms of fitness) of developmental patterns would be analogous to trying to play Scrabble without knowing how to assemble letters to form words—to spell.

Behavioral Development: Internal and External Influences

What causes animals to do certain things is a question for which there are no simple answers. When discussed in terms of nature versus nurture, rather unproductive results often are produced, because it actually is a case of nature *and* nurture in the vast majority of cases (e.g., Hailman, 1967). Analysis of the ways in which genetic predispositions and internal constraints interact with environmental factors in behavioral development is an exciting endeavor. Consideration of these issues only become counterproductive when it gums up the wheels of progress and becomes an excuse for arm chair rumination rather than empirical research. As Beck (1980, p. 136) has written, "Past excesses, oversimplifications, and a history of sociological perversions discourage use of the terms 'innate' and 'learned.' Nonetheless, their usefulness can be gauged by the frantic search for noninflammatory synonyms by many contemporary behavioral scientists."

While not ignoring the fact that behavior in most cases is a modifiable phenotype, the importance of internal constraints on behavioral ontogeny is being stressed by workers interested in diverse problems. For example, Bateson (1976, p. 402) remarks, "I believe that the relentless hunt for yet further external sources of individual variation, and the shyness of accepting substantial internal control over developmental processes, have led to a certain sterility of thinking and an incapacity to comprehend more than a small fraction of the rapidly accumulating data." Other workers also make the point that although behavior may be susceptible to modification during development, the contribution of innate motor programs to developmental processes [including embryology (Hamburger, 1980)] should not be underestimated (Marler, 1981; Marler *et al.*, 1980; Cavalli-Sforza and Feldman, 1981).

For example, neurobiological studies have shown that there are many "hard-wired," genetically specified central pattern generators responsible for coordinated movements and locomotion in various vertebrates and invertebrates (Delcomyn, 1980). Central pattern generators operate in the absence of sensory input, but may be responsive to sensory stimulation; they can influence the effect of incoming stimuli by gating the afferent signal (Grillner, 1975; Edgerton *et al.*, 1976; A. Bekoff, 1981).

The ontogeny of adaptive motor skills in humans may even depend on genetically specified central motor programs. Recently, Thelen (1981) suggested that rhythmical stereotyped motor behavior performed by human infants is a developmental manifestation of intrinsic central programs that provide the basis for coordinated movements. Infant rhythmic stereotypes probably are "phylogenetically old neuromuscular coordinations used by humans as adaptive behavior because slow cortical maturation results in long stages where voluntary movements have not fully matured" (Thelen, 1981, p. 237). As Marler *et al.* (1980) point out in their review of perceptual systems, programs of behavior are often as canalized as those responsible for the development of anatomic structures; learning tends to be species-specific and *what* is learned *when* and *how* are often rigidly specified.

While it is not our purpose to defend the frequently misinterpreted views of classical ethologists, it should be noted that they, too, would probably subscribe to the above *interactionist* view for the development of behavior. For example, Lorenz (1981, p. 257) recently wrote, "The realization of any genetic program that has evolved during phylogeny is dependent on innumerable external conditions influencing the organism during its individual development, during ontogeny." Eibl-Eibesfeldt (1979), in his review of human ethology, takes a similar position to that of Lorenz, namely, that while behavior may be *in part* preprogrammed, it is not immodifiable.

Suffice it to say, either/or views of behavioral development have generally been infertile (Hinde, 1959; Jacobs, 1981). Indeed, some of the most exciting aspects of research for scientists interested in behavioral development have involved attempts to determine how "environmentally expectant" (evolutionary) factors and "environmentally dependent" (individual experience) factors (M. Bekoff and Fox, 1972) mesh to produce adaptive behavioral phenotypes (Coss 1978, 1979).

Reproductive Success and Developmental Discontinuity

In many species, some period of time intervenes between early development and sexual maturity. Therefore, it is not always clear how variations in early behavioral ontogeny are linked to differences in individual reproductive success, because of what appears to be a discontinuous process [see Bateson (1976, 1978, 1981), Hofer (1978), and Sackett *et al.* (1981) for discussions of continuity and discontinuity in behavioral development]. However, taken as a whole, development is a continuous process, and a concentration only on adult fitness begs the question as to how individuals meet challenges at *each* stage of development, and how each stage is affected by preceding events and in turn influences succeeding stages (Galef, 1981). Longitudinal data demonstrating that different developmental pathways produce variations in fitness undoubtedly will emerge from studies in which the continuity of behavior (the formation of one continuous path assembled, perhaps, from single discontinuous systems) is highlighted.

It would be misleading to imply that in areas other than behavioral development a plethora of conclusive data exist demonstrating a causal relationship between behavior and fitness. They do not. Playing "find-the-adaptiveness" game (Klopfer, 1981) may be fun and intellectually stimulating, but the assumption that *everything* that an animal does is adaptive may be misleading (Williams, 1966; Gould and Lewontin, 1979; Cavalli-Sforza and Feldman, 1981; Reed, 1981; Rijksen, 1981; Stearns and Sage, 1980).

Variability, Plasticity, and Open Systems

Variability, due either to differences in individual responsiveness to a given set of environmental conditions or strong genetic predispositions to respond to specific stimuli in species-characteristic ways (Arnold 1981*a,b*), also is a potential bugbear in studies of behavioral ontogeny. One need not spend much time observing animals to realize how extensive are individual differences in behavior.

With respect to social behavior, it is a well-established fact, especially in birds and mammals, that variations in the social environment can drastically affect behavioral ontogeny. The demonstration of plasticity shows that behavioral systems (some more than others) are modifiable, and it is useful to ask why this is so. Selection for "open systems" in which individual experience and learning play important roles (Mayr, 1974; van der Molen, 1984) is not surprising, but there may be "selective costs" involved in the evolution of learning, such as delayed reproduction, increased juvenile vulnerability, increased parental care, greater complexity of the nervous system and genome, and developmental fallibility (Johnston, 1982). However, animals usually are born and reared in what may be characterized as species-typical environments (Boorman and Levitt, 1980; M. Bekoff, 1978*a,* 1981*b*), and the probability of "accidents of development" (Immelmann, 1975*a*) occurring is low (also see Johnston, 1982). Selection that severely limits the influences of external stimulation would be expected only in those systems in which slight modifications would have dire consequences.

In most species with open behavioral programs, maximum learning usually occurs during ontogeny (Immelmann, 1972, 1975*b*). This is especially so in slowly developing animals born into closed social groups from which emigration during early life is rare (Boorman and Levitt, 1980; M. Bekoff, 1978*a*). All that needs to be specified genetically are predispositions to respond to the broad range of stimuli that have characterized past environments; natural selection cannot imagine situations that never occur (Cavalli-Sforza and Feldman, 1981). Furthermore, information acquired during early life should remain relatively resistant to further change, in order to prevent a detrimental influence of subsequent and possibly less appropriate stimuli to which the organism will be exposed as it broadens its horizons (Immelmann, 1975*b*). Therefore, there is selection for open systems which become increasingly closed due to individual experience.

Finally, as Immelmann(1975*b*) pointed out, acquired behavioral patterns generally are more adaptable to the demands of the current environment than are innate programs. He also suggested that the amount of information that can be stored in the genome tends to be smaller than the possible amount of information that can be stored in an individual's memory.

ONTOGENY AND ECOLOGY

Habitat type (fluctuating, stable, patchy, continuous, terrestrial, arboreal, etc.) and resource availability (space, food, dens, hibernacula, social companions, etc.) can affect the development of behavior (e.g., Barash, 1974; Tschanz and Hirsbrunner-Scharf, 1975; Anderson *et al.*, 1976; Sussman, 1977; Berger, 1979, 1980; Armitage, 1981; Ferron, 1981; Fox *et al.*, 1981; Stamps, 1983). Furthermore, resource utilization has been implicated as a major selective force in the evolution of two general and contrasting suites of demographic and behavioral strategies.

Where and How?

Various methodological alternatives need to be considered in studies on the ecology of development. "Where" and "how" questions are deceptively easy to answer: given a specific question, find an animal group living in an area in which young individuals of known age can be observed, identified, and followed as they mature, using methods that allow achievement of the stated goal by amassing quantitative data that can be rigorously analyzed, perhaps using multivariate statistics (Miller *et al.*, 1977; M. Bekoff, 1978*a*; Petrinovich, 1981; Rushen, 1982). This is a tall order! Because the young of many species are difficult to observe early in life due to nesting or denning habits or due to the nature of the terrain in which they live, field studies of behavioral development are very difficult to perform. When coupled with the difficulty of reliably measuring various ecological variables that might affect development, practical difficulties often become overwhelming.

Kinship patterns also need to be detailed (M. Bekoff *et al.*, 1984); in some species this may be less of a problem than in others because of the closed nature of the group in which one litter is born at a time. However, in many species it is difficult or virtually impossible to assess reliably genealogical relationships. The use of behavioral measures to assess kinship is promising, and should be pursued more widely. For example, Walters (1981) found that when the mother of a juvenile yellow baboon (*Papio cynocephalus*) was living, she could be identified by the behavioral pattern, Presenting for Grooming.

Age determination also is important, and obvious difficulties present themselves in many cases. Knowledge of age–development relationships enables scientists to study animals of unknown birthdate and to compare data from cross-sectional studies to information gathered in longitudinal analyses (J. Altmann *et al.*, 1981). Hausfater and Gilmore (1979) found that tail carriage was a useful behavioral criterion for estimating age in anubis baboons (*Papio anubis*).

Choice of locale brings up a number of questions concerning what is a natural habitat (Bernstein, 1967; Miller, 1981). Given the variability of habitats occupied by conspecifics throughout their geographic range, defining "the natural habitat" often is difficult. As Bernstein (1967) pointed out, most field workers have an unexpressed definition of natural habitat. The selection of a study site is influenced by how accessible it is, whether the terrain is favorable, the nature of the observation conditions, the political situation, and "the availability of whatever creature comforts an individual demands" (Bernstein, 1967, p. 178). He concluded that consensus of opinion among researchers is a valid criterion for determining whether or not an area may fall into the spectrum of natural habitats. If a given group has been able to live in a particular area in the absence of human disturbance, if it arrived at the site due to normal movement patterns, and if the area contains whatever essential resources are necessary for growth, reproduction, and maintenance, then it probably is a suitable habitat.

If problems associated with a field project are insurmountable, then a study of captive animals designed with a field situation in mind is necessary. Given that many excellent studies have been performed on captive subjects, Chauvin and Muckensturm-Chauvin's (1980) declaration that it is not possible to observe a caged animal with any validity is greatly overstated. However, they are correct in stressing that restricted space can have a large influence on behavior. Laboratory studies in many instances provide the only way to analyze in detail the development of behavior of identified individuals over long periods of time, and these types of data are badly needed. Tschanz and Hirshbrunner-Scharf (1975), in their detailed analyses of chick-rearing in guillemots (*Uria aalge*) and razorbills (*Alca torda*), have clearly shown how field and laboratory studies may be used to provide a clear picture of species differences in behavior that are related to ecological conditions.

We will now discuss each of the topics listed above. The common link among them is that each may be viewed through evolutionary and/or ecological eyepieces and age-related changes in behavior or developmental mechanisms can be studied.

THE DEVELOPMENT OF COMFORT BEHAVIOR

Analyses of comfort behavior have been useful in formulating models of behavioral organization, development, and evolution (van Iersel and Bol, 1958; McKinney, 1965; Horwich, 1972; M. Bekoff, 1978b; M. Bekoff *et al.*, 1979; Richmond and Sachs, 1980; Fentress, 1978, 1981; Ferron, 1981; Ferron and Lefebvre, 1982; Vadasz *et al.*, 1983). Examples of comfort movements include yawning, stretching, rubbing, body-shaking, scratching, and preening. These motor patterns (and the way in which they become coupled sequentially) lend themselves to careful developmental studies because they (1) typically appear early in life, (2) are repeated often, and (3) can be identified as individual acts that change little if at all in form during ontogeny. Furthermore, comfort movements do not seem to differ between conspecific captive and wild populations, except perhaps for rates of occurrence.

Comfort Behavior and Its Development in Adelie Penguins

A field study of the development of comfort behavior in Adelie penguins (*Pygoscelis adelie*), a semialtricial bird, was undertaken at the Cape Crozier Rookery, Ross Island, Antarctica, in order to study the time course of age-related changes in behavior and behavioral variability; comfort activities of young chicks of different known ages were compared with those of adults (M. Bekoff, 1978*b*; M. Bekoff *et al.*, 1979). The comfort movements studied included yawning, head-shaking, wing-flapping, various forms of stretching and scratching, oiling (application and distribution of oil from the uropygial gland, located at the base of the tail, to the body), and preening directed to the breast, belly, back, side/flank, wing, shoulder, leg, cloaca, and tail-base (Fig. 1) [see Ainley (1974) and above references for descriptions]. Data were collected on the age at which actions were first observed, duration of sequences, movement rates (number of actions/minute), the relative percentage of occurrence of various actions, the temporal distribution of movements within sequences, the continuity between behaviors according to the side of the body and area to which they were directed, and two-act transitions within sequences. Sixteen variables were used in multivariate analyses to study age-related changes in comfort behavior. Because the chicks and adults were easy to observe, detailed information could be gathered under field conditions. A brief summary of the results is as follows:

1. The first two actions performed during ontogeny were yawning (day 1) and side-to-side head-shaking (day 2). Most actions appeared after the chicks were 7–9 days old (in contrast to more precocial species, in which many comfort behaviors are performed within a few days of, or even during, hatching. The earliest comfort movements were directed to body areas (breast, belly) comprising the major portion of the chick's body surface; these body areas make the most contact with the ground and thus tend to get soiled when the chick lies down. When the chick begins to become more mobile, dirt collects on other parts of the body. These changes in posture and activity appear to play a role in the change in the relative distribution of comfort behaviors during early development. By day 21, all actions except shoulder-rubbing and oiling were observed.

The last comfort activities to emerge were shoulder-rubbing (day 33) and oiling (day 35), just after the uropygial gland became functional (days 30–33) and just prior to fledging (going out to sea for the first time; approximately 6–8 weeks of age). That both oiling and shoulder-rubbing appeared at approximately the same time is not surprising, because shoulder-rubbing is the specific behavior used to distribute oil from the head to the shoulder, after the oil has been transferred to the head by wing-rubbing (Fig. 1d) (Ainley, 1974). A similar relationship was noted by Kruijt (1964) for Burmese junglefowl (*Gallus gallus spadiceus*), in which head-rubbing, an action used to release oil from the oil gland, appeared simultaneously (day 11) with the development of function of the gland. Wing-rubbing, on the other hand, a behavior that is used in other contexts as well, first appeared in the penguins very much earlier, on day 7.

The appearance of several other comfort movements in penguin chicks coincided with growth of contour feathers. At approximately 20 days of age, contour feathers began to emerge. This corresponded to a marked increase in the number of acts per sequence during nonoiling.

FIG. 1. (a) Four- to 5-week-old Adelie penguin chick back preening. Note the loose down. (b) Five-week-old chick preening its shoulder. (c) Six-week-old chick preening its shoulder. Note the patches of contour feathers on the breast, belly, and wing. (d) Seven-week-old chick performing the action "bill-to-wing-edge," during which the bird grasps the top edge of the wing between its mandibles and draws the bill along the wing's edge toward its body. This is the first step in which oil is transferred from the uropygial gland located at the base of the tail to the head. [From M. Bekoff *et al.* (1979).]

FIG. 1. (*Continued*)

A change in the distribution of preening effort also occurred during the fourth week, apparently related to feather growth. For example, down was lost from the legs at about 25 days of age (R. H. Taylor, 1962); correspondingly, the proportion of leg preening increased from 3.6% in chicks 14–20 days of age to 8.8% in chicks 21–28 days old.

2. Although highly variable, nonoiling sequences did not become more stereotyped with increasing age (oiling sequences performed by adults were more variable than those performed by chicks). These data are in agreement with other studies (Schleidt and Shalter, 1973; Wiley, 1973; M. Bekoff, 1977*d*), in which it was shown that behavior does not necessarily become more stereotyped as a result of maturation, practice, or other forms of learning. Among adults, behaviors associated with the collection and distribution of oil were tightly linked in sequence.

3. No relationship was found between the order in which actions appeared in ontogeny and the order in which they were coupled together in sequences later in life. This finding may be contrasted with observations on various rodents (Horwich, 1972; Richmond and Sachs, 1980; Fentress, 1981; Ferron and Lefebvre, 1982; Thiessen *et al.,* 1983) and bobwhite (*Colinus viginianus*) and Japanese quail (*Coturnix coturnix japonica*) (Borchelt, 1977). For example, Richmond and Sachs (1980) found that the front-to-rear progression (cephalocaudal) of grooming by adult rats paralleled the order in which particular areas were groomed during development.

Although species differences need to be reconciled and the methods of analysis compared to assure that different investigators were measuring precisely the same relationship, the striking parallel between adult organization and development found in apparently all studies of grooming in diverse species of rodents suggests that these motor programs are phylogenetically old (Baerends, 1976; Ferron and Lefevbre, 1982). In mountain beavers (*Aplodontia rufa*), the most morphologically primitive rodent, grooming bouts follow a nearly perfect anterior–posterior sequence of actions (J. A. Byers, unpublished data). Similarity between adult sequences and ontogeny also was reported by Etienne *et al.* (1982) in their study of the development of hoarding in golden hamsters (*Mesocricetus auratus*).

4. Age-related changes in comfort behavior were detected using principal components (PCA) and discriminant function analyses (DFA). Figure 2 shows an age-related progression (from left to right) along the *x* axis [principal component (PC) I] of a plot of factor scores from a PCA. PC I was related to age [gross behavioral development; see also Miller *et al.,* (1977)]. DFA also indicated age-related changes in development (Fig. 3). A comparison of 7- to 13-day-old chicks with 21- to 28-day-old chicks

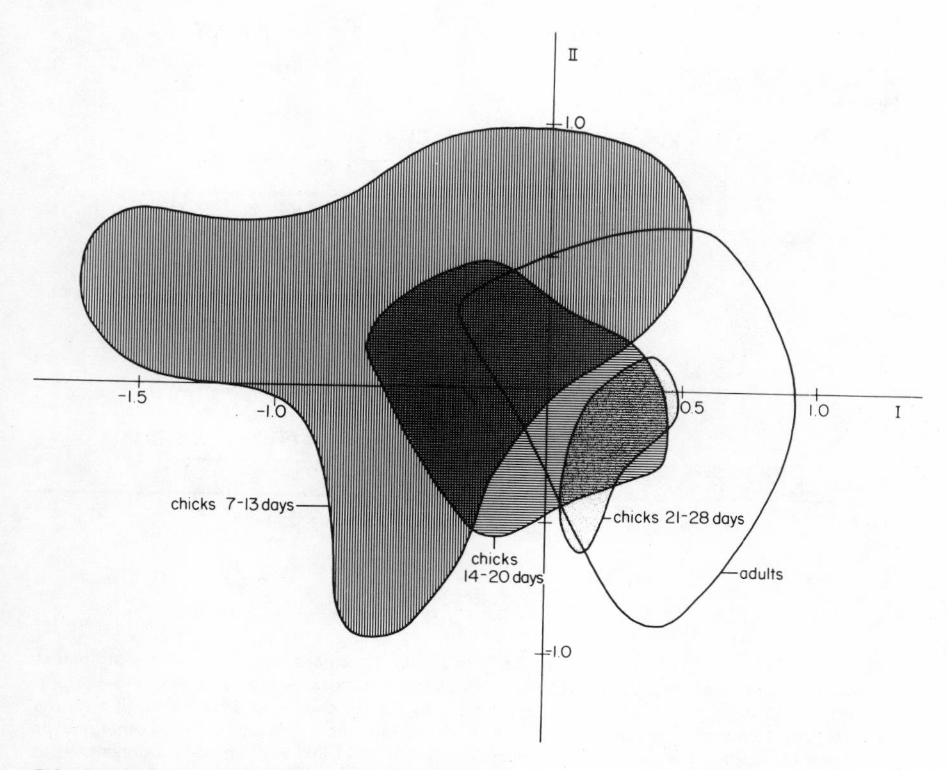

FIG. 2. A two-dimensional principal components plot of factor scores for comfort movement sequences for chicks of three different age groups and adults. Note the age progression from left to right. Factor I was related to age [gross behavioral development; see also Miller *et al.* (1977)]. [From M. Bekoff (1978*a*).]

indicated no overlap between them, and when 14- to 20-day-old chicks and adults were placed on the discriminant axis, an age progression was obvious (Fig. 3, top). Similar results were found when comparing 7- to 13-day-old chicks with adults (Fig. 3, bottom). Rate of performance of comfort behaviors and the duration of sequences were the two main variables discriminating between age groups (youngest chicks performed fewer acts/minute than did older chicks or adults; oldest chicks performed fewer acts/minute than did adults; sequence duration of youngest chicks was shorter than that of oldest chicks and adults).

The importance of rate (and duration) in discriminating the different age groups might be related to neural maturation and muscular (physical) development. Ferron and Lefebvre (1982) also suggested that there might

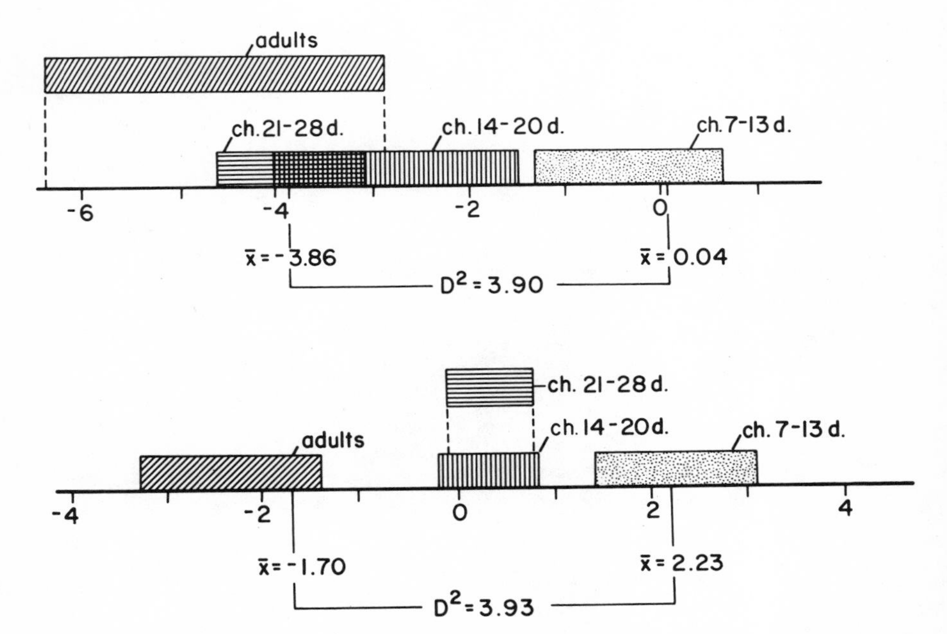

FIG. 3. Top: Linear discriminant values for 7- to 13-day-old chicks and 21- to 28-day-old chicks cast on a discriminant axis. There was no overlap between these two groups. Note the age progression (from right to left) when middle-aged chicks (14–20 days) and adults were fit onto the axis. Bottom: Linear discriminant values for 7- to 13-day-old chicks and adults. When other groups of chicks (14- to 20-day-old and 21- to 28-day-old) were fit onto this axis, an age progression also was noted. Rate of performance of comfort movements and duration of comfort sequences were the two most important variables discriminating between the youngest and oldest chicks (top) and the youngest chicks and adults (bottom). The youngest group of chicks performed fewer acts per minute and sequences were of shorter duration. [From M. Bekoff (1978*a*).]

be strong selection on the elements of the nervous system controlling rate of movement during comfort sequences in rodents (recall the discussion of central pattern generators). They too, found age-related changes in rates with which comfort movements were performed in North American squirrels (see Ecology and Evolution of Grooming). Level of maturation could be a unifying variable and might operate as an internal constraint on the behavior of young animals. It is notable that one of the most important variables (rate) responsible for age-related changes in the comfort behavior of penguins is the one that may be closely related to a relatively invariant neural process, similar to one that predisposes some species to show cephalocaudal progression of comfort movements during ontogeny and later life.

Ecology and Evolution of Grooming

The work of Ferron (1981) and Ferron and Lefebvre (1982) reflects nicely the value of detailed, comparative, developmental analyses of select behaviors. In a series of studies, five species of North American squirrels were studied; analyses of the development of locomotion, feeding, comfort movements, and elimination patterns were undertaken. Three terrestrial species [Richardson's ground squirrels (*Spermophilus richardsonii*), Columbian ground squirrels (*S. columbianus*), and golden-mantled ground squirrels (*S. lateralis*)] and two arboreal species [red squirrels (*Tamiasciurus hudsonicus*) and northern flying squirrels (*Glaucomys sabrinus*)] were compared. Ferron (1981) found a general tendency for faster behavioral development in ground squirrels as compared to tree (red squirrels) and flying squirrels. He suggested that the rate of development was closely adjusted to nest emergence, which differed among species according to the level of difficulty of moving in the environment. Later emergence was associated with more complex locomotor requirements, and all behavioral patterns associated with locomotion also were delayed. Flying squirrels showed slower rates of development than red tree squirrels. Both of these arboreal species showed delayed development when compared to the three terrestrial species, and among the terrestrial species, golden-mantled squirrels, which typically inhabit rock slides, showed slower development than Columbian ground squirrels, which usually are found in alpine and subalpine tundra.

Ferron's studies also point out that there is a close interplay between the development of different types of behavior, all of which may depend on common underlying postural, manipulative, or locomotor abilities (Chalmers, 1980*a*; Jacoby, 1980, 1983; Thelen, 1981; also see Kruijt, 1964). Similarly, motor patterns used in communication may develop from comfort and other types of general movements (Bekoff, 1977b; Barlow, 1977), such as yawning displays in primates (Hadadian, 1980) and behaviors used in conflict situations (displacement activities) (van Iersel and Bol, 1958; Baerends, 1975). Selection for open motor systems would permit organisms to use a relatively limited number of basic motor patterns in diverse behavioral context, by combining and/or modifying them according to current circumstances. Recently, Vadasz *et al.* (1983) presented evidence that three movement categories in mouse (*Mus musculus*) grooming (prenatal forelimb tremor contact, overhead stroke, and repetitive facial grooming) are inherited independently. Detailed analyses of a seemingly uninteresting phenotype such as comfort behavior may provide a key to furthering our understanding of the relationships among behavioral ontogeny, ecology, and the evolution of nervous systems.

HELPING BEHAVIOR

Comparative research on helping [alloparental (E. O. Wilson, 1975)] behavior has grown rapidly in the past few years because of its obvious relationship to theories concerned with the evolution, ecology, and development of social behavior. Helping occurs when individuals other than known parents provide care to young individuals, and a basic question is why should an individual help to rear young other than its own (see below)? General reviews of helping behavior can be found in E. O. Wilson (1975), Brown (1978, 1980), Emlen (1978, 1982*a,b*), Alexander and Tinkle (1981), Riedman (1982), and Stacey (1982). Recent data on carnivores are provided by Rood (1978, 1983), Moehlman (1981, 1983), M. Bekoff and Wells (1982), Harrington *et al.* (1983), Malcolm and Marten (1982), Macdonald and Moehlman (1982), Mills (1982), M. Bekoff *et al.* (1984), and D. D. Owens and Owens (1984). In carnivores, major helping activities include guarding the den site (babysitting) and feeding the young; prey, or parts of prey, may be brought to the den, or partially digested food may be regurgitated.

In this section we will briefly consider some aspects of helping behavior, particularly in coyotes (*Canis latrans*). Here, helping specifically refers to care-giving behavior provided to young animals by nonparents. The data are taken fom a long-term study of coyotes living in the Grand Teton National Park, outside of Jackson, Wyoming (M. Bekoff and Wells, 1980, 1981; Wells and Bekoff, 1981); additional details are provided in M. Bekoff and Wells (1982, 1985) and M. Bekoff *et al.,* (1984). Unlike other populations of wild coyotes and other carnivores, the coyotes living in our study area are readily observed, and den sites also can be watched.

An Overview of Coyote Social Ecology

Annual food and reproductive cycles are outlined in Fig. 4 [for details see M. Bekoff and Wells (1980, 1981, 1982, 1986) and Wells and Bekoff (1982)]. Coyotes on our study area depend mainly on hunter-killed elk (*Cervus canadensis*) carrion during winter months, and on small rodents

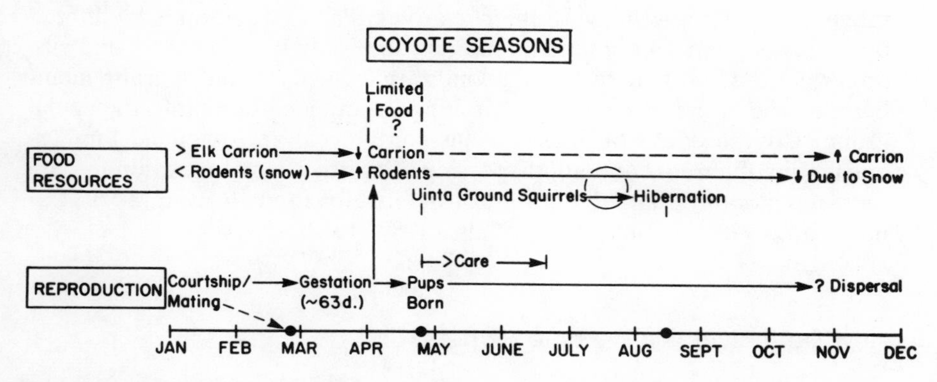

FIG. 4. A simplified scheme showing annual changes in food resources and the temporal distribution of reproductive activities for coyotes living in the Grand Teton National Park, Jackson, Wyoming [see text and M. Bekoff and Wells (1980, 1981, 1982, 1986)].

[mainly voles (*Microtus* spp.) and uinta ground squirrels (*Spermophilus armatus*)] during the rest of the year. Uinta ground squirrels are the major food resource from about mid-April to mid-August.

Coyotes begin to court in early January, and copulation usually takes place in late February. After a gestation period of about 63 days, pups are born in late April in a subterranean den. Coyotes pups are altricial, and depend heavily on maternal care during early life. After pups emerge from the den at about 3 weeks of age, they still depend on parental care; they become independent at about 4 months of age.

Between about the beginning of April and the time at which uinta ground squirrels emerge from hibernacula, food resources may be limited, because carrion, a nonrenewable resource, is depleted, and snow cover may make it difficult for coyotes to catch small rodents (Wells and Bekoff, 1982). This time period coincides with the second half of pregnancy (and possibly with maternal lactation), and could possibly be a time when pregnant females and mothers become energetically stressed. Pack-living females may be better off energetically than females living only with their mates (M. Bekoff and Wells, 1981). However, the advantage for pack-living females may only be realized in periods of extreme food shortage; there were no significant differences between the reproductive output of group-living females and females living only with their mates.

Dispersal of juveniles usually begins in early fall and continues through early winter. Individuals who spend the first 10–11 months of life with their parents and older siblings typically do not disperse; they either become helpers or remain on the periphery of their natal home range, rarely interacting with their relatives. Pack formation is facilitated by nondispersing young remaining in close proximity to parents and siblings over the first winter. Strong bonds are formed among all group members, aiding in the incorporation of nonreproducing young into the social group. Coyote packs (at least on our study area) are extended families consisting of parents and nondispersing offspring. Pack formation in coyotes appears to be an adaptation for defense of food (Bowen, 1978; Camenzind, 1978; M. Bekoff and Wells, 1980, 1982, 1986).

How Do Individuals Become Helpers?

In coyotes and many other animal species, helpers are individuals who do not disperse from their natal area before or at the age at which they become reproductively active. Typically, helpers do not breed, even though they may potentially be capable of doing so (individuals of the same age are known breeders). Some possible relationships among development, dispersal, ecology, and helping are outlined simplistically in Fig. 5. Major developmental questions include: (1) How do early social interaction patterns among siblings and between parents and offspring mesh with ecological conditions to affect the probability that an individual will either remain in its natal area or disperse (M. Bekoff, 1977*c*; Gaines and McClenaghan, 1980; Downhower and Armitage, 1981; Harcourt and Stewart, 1981)? (2) Why do some nondispersing individuals become helpers, whereas others remain within the confines of their parents' home range, but do not help? (3) What do helpers get out of helping, or, why help? Only the third question will be considered here.

Genetic Relationships among Pack Members

A study of helping demands that individuals of known genetic relatedness be observed, because of obvious implications for theories about the evolution of social behavior, especially kin selection. The pedigree for one coyote pack that was studied extensively between 1977 and 1982 is shown in Fig. 6. Though other researchers tentatively suggested that coyote packs are extended families (Bowen, 1978; Camenzind, 1978), documentation awaited the study of individuals who were identified early in life (by using ear tags, and subsequently radiocollars) (M. Bekoff and Wells, 1980, 1982, 1986). Litter size was determined when pups were 3–4 weeks of age, because denning habits make it impossible to do this

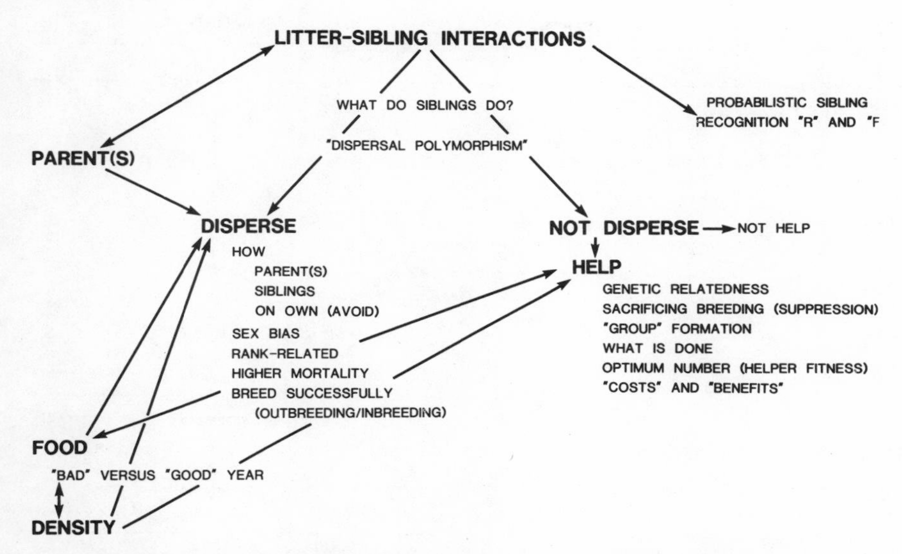

FIG. 5. A summary diagram of some possible relationships between behavioral development, dispersal, ecology, and helping. Sibling interactions can have effects on (1) future recognition of siblings, (2) patterns of dispersal, and (3) behavior of parents. Parents may also influence sibling interactions as well as dispersal of their offspring. Food availability and population density, among other ecological variables, can affect dispersal patterns and perhaps helper social biology. Under dispersal, one must consider how it comes about (i.e., the roles played by parents, siblings, and the dispersing individual itself), if there are any biases as to which sex disperses, whether dispersal is related to social (dominance) rank, if early dispersers suffer higher mortality than individuals who disperse at a later age, and if dispersers breed successfully after they leave their natal site and with whom (a relative or unrelated individual; that is, does dispersal favor outbreeding?) With respect to helping, one must consider the genetic relationship between the helper(s) and the young individuals to whom care is provided, whether helpers are actually sacrificing reproducing on their own, how helpers are incorporated into the group, what helpers do (e.g., nurse, provide other types of food, guard, carry, play), if there is an optimum number of helpers after which no additional young are successfully reared and beyond which resources may become overexploited by the presence of too many individuals, and the "costs" and "benefits" to the helper's reproductive fitness. Of course, one must also consider whether helpers actually affect survival of young to whom they provide care (see Fig. 9 and text). The alternatives presented here are rather complex, but not exhaustive. For mammals and most other species very little information is available to tie together the areas represented in this diagram, though such data would be extremely useful and important for providing insights into the ways that proximate and ultimate factors may influence behavioral patterns and "decisions." [From M. Bekoff (1981*b*).]

before pups emerge. Of 22 pups found around dens as early in life as possible, 21 were marked; five of six (83%) helpers were males.

Details concerning the pack pedigree are discussed in the legend to Fig. 6. It should be pointed out that although the present discussion centers on helpers' contributions to pup rearing, helpers also actively partake in defense of food against intruding coyotes (M. Bekoff and Wells, 1982).

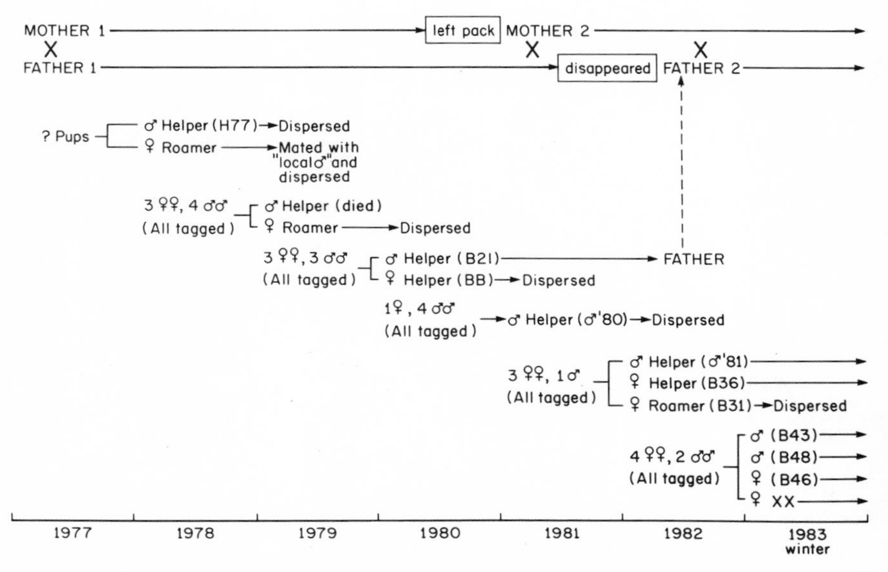

FIG. 6. Social groups of carnivores often are composed mostly of genetically related individuals (extended families). This figure presents a pedigree for a pack of coyotes observed in the Grand Teton National Park, outside of Jackson, Wyoming, from 1977 to 1983 (M. Bekoff and Wells, 1981, 1982, 1985; Wells and Bekoff, 1981, 1982). Young of each year that are accounted for either dispersed or died before they were about 9 months of age. After the original pack mother (Mother 1) left the group in late 1980, a new and unrelated female (Mother 2) joined the pack and mated with the original pack father (Father 1) in 1981. Then, after he left the pack in spring 1981, his son, male helper B21, mated with the new female in 1982 and 1983. In 1982, the help that B21 provided to male '18 and female B36 was reciprocated when B21 and the new female's pups were born. The new pack mother was the only unrelated coyote to join the pack in 6 years. [See M. Bekoff and Wells (1982, 1986) for details.]

What Do Helpers Do?

Dens were observed for 340 h from 1978 to 1981. All analyses discussed below involve four litters born in the pack (mean litter size 5.5) and three litters born to a mated resident pair (mean litter size 5.3). Although the sample size in our study (and other studies on medium to large carnivores) needs to be increased, available data permit consideration of a number of hypotheses concerning helping. Field data on helping in carnivores are difficult to collect, a fact that is responsible for the lack of data for many species in which other aspects of social behavior have been detailed.

Two major contributions that helpers potentially could make include feeding pups or the mother and guarding the den site. Only male B21 was

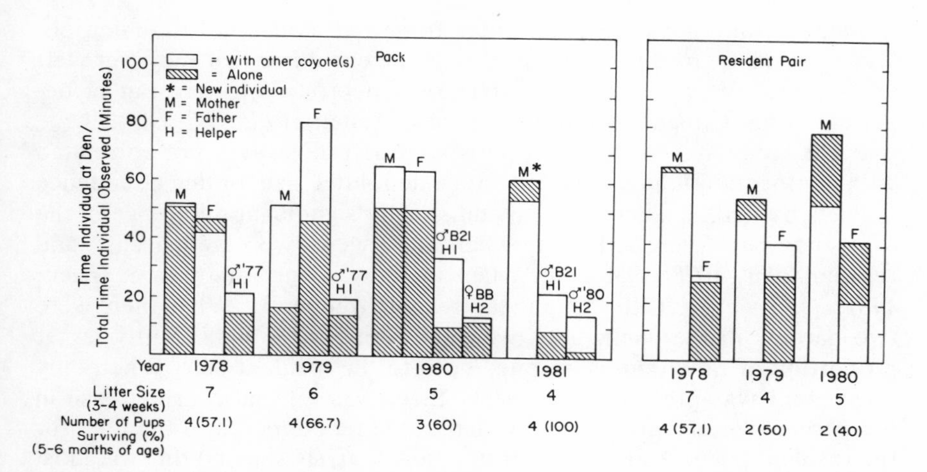

FIG. 7. The percentage of time (total time individual was observed at active den sites/total time the individual was observed anywhere) that individual coyotes budgeted to den attendance. Alone = time individual was at den alone/total time individual was at the den. For information on pack structure see Fig. 6. The same resident mated pair was observed from 1978 to 1980. There were no significant differences in average litter size between the pack and the resident pair, nor in the number of pups surviving to 5–6 months of age.

observed to feed younger siblings or half-siblings more than one time. Our data indicate that helpers do not play a significant role in food provisioning, as they do in Cape hunting dogs (*Lycaon pictus*) (Malcolm and Marten, 1982; also see Brown, 1978).

Helpers did partake in den sitting (no active defense of pups by parents or helpers was observed), but to a far lesser extent than did parents (Fig. 7). The pack father showed considerable year-to-year variability. The pack mother from 1978 to 1980 budgeted an average of 57% of her time (time observed at den/total time observed anywhere) to den attendance, while the new pack mother in 1981 budgeted 53% (pack male = pack females). Pack and resident-pair mothers budgeted about the same percentage of time to den attendance; the pack male budgeted more time to den sitting than did the resident-pair male. In the resident pair, the mother budgeted more time to den sitting than did her mate. Similar trends were noted for the percentage of time that individuals spent alone at den sites and when percentage of time spent den sitting was calculated as time the individual was observed at the den/total time the den was observed (this ratio represents the relative contributions of individuals to den attendance, given that the den was attended by anyone).

We also considered the frequency pups were left alone, the percentage of time pups were left alone (time left alone/total time den observed), and the mean time pups were alone each time that they were left unattended. We observed no pattern or synchrony for arrival at or departure from den sites by adults (see also Andelt *et al.,* 1979).

The frequency with which pups were left alone was not correlated with number of adults associated with a den, litter size, or den attendance budgets by adults. There was no statistically significant difference in the frequency pups were left alone when there were two parents (pack and resident pair, 1978–1980), but when there was only one parent (pack, 1981), pups were left alone significantly more frequently. When there were two parents in the pack, pups were left alone for a significantly lesser proportion of time than were pups born to the resident pair. When only one parent was present in the pack, there was a significant increase in the percentage of time pups were left alone (about equal to the data for the resident pair). A discriminant function analysis showed that the most important variable discriminating between the pack and the resident pair was percentage of time pups were left alone.

There was great variation in the mean number of minutes pups were left alone each time they were unattended, but there was no significant difference between the pack and resident pair. In 1980, resident-pair pups were alone for an average of almost 2 h each time they were left by their parents.

The more time pups are left alone, the more susceptible they are to harassment by coyotes and other predators. While intraspecific killing of pups by nonintruding coyotes is rare, it was observed twice by Camenzind (1978); on both occasions, the dens were unattended. Also, a case of possible predation by a gray fox (*Urocyon cinereoargenteus*) on a red fox pup (*Vulpes fulva*) was observed in upstate New York (Saggese and Tullar, 1974). The mere presence of parents and helpers at den sites may be enough to ward off any danger; the importance of pup protection provided by adults has been recognized by other researchers (Rood, 1978; D. D. Owens and Owens, 1979; Moehlman, 1979; Harrington *et al.,* 1983).

Helper Effects: Kin Selection and Reciprocity

In addition to asking what helpers do, we also need to consider what effects helpers may have on pups to whom they provide care. With respect to possible genetic advantages that helpers may gain by providing aid to close relatives with whom they share a certain proportion of common genes (*r,* Wright's coefficient of genetic relationship, averages $\frac{1}{2}$ for diploid full siblings, equals $\frac{1}{2}$ for parents and offspring, equals $\frac{1}{8}$ for first cousins,

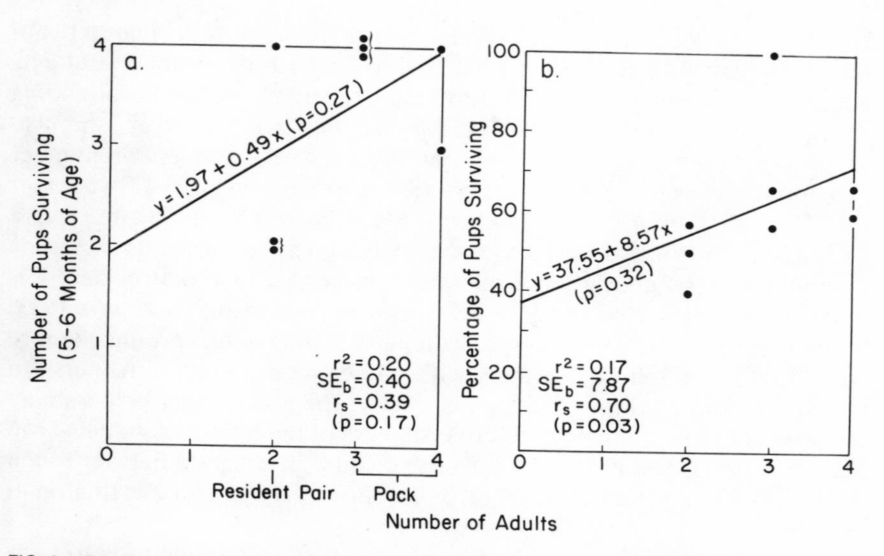

FIG. 8. Regression analyses for the number of adult coyotes present at dens (number of helpers = number of adults − 2) versus (a) number and (b) percentage of pups surviving until at least 5–6 months of age.

etc.), helper effects on pup survival need to be studied. Detailed discussions of the evolution of behavior via kin selection, based on the fact that genetic relatives share a certain proportion of identical genetic material, can be found in Hamilton (1964), E. O. Wilson (1975), and Boorman and Levitt (1980).

In the present analysis, the relationship between number of pups surviving until 5–6 months of age and number of adults attending a den was not statistically significant (Fig. 8a), but there was a positive correlation ($r_s = +0.39$) between these two variables. Perhaps the addition of more data will yield a significant result. The low value of r^2, the coefficient of determination, suggests that variables other than number of adults affect pup survival. Litter size was not significantly correlated with number of adults, number of pups surviving, or percentage of pups surviving. It should be noted that two parents reared an average of 2.7 pups, but each helper only added 0.49 pups. Therefore, it seems as if it would be better for individual coyotes to breed and rear their own pups if possible. The major payoff to a helper, from a genetic point of view, would be an increase in number of surviving kin as a result of helping (and foregoing mating).

Helpers' presence did increase the percentage of pups surviving (Fig.

8b). Moehlman (1981) [for additional information see Moehlman (1983)] reported a significant positive relationship for both the number and percentage of pups surviving to 3.5 months and the number of adults attending a den for black-backed jackals (*Canis mesomelas*). However, she also found significant correlations between litter size at 3 weeks and number of adults, and litter size and number of pups surviving to 14 weeks of age. Excluding total breeding failures, Malcolm and Marten (1982) found a significant positive relationship between the number of adults attending a den and the number of African wild dogs reared to 1 year of age. Ecological variables also need to be considered. Harrington *et al.* (1983) found that when prey was scarce, wolf pairs (*Canis lupus*) produced more surviving young than did groups with one or more potential helpers. In red foxes (*Vulpes vulpes*), if food is scarce, the presence of helpers may temporarily reduce the reproductive success of breeding individuals (von Schantz, 1981). Detailed comparisons of canid helping studies are found in M. Bekoff and Wells (1982), M Bekoff *et al.* (1984), and Harrington *et al.* (1983).

As mentioned above, helpers were observed to provide food to pups only rarely. A nonsignificant relationship was found between pup weight at 5–6 months of age and number of adults attending the den(s). Harrington *et al.* (1983) also noted that the presence of helpers in wolves had no effect on juvenile weight.

Why Help?

There are many reasons why an individual might forego mating and help to rear young. If the pups are closely related (full-siblings, half-siblings) and helper presence increases pup survival, helping could evolve (Moehlman, 1981, 1983; M. Bekoff and Wells, 1982) [for comparative data on birds see Brown (1978, 1980), Emlen (1980, 1982*a,b*), and Ligon (1981)]. However, a correlation between helper presence and survival of young may be confounded by habitat quality (Brown and Balda, 1977). Young individuals may be more likely to delay dispersal if they live in an area where vital resources, especially food, are abundant. If helpers do not contribute to feeding young, but survival rate is higher in high-quality habitats, pup survival may be linked to the high quality of the natal area and not to the presence of helpers.

Although kin selection is useful for explaining some examples of helping, there are exceptions that may be explained by reciprocal altruism (Trivers, 1971) or simple reciprocity (Rood, 1978; M. Bekoff and Wells, 1982). For example, Rood (1978) found that unrelated dwarf mongooses

(*Helogale parvula*) helped to raise pack young; he argued that helpers might benefit from being part of the group, which offered better protection against predators. Also, helpers that subsequently breed might benefit from receiving help from young that they had previously helped or from other group members. Furthermore, the helper(s) may benefit from remaining in an area where there is adequate food and inheriting breeding space when one of its parents dies or disperses (Stallcup and Woolfenden, 1978; Rowley, 1981). In the present analysis of helping in coyotes, male B21 not only increased his inclusive fitness by helping to rear younger siblings (1980) and half-siblings (1981), but he also inherited a breeding area, paired and mated with a female with whom he was unrelated (1982 and 1983), and received help from individuals to whom he previously provided care (male '81, female B36; see Fig. 6).

Analyses of helping behavior certainly will provide useful information for tying together evolution, ecology, and behavioral development. Long-term studies of identified individuals of known genetic relatedness need to be conducted under conditions that allow unobstructed observation. Data from all animal groups are welcome. Especially needed are studies of helping behavior in conspecifics living in different habitats in which food (and other) resources vary (e.g., M. Bekoff and Wells, 1982; Harrington *et al.*, 1983).

DEVELOPMENT OF AGONISTIC BEHAVIOR

Analyses of agonistic behavior have provided insights into many aspects of behavioral development. As used here, *agonistic behavior* refers to motor patterns involved in aggression (fighting and various types of threatening) and submission (Scott and Fredericson 1951). The term "agonistic" accounts for behavioral patterns of all interactants, not only those individuals behaving in an offensive or competitive manner. *Descriptive* definitions of agonism often are difficult to provide in developmental studies due to the dynamic nature of development itself. We may indeed be dealing with a "different" organism at different stages of development (Lehrman, 1953). For example, agonistic behavior may change in form over time due to a combination of neuromuscular maturation, hormonal effects, and/or learning (Collias, 1943; Guhl, 1958; Tooker and Miller, 1980). Also, different behavioral patterns may be used to communicate agonism at different ages (Kruijt, 1964; K. S. Cole and Noakes, 1980). Sequences of behavior may also be developmentally labile (M. Bekoff, 1978*b*; M. Bekoff *et al.*, 1981*a*; Jacoby 1980, 1983).

Functional definitions of agonism are difficult to develop regardless of whether an ontogenetic analysis of the phenomenon is undertaken. When the developmental factor is added, functional definitions usually are rather limited in scope, mainly because functions of behavior may change during development (Kruijt, 1964), and there may be marked species and individual differences in the immediate and future consequences (functions) of engaging in agonistic encounters (M. Bekoff, 1977*c*). Nevertheless, the immediate effects of prior engagement in agonistic encounters do not appear to be very different for young or adult animals. These include priority of access to some resource, the establishment of social hierarchies, and associated changes in behavior such as increased freedom of movement within the social group by more dominant individuals (Scott, 1962; M. Bekoff, 1977*c*; M. Bekoff *et al.,* 1981*a*).

Sibling competition may also affect litter sex ratios (Clark, 1978), in addition to the possible role that parents may play (Trivers and Willard, 1973; Maynard Smith, 1980; McClure, 1981). Further, it has been suggested that an infant also may compete with as yet unborn siblings by postponing the date when its mother next conceives (Simpson *et al.,* 1981)! However, the ways in which immediate consequences are translated into future effects via the cumulative manner in which development appears to work is another problem, one for which there are basically no general tales, but some intriguing just-so stories.

Most studies of the development of agonistic behavior have been performed in the absence of an ecological/evolutionary perspective. Causal relationships between variations in dominance status early in life and differences in reproductive success have not yet been demonstrated, but Kikkawa (1980) found that silvereyes (*Zosterops lateralis,* Aves) born early in the breeding season dominated later-born young, and dominant birds also had better chances of survival. In pronghorn (*Antilocapra americana*), fawns contest dominance status as early as 7 days of age (Byers, 1984*a,b*) and individuals born early in the spring dominate smaller, later-born, individuals (Byers, 1984*a,b*). For females, the percentage of dominance encounters won during the first 3 months of life correlates very strongly with the proportion of dominance interactions won as a yearling and as a 2-year-old (Byers, 1984*a,b*). Pronghorn, especially females, seem to be set on divergent developmental pathways as a consequence of birth-order effects.

Ontogenetic pathways [or trajectories (Wiley, 1981)] followed by "winners" *and* "losers" of early dominance interactions need to be traced longitudinally, as Byers is doing for pronghorn living on the National Bison Range, Montana. Although in some respects it may not be disadvantageous to be a subordinate animal in a social group (Alexander,

1974; Rohwer and Ewald, 1981), subordinate individuals may show decreased resistance to disease, delayed maturation, decreases in the size of seminal vesicles, and drastic weight loss when compared to more dominant individuals (Barnett, 1958; Geist, 1978; Rasa, 1979).

Subordinate individuals may also show bioenergetic disadvantages when compared to more dominant group members. When population density is high, subordinate adult deermice (*Peromyscus maniculatus*) suffer bioenergetically, while more dominant animals do not (Farr and Andrews, 1978*a,b*). Golightly (1981) found that dominant and subordinate captive adult coyotes in one group had higher daily metabolic rates than did middle-ranking individuals. He suggested that the energetic differences might have been due to the regular challenges to which dominant coyotes were subjected and the harassment suffered by "scapegoat" subordinate individuals. He also noted that dispersing individuals of different rank might benefit energetically by leaving their group.

Although corroborative developmental energetic data do not exist, dominance hierarchies among very young coyote littermates are well-pronounced, and scapegoat individuals have been observed in coyotes and other species (M. Bekoff *et al.*, 1981*a*). Furthermore, it has been suggested that dispersal may be influenced by early social interaction patterns [as well as by ecological factors (M. Bekoff, 1977*c*)]. The independent predictions that dispersal may be related to dominance rank and that high- and low-ranking animals may leave their natal group for *different* behavioral reasons but for *similar* energetic reasons (M. Bekoff, 1977*c*; Golightly, 1981) requires field verification.

Ontogeny of Agonistic Skills

Origins of Agonistic Motor Patterns

In some mammals it appears that rudiments of "real" aggression first appear during "play-fighting" (T. B. Poole, 1966; Ewer, 1968; N. W. Owens, 1975). While the differences between play-fighting and true aggressive behavior have only been studied in a few mammals, detailed analyses indicate that these two behavioral patterns can be distinguished (M. Bekoff and Byers, 1981). T. B. Poole (1966) found that of 13 motor patterns commonly observed during aggression in adult polecats (*Putorious putorious putorious* L.), nine appeared during aggressive play in young polecats by 8 weeks of age. The four patterns of aggression that were absent from play-fighting were those used to intimidate an opponent during real fights. In Australian fur seals (*Arctocephalus forsteri*), full neck displays that are used by adults when threatening another seal first appear during play-fighting between 3 and 4 weeks of age (Stirling 1970).

In some species, however, fighting typically occurs before or concurrently with the onset of play-fighting (M. Bekoff *et al.*, 1981*a*; Byers, 1983; Henry, 1985), while in others, isolation-reared animals fight although they have had no prior social experience (Cairns, 1973). Therefore, play-fighting experience cannot be considered a necessary step in the development of fighting skills (Cairns, 1973). Also, in some species, actions used during aggression emerge from motor patterns used during simple locomotion (Kruijt, 1964; Tooker and Miller, 1980) (see discussion of comfort behavior).

There are few data concerning the ontogeny of submissive signals. In canids, two forms of submission occur which can be traced to behaviors that were either directed toward infants by mothers (anogenital stimulation to elicit urination or defecation) or performed by infants during food-begging (Schenkel, 1967). Active submission, which appears to be derived from infantile food-begging, involves muzzle-nudging, often accompanied by whining; passive submission appears to be derived from the posture that pups assume when they are given anogenital stimulation. The animal lies on its back, often with its hindlegs spread, and remains immobile. Similarly, in primates, some submissive behaviors, such as "pucker-lip" expressions and lip-smacking, appear to be derived from infantile food-getting responses (Chevalier-Skolnikoff, 1974).

Communication of Aggression and Submission

Developmental analyses in this area must provide basic information concerning: (1) when during ontogeny signals of aggression and submission first appear; (2) the stimuli that first evoke them; (3) the response that is first elicited in the recipient of the defined signal(s); and (4) ontogenetic changes that occur as individuals become socialized. The severity of early fights in coyotes and the fact that after only a few fights behavioral controls are evident (M. Bekoff, 1977*a*) suggest that young animals are learning something about the nature of agonistic interactions and are developing skills necessary for the exchange of information very rapidly. What actually reinforces agonistic behavior during development is still not clear; some of the ideas developed by G. T. Taylor (1979) in his discussion of reinforcement and adult aggressive behavior may be relevant. As stated by Cairns (1973, p. 66):

> Aggression as a construct has been useful to pointing to problems but not very helpful in yielding solutions. A basic difficulty has been that the experimental procedures that it supports have not facilitated attention to the details of the

> relevant interactions or to the ways that the behaviors of *each individual* [my emphasis] in a sequence serve to elicit, maintain, or inhibit noxious and painful actions of the other individual . . . an essential concern of a developmental analysis is the identification of the particular ways in which maturational changes are bidirectionally interlaced with experiential events.

Agonistic Behavior, Development, and Assessment of Self and Others

Another set of developmental questions concerns the ways in which developing animals learn to assess their own fighting prowess and the probability of winning an encounter with another familiar or strange individual. By making the correct assessment of a situation, energy that might otherwise be harnessed into growth is not wasted and unnecessary risks are avoided (Popp and DeVore, 1979). It is known that in red jungle fowl (*Gallus gallus*) certain postures are associated with the subsequent winning or losing of an encounter (R. H. Wilson, 1974), and it remains possible that a similar relationship exists in a wide variety of species. Therefore, one way in which an individual might assess the outcome of an impending interaction would be to "read" its opponent's initial actions before an encounter and change its own behavior accordingly. It would not be difficult to imagine that this assessment skill could develop after only a few trials.

Physical characteristics such as size and weight, as well as the location of an encounter, might also be important variables that could influence the outcome of an agonistic interaction. For example, based on past experience, an individual might learn that if another animal is "this much" larger than itself (it has to raise its neck up a certain amount), it has a very small chance of winning a fight with an animal of such a size. Therefore, it should not fight with individuals more than a certain degree larger than itself.

There is little information concerning the development of assessment skills. However, data for adults suggest that such relative interindividual assessment is possible (Parker, 1974; Wachtel *et al.*, 1978; Clutton-Brock and Albon, 1979; Berger, 1981). It is possible that the same processes are responsible for the development of the perceptual abilities involved in assessment in young and adult animals. Basically, individuals must have the ability to match their own resource holding potentials (RHPs) (Parker, 1974) against those of their opponent(s). However, in some species *individual recognition* might override *character recognition* (Breed and Bekoff, 1981), in which a particular character, rather than an identifiable individual, is associated with fighting skill. In altricial mammals, for example, character recognition may be less important than individual recognition, given the nature of the developmental environment in which strange individuals are rarely encountered early in life.

An important problem in developmental studies of assessment is that individuals change as they develop and characters are in a state of relative flux; animals grow larger and often change in color (Fernald and Hirata, 1979). Intriguing questions for future research include: (1) what characters are used to assess relative resource holding potentials; (2) are these characters developmentally stable; and (3) does an individual learn to recognize characters or individuals or both? That is, do all individuals of a given size elicit withdrawal or submission, or only those individuals with whom an animal has had previous contact? These types of questions lend themselves to experimental analysis, but unfortunately they have not been pursued.

One way of studying the notion of assessment involves analyzing patterns of escalation (Clutton-Brock and Albon, 1979; M. Bekoff *et al.*, 1981*a*). The basic assumption is that by performing the first unprovoked, potentially dangerous act (escalating) (Maynard Smith and Price, 1973), an individual is "stating its intent" to engage in a serious fight from that point onward, depending on the response of its opponent. We have found that young coyotes that are familiar with one another apparently have the ability to assess their chances of winning an aggressive encounter before they engage in a fight, as well as during a fight (M. Bekoff *et al.*, 1981*a*). Escalation was used to "test" an opponent during a fight, and escalation patterns were found to be rank-related. Young coyotes tended to escalate with individuals closest in rank, with whom relative social standings were not well defined.

Ecology and Development of Agonistic Behavior

Species differences in the development of agonistic behavior may be related to species-typical social organization (King, 1973; Barash, 1974; Happold, 1976; M. Bekoff, 1977*c*, 1981*a,b*; Farr, 1980; Armitage, 1981; Byers, 1983; Lott, 1984). In general, many species that are characterized as being highly social show a delay in the appearance of rank-related, injurious agonistic behavior when compared with closely related species in which sociality is not as pronounced. However, significant variability in "species-typical" social organization has been documented for diverse species (Lott, 1984), and in many animal groups, environmental resources such as food appear to exert a major influence on sociality. For example, there are large variations in social organizational patterns displayed by coyotes, which seem to be influenced greatly by winter food resources

(Bowen, 1978; Camenzind, 1978; M. Bekoff and Wells, 1980, 1982, 1985). Observations of captive coyotes reared differently (M. Bekoff, 1978*c*; Knight, 1978) and limited field observations (Ryden, 1975; Camenzind, 1978) indicate that regardless of the environment in which coyotes are raised, rank-related, intense agonistic behavior appears very early in life. Zimen (1975) also noted that the development of aggressive behavior in wolves (*Canis lupus*) was similar, regardless of the different conditions in which litters were reared.

These data and other information (e.g., Berger, 1979) suggest that a question that needs careful attention in future research is if, and how, similar patterns of development can be overridden by ecological conditions. As with expressions of phenotype other than behavior, selection may predispose some species to display characteristic developmental patterns, which subsequently may be modified by proximate environmental conditions (Jacoby, 1983).

Hatching Asynchrony in Birds

Some very good examples of the way in which ecological variables can influence the development of agonistic behavior come from studies of hatching asynchrony in birds (O'Connor, 1978; Stinson, 1979; Ryden and Bengtsson, 1980; Bengtsson and Ryden, 1981; Clark and Wilson, 1981, 1985; Mock, 1981; Nuechterlein, 1981; Safriel, 1981; Richter, 1982). Here we consider the effects of hatching asynchrony on sibling interactions; asynchronous hatching leads to sibling competition in a wide variety of birds.

In general, scarcity of food seems to be the proximate stimulus that facilitates the occurrence of sibling aggression (siblicide) (Mock, 1981) in asychronously hatched birds, especially altricial species (A. Poole, 1979; Ryden and Bengtsson, 1980; Bengtsson and Ryden, 1981; Mock, 1981; Nuechterlein, 1981; Safriel, 1981). First-born individuals typically are heavier than, and can out-compete, later-born siblings when food is in limited supply. Early sibling competition, via the formation of dominance hierarchies among nestmates, can have significant effects on the survival of late-born young, even in precocial species. Safriel (1981) observed that first-born young oystercatchers (*Haematopus ostralegus*; a precocial species) dominated later-hatched siblings, and that the social hierarchy determined the way in which food was partitioned among them. When food was scarce, food was distributed unequally by the parents to the more dominant chicks. The growth rate of subordinate chicks was impaired; they also suffered higher predation.

In an experimental study of hatching asynchrony in western grebes (*Aechmophorus occidentalis*), Nuechterlein (1981) found that the older chicks suppressed the response of younger siblings to playbacks of the parental food call. Subordinate individuals were pecked at when they begged simultaneously with older siblings, and they soon refrained from begging until older chicks were satiated. Chick dominance relationships were evident even in situations unrelated to food. When space became limited, dominant chicks gained access to space in which they were more protected from violent windstorms, a major source of mortality in young grebes.

Obviously, death reduces fitness gains due to an individual's own reproduction to zero. If late-hatched chicks die, there is a clear advantage to being the first-hatched chick in an asynchronously hatched brood. Long-term studies should tell us whether there are any differences in fitness between first- and later-hatched birds when all survive periods of food shortage. When food is not limiting, asynchrony may produce little fitness decrement to late-hatched chicks (Nuechterlein, 1981).

The results of studies of hatching asynchrony, sibling competition, and food availability in birds seem robust. Similar relationships among development, food availability, body size, and dominance have been observed in domestic pigs and in fish (Magnuson, 1962). When nursing orders are formed in very young domestic pigs, feeding usually takes place with less interruption. It is possible that the more dominant animals also gain access to teats providing more milk, or are able to ingest large amounts of high-quality milk (Hartsock and Graves, 1976; Hartsock *et al.,* 1977; Fraser *et al.,* 1979, and references therein). In one study, the more successful fighters among piglets were heavier at birth and claimed and defended a specific teat earlier in life than did less successful fighters (Hartsock *et al.,* 1977). Furthermore, earlier-born piglets were heavier at birth, enabling them to fight more successfully than later-born siblings; the former individuals also suckled more frequently (Hartsock and Graves, 1976) and showed significantly greater chances of surviving.

Research into the ways in which ecological conditions can affect the development of agonistic behavior and later behavior are needed on a wider variety of animals. Careful measurements of resource availability are necessary, and identified animals have to be studied closely over long periods of time. There is no reason why laboratory studies should not be performed when field observations and measurements, or manipulations of resource availability, are not possible (e.g., Baldwin and Baldwin, 1976). The question with which this section started, concerning the modifiability of species-characteristic developmental patterns by ecological factors, could be approached in this way.

SOCIAL PLAY: THE COMING OF AGE OF A BEHAVIORAL ENIGMA

For many years, studies of animal social play behavior produced hard qualitative data consisting mainly of vague descriptions of a phenomenon that everyone saw, but few took seriously. As Byers (1981, p. 1493) noted, "Study of animal play historically has languished at the fringes of behavior research. Suspicions that play did not really exist or that it was a trivial phenomenon kept many investigators away. However, like an unwelcome guest, play continued to present itself to fieldworkers." Early observations of play usually were accompanied by grandiose, eponymous theories, explaining everything, and in some cases, nothing (M. Bekoff, 1976*a*). Perhaps if it was as much fun to study play as it is to engage in the activity, more would be known about it. This section is a modified and updated version of a recent review of ecological and evolutionary aspects of mammalian social play (M. Bekoff, 1984).

Although it has been a long time in coming, recent reviews of social play behavior (M. Bekoff, 1984, and Martin and Caro, 1985, and references therein) clearly show that the study of play is coming of age. One of the main reasons behind the increased vigor and rigor with which play research is being conducted stems from broad multidisciplinary interest in the activity by neurobiologists, population biologists, demographers, exercise physiologists, pharmacologists, ecologists, and evolutionary biologists (Welker, 1971; Fagen, 1977, 1981; M. Bekoff *et al.,* 1980; Panksepp, 1979; Beatty *et al.,* 1981, 1982; Meaney *et al.,* 1981, 1985; M. Bekoff and Byers, 1981; Martin, 1982, 1984; Festa-Bianchet and King, 1984). Scientists who otherwise wrote play off as a topic not worthy of serious study came to realize that detailed analyses of play could further our understanding of the evolution of social behavior, developmental processes (neural and behavioral), and learning. Indeed, E. O. Wilson (1977) lists play as one of five major areas of sociobiology warranting detailed explanation, the others being kin selection, parent–offspring conflict, the economic role of territory, and homosexuality.

What Is Play?

Play is a vague word that is used to describe a wide variety of motor patterns. Although most people have little trouble recognizing play, it has been difficult to develop a general, comprehensive definition [see Fagen (1981, Appendix I), for a list of definitions]. As more and diverse species are studied, the definitional problems may become increasingly difficult, but a definition also may be more necessary. The possibility remains that "play" may be broken down into finer categories, which can be more easily characterized.

Definitions of play based on structure (what individuals do when they play) are problematic, because play assumes many different forms in different mammalian orders. General definitions based on function are impossible to develop at this time because the function(s) of play largely remain a mystery. The enormity of the problem of definition is evidenced by the fact that the word "play" itself often is used in definitions of play.

Some defining characteristics of play include: (1) activities from a variety of contexts are linked together sequentially; (2) specific sets of signals (visual, vocal, chemical, tactile), including gestures, postures, facial expressions, and gaits, are important in its initiation; (3) certain behaviors, such as threat and submission, are absent or occur infrequently; (4) there are breakdowns in dominance relationships, role reversals, changes in chase–flee relationships, and contact time, and individuals engage in self-handicapping; and (5) there are detectable changes in individual motor acts and differences in sequencing when compared to nonplay situations [see M. Bekoff and Byers (1981, Table 11.1) for references, and also Chalmers (1980*b*), Fagen (1981), and Rasa (1984)]. Not all of the characteristics apply to all species, nor is it presently possible to say that at least two or three must apply in order for the activity to be called play.

While attending a conference dealing with behavioral development (Immelmann *et al.,* 1981), M. Bekoff stubbornly refused to offer a definition of play, because he thought that it would be impossible to suggest anything that would be applicable to even a small number of the few species in which play has been described. However, it became apparent that regardless of what definition was offered in different studies, a common component was that virtually all postnatal activities (performed in nondeprived settings) that *appear* functionless are usually referred to as play. Furthermore, striking similarities between prenatal activity (motility) (Hamburger, 1963; Oppenheim, 1974; A. Bekoff, 1976, 1978) and postnatal play became apparent. As Chalmers (1980*a*) and Thelen (1981) pointed out, and as was discussed in the section on comfort behavior, diverse behaviors may have common ontogenetic (and phylogenetic) precursors. [Some of the following discussion comes from M. Bekoff and Byers (1981); the ideas presented were developed as a joint effort among these authors and A. Bekoff (M. Bekoff *et al.,* 1980).]

Prenatal Motility

One of the most interesting findings in this field of research, and one that is relevant to the present discussion, is that although overt motility at different stages of chick embryonic development does not resemble later perinatal or postnatal behaviors such as hatching or walking, it has been found that there is a strong resemblance in the covert patterns of muscle activation and coordination between early and later motility and between hatching and walking, as demonstrated by electromyographic (EMG) recordings (A. Bekoff, 1976, 1978, 1981). These findings do not necessarily mean that motility and later behaviors are functionally isomorphic, nor do they attend to the important issue of whether or not the later (ontogenetically speaking) behavior patterns are really the same as ones that appeared earlier in development (Bateson, 1978). Rather, the discovery that the underlying neuromuscular patterning of both prenatal and postnatal behaviors may be the same, coupled with experimental documentation that prenatal movements are spontaneous or endogenous, provides us with a new perspective from which we can view postnatal play behavior.

Postnatal Play

The perspective provided by consideration of prenatal development is that outwardly purposeless motor activity is a regular feature of vertebrate ontogeny that seems to have reached the peak of its expression in mammals, in which it is continued, long after birth, as play. It seems as if early activation of the developing neuromuscular system was selected for at some point in vertebrate evolution, and that this requirement for activity became more pronounced, elaborate, and longer lived (ontogenetically speaking) as the vertebrate neuromotor system became more complex.

Because prenatal life is basically nonsocial and opportunities for motor activity are limited, likely benefits of prenatal activity include the facilitation of nueromuscular development. Therefore, the phylogenetically oldest function of postnatal outwardly purposeless motor behavior (play) was probably the facilitation of neuromuscular development.

At this point, natural selection probably began to mold this postnatal behavior in response to a number of selection pressures that were most likely consequences of a species adaptive syndrome (e.g., its size, habitat, and food habits) and its degree of adult neuromuscular and behavioral complexity. The result, as we observe it today, is play. As more descriptions of play in different mammalian orders become available, we should be able to fill out a phylogenetic tree of play, and thus may be able to deduce its evolutionary history more precisely.

Therefore, there are two possible avenues of continuity between prenatal and postnatal motor activities. The first is phylogenetic, in that embryonic motility may be ancestral in evolution to postnatal play. Many vertebrates that show prenatal motility typically do not play (e.g., fish, amphibians, reptiles, many birds, and perhaps a large number of mammals). The second is ontogenetic, in that there may be developmental continuity between prenatal motility and postnatal play.

Why Do Only Endotherms Play?

Prenatal motility is ubiquitous in the vertebrates, but its apparent postnatal counterpart, play, is confined to the two endothermic classes, Aves and Mammalia. Why is this so? Byers (1985) suggested the following evolutionary explanation. With the evolution of endothermy and its attendant energetic costs, selection for energy conservation likely became more intense. As a result, the endotherms enhanced the use dependence of musculoskeletal systems, in which individual bones hypertrophied, atrophied, and became remodeled in response to the specific stresses placed on them, and individual muscles hypertrophied, atrophied, and showed cellular changes (Holloszy, 1967; Edgerton, 1978) in response to the amount they were worked.

The general hypothesis is that the endothermic musculoskeletal system became more sensitive to use or disuse of its specific components, to conserve energy by maintaining only the biomass that appeared (based on experience) to be necessary. If such use-dependent maintenance, present to some extent in most vertebrates, became more pronounced as a result of the evolution of endothermy, it can be viewed as a preadaptation necessary for the emergence of play. Juveniles with use-dependent musculoskeletal systems could prepare the systems for motor tasks (e.g, prey capture, fighting, flight) likely to be closely linked to survival and reproductive success before the tasks were encountered for the first time. In other words, the ancestral function of play was motor training, and the necessary precondition for this, highly sensitive use-dependence in musculoskeletal systems, had already emerged as a consequence of the evolution of endothermy. Byers' (1985) hypothesis predicts that ectotherms should not show the use-dependent musculoskeletal plasticity that endotherms do. Information on physiological adaptation to exercise in ectothermic vertebrates is very limited, and does not yet offer a clear answer to this prediction.

A Working Definition of Play

Similarities between prenatal motility and postnatal play are obvious: both *appear* to be functionless, and in many cases, play, like motility, appears to be spontaneous. One characteristic of locomotor play is that of sudden, persistent, frantic motor activity (e.g., locomotor–rotational movements) (S. C. Wilson and Kleiman, 1974) that may include undirected flight.

Having clarified these points, we will define as it is seen in natural conditions or in settings that closely simulate them. *Play is all motor activity performed postnatally that appears to be purposeless, in which motor patterns from other contexts may often be used in modified forms and altered temporal sequencing* (M. Bekoff and Byers, 1981, p. 300). [Whether or not an activity is deemed to be "purposeless" may depend on the inventiveness of the observer (G. R. Michener, personal communication)]. If the activity is directed toward another living individual, it is called social play; if it is directed toward an inanimate object(s), it is called object play; if the activity carries the individual in a seemingly frantic flight about its environment, it is called locomotor play. Both social and object play can be locomotory in nature, and what has been referred to as self-play could include object and locomotor play. In some cases, play may be "involuntary" (Ewer, 1968). Many of the first interactions of young mammals, either with littermates, parents, or objects, seem to develop almost reflexively from fidgeting and mouthing. For example, early rooting and mouthing movements used in nursing, when transferred to the leg or face of a littermate, are usually called play. The emergence of vigorous play from more subtle forms of interaction suggests that the form of play that is observed at a given moment is dependent on the motor and sensory abilities of the developing individual.

To summarize, the apparent lack of purpose of play activity, its spontaneity, and its emergence from early fidgeting movements suggest that there may be some continuity between prenatal motility and postnatal play. Whether the two activities simply look discontinuous because of the differences between prenatal and postnatal environments is open to speculation. The outward purposelessness of both suggests that there has been selection for early rehearsal of certain motor activities. Although the apparent lack of purpose of play is an important defining characteristic, it really is not functionless at all.

Ecology of Play Behavior

Habitat type, resource availability, and the social environment can affect the performance (type and amount) of play activities (S. C. Wilson, 1973; Baldwin and Baldwin, 1973, 1976; Sussman, 1977; Symons, 1978*a*; Berger, 1979; Nowicki and Armitage, 1979; Dubost and Feer, 1981;

Fagen, 1981; Thomas and Taber, 1984). However, even when the potential cost of play is high due to energy constraints or risks [see Fagen (1981), p. 276, for a list of possible risks], play usually is not totally extinguishable (Müller-Schwarze *et al.,* 1982) unless the animals are temporarily out of energy, are moribund, or run out of time in which to play [e.g., large amounts of time have to be spent gathering food (Baldwin and Baldwin, 1976; Oakley and Reynolds, 1976)]. Martin (1984) found that 10- to 12-week-old cats devoted 4–9% of their total daily energy expenditure (excluding growth) to play.

Field studies of the same species living in different habitats provide useful information concerning the relationship between proximate environmental conditions and the development of play. For example, Berger (1979, 1980) found that social development differed in natural populations of bighorn sheep (*Ovis canadensis*) living in different habitats. Desert sheep living in the Santa Rosa Mountains of southern California, where the presence of cholla cactus (*Opuntia* spp.) made play a risky affair, played significantly less than sheep living in the Chilcotin area of British Columbia, Canada, where there were more suitable play areas such as grassy fields and sand bowls (M. Bekoff and Byers, 1981, Fig. 11.1). The structure of play also differed; desert yearlings never were observed to engage in rapid locomotor play, possibly because of the risk of running into cacti. Berger also found that sheep in British Columbia lived in larger groups and grew up in a more diverse social environment. They had more playmates and showed greater diversity in play patterns.

The necessity for studying allopatric conspecifics is clearly exemplified in this study. Speaking about the structure and function of play in *the* bighorn sheep would be misleading because of the influence of habitat and the social environment on the development and elaboration of play activities.

Evolution, Life-History Patterns, and Play

> Social play reflects biological adaptation. Like other aspects of social behavior, it has been selected to adjust to a broad range of environmental conditions in the service of inclusive fitness. (Fagen 1981, p. 387)

Evolutionary considerations of play have become very popular recently, and have helped to clarify various aspects concerning the development of the activity, its structure, how its nature may vary with particular partners, and its functions (Fagen, 1977, 1978, 1981; M. Bekoff, 1978*a*; Symons, 1978*a,b*; M. Bekoff and Byers, 1981). The application of various evolutionary models to play behavior [such as viewing social play as an evolutionary stable strategy, ESS (Maynard Smith and Price, 1973)] provides interesting suggestions concerning how animals might play optimally, but theoretical predictions remain just that, *predictions,* until play is quantitatively analyzed and testable hypotheses are entertained. Facile ultimate explanations (Byers, 1981) of play or other behavioral patterns tend to imply that a lot more is known about the activity than actually is. Furthermore, as discussed above, one of the most exciting aspects of developmental research involves the dissection and reconstruction of how behavior unfolds throughout life. Using inclusive fitness as the ultimate (in both senses of the word) reason to explain why animals play tends to remove attention from the fact that selection operates at all stages of development (Williams, 1966) and that individuals need to meet successfully challenges at all ages in order to make it to the next stage (Galef, 1981).

Play appears to have evolved mainly in species (especially mammals) in which there is a prolonged period of immaturity, dependence on adult care-giving, and a high encephalization quotient [EQ, ratio between the observed brain weight and the expected brain weight for a defined body weight (Jerison, 1973; Eisenberg, 1981)]. However, it must be remembered that only very few species actually have been studied and that the distribution of play among mammals still is unknown.

In species in which play has been observed, young individuals typically play more than adults; there appears to have been strong selection for the appearance of play early in life (Fagen, 1977). However, playlike activities may continue after sexual maturity. For example, in many polygynous ungulate species sexually mature, prereproductive males engage in much "sparring," a form of "mock fighting" (Reinhardt and Reinhardt, 1982), or low-intensity fighting among the members of all-male groups (Byers, 1984*c*). Apparently there was selection for sustained, postjuvenile repetition of these motorically complex tasks because they are closely linked to adult male reproductive success. Even if food is limited, due perhaps to early weaning (naturally or experimentally induced), young individuals may show a facultative response to decreased food and pack the experience into a shorter period of time (Bateson and Young, 1981; Bateson *et al.,* 1981).

It also has been suggested that the benefits derived from play may decrease with age (Caro, 1981; Chalmers and Locke-Hayden, 1981) [organisms are not infinitely plastic (Fagen, 1981)]. However, do the benefits derived from play decrease because animals spend less time playing, or do animals spend less time playing because they do not get as much out of it? As individuals get older and more independent, there are more things that they need to do of their own accord, and time and energy constraints

TABLE I. The number of Play Facilitation Bouts in Which a Single Female's Own, Closely Related, or Distantly Related Kids Were Involved[a]

Own	Closely related	Distantly related
21 (15/6)	2 (0/2)	2 (1/1)

[a] The distribution is clearly uneven ($\chi^2 = 28.9$, df $= 2$, $p < 0.001$). In parentheses to the left of the slash is the number of bouts in which a single kid participated; to the right of the slash is the number of bouts in which 1–3 additional kids were involved.

because of competing activities may also affect the observed developmental time course for play (Fossey, 1979; Sharatchandra and Gadgil, 1980; Fagen, 1981; S. C. Wilson, 1982). As Fagen (1981) noted, animals have limited time and energy to devote to survival, growth, and reproduction, the important components of life histories; the amount of time devoted to a specific activity depends on how much goes to others.

If there is an optimal age scheduling of play, mothers sometimes might be expected to stimulate play in their own or closely related offspring, to ensure an appropriate rate of play at the proper ages. Byers observed this in Siberian ibex (*Capra ibex*) housed in a naturalistic exhibit at the Brookfield Zoo in Chicago. During a study of play (May–August 1977) 36 instances in which an adult female showed locomotor–rotational play motor patterns while she was close to ibex kids (ages 4–13 weeks) were observed. Eleven of the 12 mothers were observed to do this at least once. In the Brookfield ibex, play was almost exclusively confined to discrete dawn and dusk sessions, in which kids first showed social play (sparring and related movements), then locomotor–rotational (LR) play (running across steep slopes with leaps accompanied by rotational movements) (Byers, 1977). The 36 instances in which adult females showed LR movements all occurred during dawn or dusk play sessions in which some or all kids began to recline after social play ended. In 33 instances, the female movements immediately were followed by LR play in nearby kids. For 26 of the 33 (79%) instances, reliable data were available on whether the kids involved had shown any LR play earlier in that session; in 22 instances (84%) they had not. Of the 33 instances, 25 (75%) involved a single adult female. Table I shows the frequencies with which the females's own, closely related (offspring of mother or daughter), or distantly related (not offspring of mother or daughter) kids were involved in these bouts. The uneven distribution shows a clear tendency for these bouts to involve mother–offspring pairs.

Females did not continue to play after kids began, and were not observed to perform LR motor patterns at other times. Adult females performing these acts were so conspicuous that it was very likely that all such events that occurred in view of the observer were recorded. It seems reasonable to conclude that ibex females occasionally behave to increase the likelihood of play in their own or closely related offspring. This finding indirectly supports the notion of optimal age scheduling of play, and suggests that, for ibex, LR play has a strong benefit—strong enough to act indirectly through offspring fitness on maternal behavior.

Why Play: Functional Considerations

Study of the function of play is a study of the process through which natural selection shaped this aspect of behavioral phenotype. Therefore, it is different from the study of proximate causation, which seeks to define the immediate stimuli and physiological mechanisms that control play. It also needs to be stressed that not all of the beneficial consequences of a behavioral pattern constitute its function (Williams, 1966; Hinde, 1975; Symons, 1978b). *By function, we mean the specific consequences of a behavioral pattern that have resulted in its fixation in a species' repertoire by natural selection* (M. Bekoff and Byers, 1981, p. 312). The possible functions of play must be studied with respect to both the young individuals that are playing at a given age (immediate benefits) and to the delayed, cumulative consequences possibly affecting fitness.

Ideas about the functions of play [reviewed in Symons (1978*a,b*), M. Bekoff and Byers (1981), Fagen (1981), Pagés (1982, 1983), Smith (1982), Fabri (1983), Chalmers and Locke-Haydon (1984)] must be taken as what they are—*suggestions*. While it may be all right to favor one hypothesis over others, it would seem unwise to discount a reasonable suggestion as being inapplicable, given the enormous variability of play (even among conspecifics). Most authors conclude that there is no one function of play. However, Byers (1984*c*) suggested that there likely *was* one ancestral function (motor training) for play, and that the appropriate course of investigation is to deduce how play has been modified to serve new functions in the mammalian orders where it is apparent that this has occurred. Data from wholesale deprivation studies (M. Bekoff, 1976*a,b*; DeGhett, 1977) have provided little insight into the question of why animals play.

Martin (1984), after estimating from rough measurement that 10- to 12-week-old cats devote 4–9% of their energy budgets to play, suggested that this was a minor cost and that therefore play might have only minor benefits. However, as Martin himself noted, a minor cost does *not* necessarily imply a minor benefit. Also, it seems to us that 4–9% of a young cat's energy budget may not necessarily be "minor" at all. Cost and benefit, to be equated as Martin suggests, must be expressed in common units, such as altered expectation of survival or reproductive success. Martin measured an energy cost of play and then made the apparently intuitive conclusion than an energy expenditure of this size would not have major effects on survival or reproductive success. Martin's work is aimed at the worthwhile goal of assigning a selection coefficient to play; this is different from ascertaining its functions.

In his consideration of the beneficial aspects of play, Fagen (1981, p. 271) lists six overlapping hypotheses: (1) play develops physical strength, endurance, and skill; (2) play regulates developmental rates; (3) play experience yields specific information; (4) play develops cognitive skills necessary for behavioral adaptability, flexibility, inventiveness, or versatility; (5) play is a set of damaging behavioral tactics used in intraspecific competititon; and (6) play establishes or strengthens social bonds in a dyad or social cohesion in a group. M. Bekoff and Byers (1981) idealize the possible functions as involving (1) motor training, (2) socialization, and (3) cognitive training. Here we will consider only motor training and social cohesion hypotheses.

Motor Training

Different motor patterns from various contexts are used in play, and repetition of these motor patterns along with vigorous exercise may be important in the development of both general physical fitness particular motor skills (Chalmers and Locke-Haydon, 1984). Smith (1982) concluded that play primarily affords practice for certain socially competitive skills [especially agonistic and predatory; see also Symons (1978*a*)], and that the development of social skills that also appear to be associated with play experience are actually only incidental benefits. However, Smith admits that at present the evidence is not available to support this notion.

Structural analyses of play support the training hypothesis (Byers, 1977, 1980, 1984*c*; Markstein and Lehner, 1980), but Fagen (1981) believes that they do not explain adult play, courtship play, and social play. If skill development is considered to be important throughout life, then it would be entirely possible that an otherwise experienced adult could be inexperienced when it comes to the skills needed in courtship and mating.

Fine-Tuning. Associated with the training hypothesis and the common observation that play borrows actions from various contexts [except, perhaps, those used in the communication of play intention (M. Bekoff and Byers, 1981, and references therein)] is the notion that while playing an individual may be looking for the optimal goodness of fit (Klopfer, 1970) of individual behavioral patterns in situations in which it does not have to pay severe consequences for mistakes; it tries different combinations of behavior that would never be tried under pressure (Bruner, 1974). Animals also may be learning to make fine motor movements from more global actions involving many muscles and body parts.

Fine-tuning provides behavioral flexibility, which is useful when conditions (social and otherwise) change. The ability to match behavioral responses with changing situations may be a skill that is necessary throughout life; it could facilitate the social integration of individuals into a group under different environmental conditions (Nowicki and Armitage, 1979) and increase success in catching different prey.

Support for the ''fine tuning through repetition and training'' hypothesis comes from a wide variety of studies (Bruner, 1974; Simpson, 1976; Mason, 1979; Nowicki and Armitage, 1979; Baerends-van Roon and Baerends, 1979; Berger, 1980; Chalmers, 1980*b*; Latour, 1981*a*; Pellis, 1981; but see Symons, 1978*b*). As mentioned above in the discussion of comfort behavior, there might be strong selection against the continued addition of new motor patterns whenever something new is confronted. Fine-tuning could evolve as an adaptive skill via selection for the ability to modify slightly already existing motor patterns and the way in which they are linked sequentially in order to deal with changing situations; selection for open systems is apparent once again. A major question is whether the ability to fine-tune responses to proximate conditions conferred a selective advantage that led to the evolution of play in animals (M. Bekoff and Byers, 1981).

Social Cohesion

The social cohesion hypothesis basically suggests that play may be a mechanism through which social attachments are formed and social bonds strengthened and maintained (S. C. Wilson, 1973; Happold, 1976; M. Bekoff, 1977*c*; Latour, 1981*a*; Panksepp, 1981). Of course, social bonds also may be formed through activities other than play, including mere exposure, but here we consider only play behavior. In some species in which play occurs in high frequencies among young individuals there may be a cumulative effect that might reduce and delay the tendency to disperse (Barash, 1974; Happold, 1976; M. Bekoff, 1977*c*). Delayed dispersal may be associated with the existence of cohesive social groups in which individuals of different ages live (Barash, 1974; M. Bekoff, 1977*c*; Armitage, 1981; M. Bekoff *et al.*, 1981*b*).

Smith (1982) is critical of the cohesion hypothesis. He wrote that if play is important in social bonding, "it would be expected to occur most in species with social groups of moderate size, and not in solitary species" (Smith, 1982, p. 145). The reason for this assumption is not clear. Theories relating play to social bonding and dispersal attempt to account for both intra- and interspecific differences in the developmental time course of play (and other behaviors). They stress that *differential* play experience may affect the *probability distribution* of individuals leaving their natal group. Smith's (1982, p. 145) statement, "if social play serves a bonding function it would not be expected to occur if offspring leave their natal group to join another after the juvenile period," ignores the differential effects that play may have on individuals growing up in the same group. The question is not whether or not they play, but how much.

Fagen (1981) is critical of the cohesion hypothesis mainly because he can find exceptions to the "rule." However, he recognizes the cohesion hypothesis as one of three that deserves consideration in evolutionary analyses of play: "Without a doubt, each hypothesis will prove valid in particular instances" (Fagen, 1981, p. 358).

Both Smith's and Fagen's objections are answered by Byers' (1984*c*) suggestion that the proper course of research is to trace the evolutionary history of the function of play, from a common mammalian function to possibly new, specialized functions, accompanying anatomic and behavioral specializations in different mammalian orders. For example, Byers (1984*c*) reviewed play in ungulates and showed that the common distribution of many motor patterns such as running strongly supports an ancestral motor training function, but that the group play, including adults, of collared peccaries suggests modification to serve a social cohesion function. Phylogenetically based play research will avoid the either–or, my example versus your example search for functions, which clearly moves in circles. As Byers (1984*c*) noted, no one is foolish enough to claim a unitary function for the mammalian hand, so why do so for play?

Social Cohesion and Fine-Tuning. A combination of these two hypotheses may be important in understanding both the act of dispersal and how dispersing individuals respond to new environmental conditions. Geist (1978) suggested that play facilitated development of flexibility and prepared an individual to be a successful disperser or colonizer. The dispersal phenotype would be characterized by an adaptable animal capable of tolerating diversity (Fagen, 1981).

Fagen (1981) stated that the flexibility hypothesis and the social cohesion hypothesis made opposing predictions about dispersal. We believe that both may work hand-in-hand. As mentioned above, the cohesion hypothesis does not state that some individuals play and others do not. It suggests that differential early play (and other) experience may predispose some individuals to remain in their natal area and others to leave. With few exceptions, all individuals probably partake in some play early in life [although the age distribution may vary according to food resources (Peterson and Young, 1981) or dominance rank], and therefore all individuals would be primed to be able to deal with novel problems both within and outside of their natal home range. The question that needs to be answered now is how much (little) play is necessary to produce an individual who can fine-tune responses to various external conditions, and if this amount of play is insufficient for the formation of social bonds strong enough to delay dispersal. That is, *if* play is important in the formation of social bonds, and *if* individuals who have not played enough to form bonds leave their natal group, have they still had enough play experience to be able to deal with the novel situations to which they may be exposed during dispersal? We agree with Geist's suggestion about the preparatory role that play may serve for dispersal, and think that it is compatible with suggestions relating differential development play to dispersal.

Future research is filled with exciting possibilities, such as: (1) Why do different functions apply to different species? (2) Which functions may be widespread phylogenetically? (3) Do (and why do) different functions apply to conspecifics living in different habitats? If critical reviews are what it will take to get the ball rolling, one need not search far. Understandably, all recent reviews concerning the possible functions of play have been critical, but useful. That unwelcome guest is here to stay, like it or not!

LIFE-HISTORY TACTICS, DEMOGRAPHY, AND DEVELOPMENT

A life-history tactic is "a set of coadapted traits designed, by natural selection, to solve particular ecological problems" (Stearns, 1976, p. 4). Recent reviews indicated that important associations exist among various traits such as body size, patterns of reproductive behavior, behavioral and physical development, and social behavior (Fig. 9) (Wiley, 1974; Stearns, 1976, 1977, 1980; Gould, 1977; Blueweiss *et al.*, 1978; Fleming, 1979; Mace, 1979; Western, 1979; Warner, 1980; Armitage, 1981; M. Bekoff *et al.*, 1981b; Eisenberg, 1981; Fagen, 1981; Michener, 1983); body

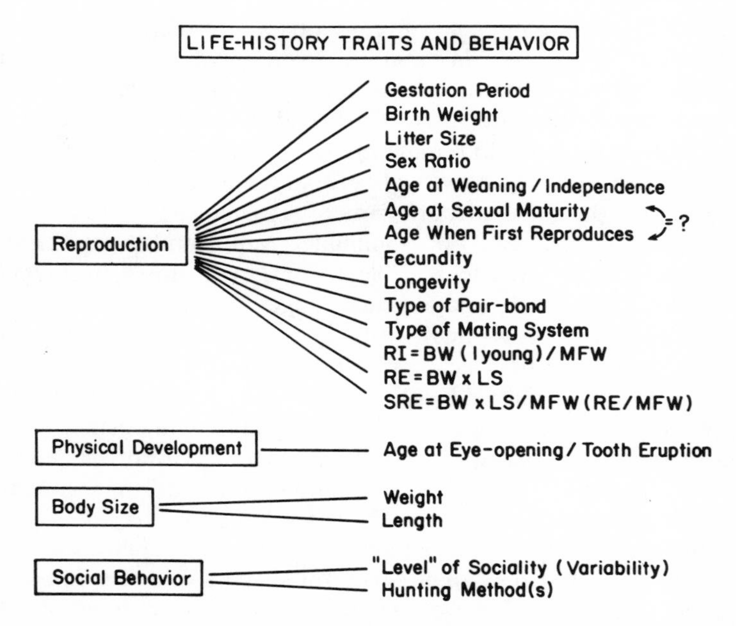

FIG. 9. Studies in which the relationship between life-history traits and behavior are analyzed consider a large number of variables associated with reproduction, physical development, body size, and social behavior. In some cases, animals do not mate when they first reach sexual maturity (see text). RI, Reproductive index; RE, reproductive effort; SRE, specific reproductive effort; BW, birth weight; LS, litter size; MFW, minimum female weight. RI represents the effort by a single female to bring one young to term relative to her weight, RE represents the total resources allocated to bringing a litter to term irrespective of female weight, and SRE represents the total resources allocated to reproduction relative to female weight (Armitage, 1981).

size frequently is thought to be the major variable linking other traits. Life-history and demographic analyses of behavioral development involve consideration of reproductive strategies, relative degree of maturity of newborn animals, and environmental variations (Portmann, cited in Gould, 1977; Gould, 1977; Mason, 1979; Eisenberg, 1981).

Reproductive Strategies and Developmental Patterns

The relationships among life-history, demographic, reproductive, and behavioral characters usually are discussed with respect to *r*- and *K*-selection (MacArthur and Wilson, 1967). Although these terms are considered to be barbaric (Horn, 1978), a careful consideration of the general suites of characters that are associated with each type of selection (E. O.

Wilson, 1975; Stearns, 1976, 1977; Horn, 1978; Fleming, 1979; Parry, 1981) is useful for developing general models and testable hypotheses about the relationship between life-history tactics and behavior. The terms *r* and *K* do not represent ends of a one-dimensional continuum (Horn, 1978). Also, a comparison of different animals may indicate that a given species is *r*-selected with respect to some species and *K*-selected with respect to others (Western, 1979).

In general, *r*-selected species tend to be small animals that evolved in fluctuating environments, in which large litters of precocial individuals (large prenatal reproductive effort) are produced, and into whom parents invest little postnatally. These species show early maturation and reproduction. *K*-Selected species tend to be larger animals that evolved in more stable habitats, in which smaller litters of more altricial individuals (low prenatal reproductive effort) are produced, and into whom parents invest more postnatally. They show slower development and first reproduce later in life; there is an increase in the interbirth interval. Demographic predictions can change due to bet-hedging (Stearns, 1976), in which reproductive risks are minimized rather than offspring production being maximized; fluctuations in mortality and fecundity are considered (Table II).

The immaturity syndrome shown by *K*-selected species most likely evolved under conditions that placed a premium on behavioral flexibility, which created openess in information processing and encouraged the intensification of curiosity, play, and exploration (Mason, 1979). Though open systems may be susceptible to developmental accidents (see above), the newborn typically enters an environment that has been shaped by natural selection. Because the infant is "prepared" to encounter a species-characteristic postnatal environment, atypical outcomes are minimized (Mason, 1979).

K-Selected species also tend to have larger brains (relative to body size) than *r*-selected species, but there is a good deal of scatter within Families when encephalization quotients (EQs) are calculated (Eisenberg, 1981, Chapter 21). In general, high EQ values are strongly associated with long potential life span, delayed sexual maturity, and an extreme iteroparous reproductive strategy (Eisenberg, 1981). Eisenberg's (1981, p. 283) list of life-history correlates of EQs also shows that higher EQs tend to be associated with information storage and retrieval based on individual experience (which is superimposed on preprogrammed information), a greater dependence on multimodal sensory input, more active antipredator strategies, and a long period of total life spent in learning situations. Life-history analyses of behavioral development lead to interesting theories and testable hypotheses about the way in which various develop-

TABLE II. Demographic Predictions Resulting from *r*- and *K*-Selection and Bet-Hedging Strategies, in Which Reproductive Risks Are Minimized Rather Than Offspring Production Being Maximized[a]

Stable habitats	Fluctuating habitats
r- and *K*-selection and bet-hedging with adult mortality	
Slow development; late maturity	Rapid development; early maturity
Interoparity	Semelparity
Small reproductive effort	Larger reproductive effort
Few young	More young
Long life	Shorter life
Bet-hedging with juvenile mortality	
Early maturity	Later maturity
Iteroparity	Semelparity
Large reproductive effort	Smaller reproductive effort
Shorter life	Longer life
More young	Fewer young
Fewer broods or litters	More broods or litters

[a] Adapted from Fleming (1979). Bet-hedging attempts to deal with fluctuations in mortality and fecundity schedules (Stearns, 1976). Two basic suites of demographic characteristics are evident: (1) early maturation, large reproductive effort (in terms of number of offspring produced per brood), and short life span; (2) delayed maturity, small reproductive effort per reproduction, and long life span.

mental strategies evolved and how susceptible they are to modification by environmental variables.

Age at First Reproduction

It is obvious that many of the general characteristics associated with *r*- and *K*-selection syndromes can have large effects on behavioral development [litter size, degree of maturity at birth, availability of siblings other than littermates (M.Bekoff, 1981*b*)]. The life-history trait that we will consider here is age at first reproduction, which may or may not be the same as the age at which sexual maturity is reached.

Age at first reproduction is considered to be one of the most significant demographic characteristics of a species (L. C. Cole, 1954; see also Thorne, 1981). The age at which an individual first reproduces can have direct effects on fitness, and it may be influenced by proximate factors such as the social environment in which an individual grows up [for reviews see Wiley (1974, 1981), Kleiman (1977), Packard (1980), M. Bekoff

(1981*b*), Armitage (1981), Stearns (1980)]. In a given species there also may be selection for sexual bimaturism, or sex differences in age at maturation (Wiley, 1974). Stearns and Crandall (1981) found that for salamanders and lizards, a fecundity gain or improvement in juvenile survival of offspring yielded predictions of a delay in maturity.

In most mammals that show obligate monogamy [a lone female cannot rear a litter without aid from conspecifics, but only one female can breed in a group because the carrying capacity of the habitat cannot support more than one breeding female (Kleiman, 1977)] and helping, individuals may show suppressed maturation and/or reproduction in the presence of their parents. In members of the family Canidae, there appears to be a close relationship among delayed dispersal, delayed breeding, and increased sociality (M. Bekoff *et al.*, 1981*b*). The young of more social (and larger) species tend to become independent and disperse at a later age than offspring produced by smaller species; delayed independence and dispersal is associated with a postponement in the age at which both males and females first breed.

In an analysis of sociality as a life-history tactic in 18 species of ground squirrels, Armitage (1981) found that variation in sociality was best explained by the age of first reproduction and age at which adult weight was reached. Sociality occurs in species whose large size associated with a short growing season delay reproductive maturity beyond 1 year of life. Sociality in these species may have evolved by the selective retention of females within maternal home ranges [see also Michener (1981, 1983) and discussion below].

Barash's (1974) analysis of development and reproduction in three species of marmots also showed that ecological variables were associated with different degrees of sociality in these animals. He found negative correlations between the length of the growing season and age at dispersal and between the length of the growing season and social tolerance. Barash suggested that greater maturity may be needed for success in more severe environments, and that there was selection for increased social tolerance of young until they have nearly reached adult size. Anderson *et al.* (1976) offered an alternative explanation. They suggested that hibernacula density may influence marmot sociality. It appears that in more severe environments, there is a lower density of hibernacula and a higher degree of saturation of this resource, and this would favor the production of fewer (if reproductive rates vary inversely with adult survival rates), but higher quality young (larger size at dispersal), and greater social tolerance between adults and young individuals until the optimal dispersal size was reached. Demographic differences among the three species lend credence to this explanation (Fleming, 1979).

In at least select ground squirrels and canids, it appears that sociality evolved through delayed dispersal associated with a postponement in reproduction. This trend also may apply to other taxa. Life-history and demographic analyses offer an attractive way in which to study how strategies of behavioral development have been selected along with other traits.

Reproductive Value, Eligibility, and Generosity

Another life-history attribute likely to have important effects on the evolution of behavioral development is reproductive value (Trivers, 1972; Gadgil, 1982; Rubenstein, 1982). These authors and others have pointed out that the cost–benefit ratio for any aid-giving act that one individual performs for another will vary with ages of the interactants. This is because reproductive value (especially in *K*-selected species) rises, then falls with age. The ability to transform social help into augmented reproductive value (Gadgil's "eligibility") is likely to follow a similar, but earlier peaking course. In contrast, the amount of social help an individual can give per unit reduction of its own reproductive value (Gadgil's "generosity") is likely to begin near zero, then increase with age. For any social interaction in which one individual aids another, the cost–benefit ratio will depend on the age-related reproductive value, eligibility, and generosity of the interactants. Gadgil (1982) presented a matrix of age dyads in which he qualitatively predicted directionality of helping and competitive interactions. Rubenstein (1982) developed similar predictions. The most intuitively obvious predictions [for instance, parents should help offspring more than young siblings should help each other] are, of course already abundantly confirmed, but other predictions are less obvious and deserve attention.

BEHAVIORAL DEVELOPMENT, EARLY SOCIAL EXPERIENCE, AND SOCIAL ORGANIZATION

Species-typical patterns of behavioral development and early experience may be closely related to modal social systems, especially in mammals. Basic questions here deal with how young individuals become integrated into already existing social groups (Bertram, 1976; M. Bekoff, 1977*a*, 1981*a,b*; Biben, 1983; Stamps, 1978; Burgess and Shaw, 1979; Johns and Armitage, 1979; Altmann and Scheel, 1980; Greenwood, 1980;

Byers, 1980, 1983; Byers and Bekoff, 1981; Downhower and Armitage, 1981; Harcourt and Stewart, 1981; van Honk and Hogeweg, 1981; Jarvis, 1981; Stein and Stacey, 1981; Sunquist, 1981; Wiley, 1981; Berman, 1982*a,b*; S. C. Wilson, 1982; Ikeda, 1983), or how do successive generations of individuals come to adopt similar social organizations (Wiley, 1981), and what role may early experience effects play when evolutionary shifts in social organization occur (Byers, 1983)?

It also is necessary to consider if and how development may be affected by the habitat or by the composition (age, sex) of the social group (or both) into which young are born. Intraspecific variability in social structure associated with proximate ecological conditions is well-documented in many animal groups, even within local populations (S. C. Wilson, 1973; M. Bekoff and Wells, 1980; Fox *et al.*, 1981; Latour, 1981*b*). Whether or not conspecifics show differential development depending on environmental (social and/or habitat) conditions needs field verification in almost all species in which variability in social structure has been documented. Fox *et al.* (1981) showed that the development of dominance in juvenile lizards (*Uta stansburiana*) was related to habitat. Dominant individuals originated from larger, more diverse, and higher quality home ranges than did subordinate animals. Also, the home ranges of dominant individuals resembled those of oversummer survivors, whereas home ranges of subordinate lizards resembled those of nonsurvivors. It would be interesting to determine if the reproductive output of surviving subordinate individuals is different from that of surviving dominant lizards.

Ontogenetic Trajectories

Wiley's (1981) concept of ontogenetic trajectories, which describe movements of individuals through social positions, may be useful both for longitudinally tracing how individuals become integrated into social groups and for determining if variations in the natal environment affect development [it may also be important to consider prenatal environments (von Saal, 1981)]. He related the evolution of different mating and social systems to ontogenies differing in the age at which individuals first breed, and pointed out that we should not consider delayed reproduction to be a secondary consequence of social structure.

Wiley (1981) stressed that the evolution of complex social structures is inextricably linked with the evolution of an optimal allocation of effort to reproduction, growth, and maintenance throughout the life of an individual. The model of ontogenetic trajectories makes us think about the complex tradeoffs among demographic variables such as age at first reproduction, fecundity, mortality, and developmental pathways (also see Schaffer, 1974; Charlesworth, 1980; Thorne, 1981). Clearly, life-history and demographic analyses are important in studying the relationship between behavioral development and the evolution of sociality (Armitage, 1981; M. Bekoff *et al.*, 1981*b*; Fagen, 1981).

Squirrels, Peccaries, and Gorillas

Comprehensive studies focusing on possible relationships between social development and social organization need to account *at least* for the complex and interrelated roles that genetic relationships, familiarity, recognition abilities, individual early experience, spatial associations, patterns of emigration and group transfer, and ecological conditions play, both during early behavioral ontogeny and throughout the life of identified individuals of known genetic relatedness. Douglas-Hamilton and Douglas-Hamilton (1975) pointed out that in African elephants (*Loxodonta africana*), kin relationships may extend over 100 years. Meikle and Vessey (1981) found that after male rhesus monkeys (*Macaca mulatta*) left their natal group they frequently transferred into the same social group as their older, maternally related brothers. Subsequently, brothers spent more time close to each other than to other males and formed alliances with brothers more frequently than with nonbrothers. The necessity for tracing the ontogenetic trajectories of identified individuals is obvious.

Richardson's Ground Squirrels

The work of Michener (1979, 1980, 1981, 1983) on the development of behavior and social organization in Richardson's ground squirrels provides a solid example of the value of detailed field work on identified individuals. For the purpose of this discussion, her analysis of the ontogeny of spatial relationships and social behavior is relevant.

Michener (1981) found that after juvenile squirrels emerged from their natal burrows at about $4\frac{1}{2}$ weeks of age, they remained in close spatial proximity to their mother and littermates for about 4 weeks. There was extensive overlap of sibling home ranges after young squirrels first emerged. Between about 7 and 10 weeks of age, young squirrels became increasingly independent from the family unit and established their own spatially distinct areas. There was a decline in the extent of spatial overlap. Juveniles usually remained physically closer to (and more amicable with) their littermates and mother than with neighboring conspecifics.

These kin clusters behaved agonistically toward members of adjacent kin groups.

By about 9–10 weeks of age, juvenile squirrels resembled adults with respect to spatial and social patterns. Daughters were more likely than sons to continue to live near kin as adults (Michener, 1980). Interactions between juveniles and other squirrels differed depending on kinship [which is related to familiarity (Michener, 1983)]. Kin interactions were predominantly amicable (110/124 = 89%), whereas only 19% (24/126) of the interactions between non-kin were amicable.

By about 3 months of age, juveniles were behaviorally indistinguishable from adults both in their use of space and in their social interactions. Comparative analyses indicate that the rapid development of adult-like spatial and social patterns appear to be a characteristic of ground-dwelling sciurids that are obligate hibernators and who reproduce as yearlings.

Collared Peccaries

Collared peccaries (*Tayassu tajacu*) are very social ungulates that live in permanent, mixed-sex, cohesive herds (about 12–15 individuals in the American Southwest), in which cooperative and amicable behavior predominate, and in which overt conflict is rare (Byers and Bekoff, 1981). Agonistic behavior, when it occurs, usually does not involve physical contact. Cooperative nursing, predator defense, and feeding are common, and apparently all adults are tolerant of young animals. The absence of sexual dimorphism in body size, canine size, and other characters that might influence fighting ability, the 1:1 sex ratio within herds, and the production of small litters of precocial young (Fig. 10) suggest an evolutionary history of marked sociality. Unlike other suoids, subordinate animals are not driven away from the social group either by same-aged individuals or by adults, and it is likely that group members are closely related to one another. Kin selection probably played an important role in the evolution of social behavior in collared peccaries, but detailed observations of identified individuals need to be done.

Byers (1983) was able to test a hypothesis concerning the relationship between behavioral development and social organization by directly observing animals in the field (the precocity of peccaries facilitated the observation of young individuals). Specifically, he tested the social cohesion hypothesis that suggests that amicable interactions, frequent olfactory contact, and a delay, with respect to amicable behavior, in the emergence of agonistic behavior among juveniles are the developmental mechanisms by which stable social groups are achieved (S. C. Wilson, 1973; S. C. Wilson and Kleiman, 1974; Barash, 1976; Happold, 1976; Bekoff, 1977*c*).

FIG. 10. A 1-week-old collared peccary with an adult. Infant and juvenile peccaries maintain close spacing with adults for about the first 3 months of life. Peccaries are highly social, and the way in which precocial young are incorporated into cohesive herds was studied by Byers (1980, 1983) (see text). (Photograph by John A. Byers.)

The results of Byers' study did not fully support the social cohesion hypothesis when applied only to same-aged individuals. Rather, Byers found that juveniles interacted at very low rates, showed no delay in the expression of agonistic behavior among themselves, and had most of their social experience and amicable contact with adults. The only amicable interactions in which juveniles partook during the first 3 months of life were playful, and this experience may have facilitated the development of bonds among juveniles. Play and all other forms of amicable interaction are developmentally segregated in peccaries.

In collared peccaries, it appears that both juvenile–juvenile and juvenile–adult amicable interactions are important for the maintenance of cohesive social organization. In species in which there is pronounced sociality and cooperation among individuals of different ages, amicable interactions both within and between age classes, probably facilitate continued maintenance of group cohesion when young born into the group need to be incorporated into the existing social system.

Byers' (1983) results also illustrated that ideal rules concerning the relationship between social organization and behavioral development often may be blurred by the specific evolutionary path by which a species achieves a specialized social trait. Juvenile peccaries show no delay in the emergence of agonistic behavior among themselves, and this seems to be largely a consequence of female passivity in nursing. Females stand while juveniles suckle from between the hind legs, and females make no observable attempt to check the young's identity. This behavior by females results in communal nursing, but also results in (sometimes severe) competition among juveniles for access to teats. Female passivity in nursing also is practiced by pigs (*Sus scrofa*) in which there are large litters of altricial young that fight among themselves for teats. Maternal passivity likely was present in the common ancestors of pigs and peccaries. In peccaries it was retained, despite reduction of litter size to two precocial young, apparently in response to selection for communal nursing. Thus, it seems that the evolutionary pathway by which peccaries achieved communal nursing precluded modification of developmental schedules to produce a delay in the emergence of agonistic behavior among young.

Gorillas

Patterns of emigration and group transfer in primates suggest that ontogenetic factors may influence individual dispersal and also may play a role in observed sex differences in movement away from the natal group (Harcourt, 1978). Harcourt and Stewart (1981) were able to test how differences in ontogeny might be related to variations in reproductive tactics in male gorillas (*Gorilla gorilla*). Young male gorillas either leave their natal group to find mates or remain to inherit leadership and mating rights within it [usually not involving inbreeding (Harcourt *et al.*, 1981)]. Harcourt and Stewart found that a close social relationship with the leading male during infancy and adolescence could predispose a young male to stay in his natal group. They concluded that future studies of intraspecific variation in reproductive tactics will be incomplete without accompanying investigation of associated variation in the course of development.

GENERIC APPROACHES TO FUTURE RESEARCH: MODELS, JARGON, AND THEORY

> Occasionally, the use of a term spreads through the scientific literature with amazing swiftness but without a valid birth certificate, that is, without having

> been defined in some precise manner. Actually, the swifter the spreading, the greater is everyone's confidence that the meaning of the term is perfectly clear and well understood by all. (Georgescu-Roegen 1971, p. 211)
>
> Especially in ethology, it is difficult to avoid the unprofitable extremes of blinding skepticism and crippling romanticism. (Gould 1975, p. 692)

The anticipated statement that more data are needed applies here: they are! Multidisciplinary approaches to problems in behavioral development are necessary, and they have proven again and again to be very useful. Furthermore, animals are biological systems, behavioral characteristics comprise phenotypes, and general principles of organic evolution apply to the study of behavior. For the purpose of this discussion, the word "generic" is used simply to stress the points that in the field of behavioral development, among others, raw, empirical data are needed that inform us about the things that animals do, and that models and jargon, despite their utility and appeal, must be used cautiously lest we conclude erroneously that we know more than we actually do.

Models attempting to explain various aspects of the evolution of social behavior have become very popular. Only empirical research will determine how robust are models based on restrictive assumptions. As Fleming (1979) and J. Altmann (1980) point out, many models dealing with the evolution of social behavior are not directly applicable to a large number of species that are studied intensively, including primates and large mammals. Highly restrictive assumptions about litter size, the lack of overlap between sibships, mating systems (semelparity, asexual reproduction), and the way in which genes with specific personalities operate need to be taken into account when assessing the applicability of a given model to one's own research.

We also need to be more rigorous in our choice of descriptive terms (Estep and Bruce, 1981; Leach, 1981; Gowaty, 1982). If catchy jargon is used, it should serve to attract attention to theories with strong supportive evidence, or at least testable ideas, and not to unfalsifiable claims that tend to demean the field of social biology.

There also needs to be a rapprochement between theorizing about the way things should be and collecting data that will inform us about the way things are. However, one cannot completely discount the utility of seemingly outrageous ideas (Dunbar, 1980); what seems to be a ludicrous suggestion today may become the standard explanation of tomorrow. We should be concerned with coming to an understanding of why an individual behaves the way it does, and not make value judgments about what an animal should do (Emlen, 1980).

We have stressed that in most instances nature and nurture interact to produce behavioral phenotypes. However, internal constraints imposed by nervous systems and/or social environments are very important to consider in our attempts both to understand ontogenetic processes and to generate comprehensive explanations for common patterns of development and behavior in diverse species.

There obviously are holes in the available data base for the topics considered herein and for those important areas that were ignored. Also, there is no denying the absolute importance of furthering our understanding of development; the absence of ontogenetic information invariably means that research on a given topic is incomplete. Those of us interested in the development of behavior should consider ourselves lucky to be involved in one of the most exciting research areas in the behavioral sciences.

Acknowledgments

Various aspects of our research have generously been supported by the National Science Foundation and the Public Health Service. M. B.'s research program also was supported by a grant from the Harry Frank Guggenheim Foundation, a Faculty Fellowship from the University of Colorado, and a fellowship from the John Simon Guggenheim Memorial Foundation. J. B.'s work on peccaries also was supported by Sigma Xi and University of Colorado Graduate Grants. Joel Berger, Andrew Blausteain, Gordon Burghardt, Douglas Conner, Thomas Daniels, Les Greenberg, Gail Michener, and Michael Wells provided useful comments on this chapter, even if we did not always incorporate their suggestions into the final piece. Jeanie Cavanagh miraculously organized and typed, in record time, a multicolored, follow-the-arrow, surgically reconstructed draft, and her and Mary Marcotte's patience are deeply appreciated and warmly acknowledged. Portions of Section 5 appeared previously (M. Bekoff, 1981*a*), and are reprinted here with permission of the publisher. Portions of Section 6 were also originally published elsewhere (M. Bekoff, 1984) and are reprinted here with permission of the publisher.

REFERENCES

Ainley, D. G., 1974, The comfort behaviour of Adelie and other penguins, *Behaviour* **50:**16–51.

Alexander, R. D., 1974, The evolution of social behavior, *Annu. Rev. Ecol. Syst.* **4**:325–383.
Alexander, R. D., and Tinkle, D. W., eds., 1981, *Natural Selection and Social Behavior: Recent Research and New Theory,* Chiron, New York.
Altmann, D., and Scheel, H. G., 1980, Beginn des Sozialverhaltens und erstes Lernen beim Milu, *Elaphurus davidianus, Milu (Berl.)* **5**:146–156.
Altmann, J., 1980, *Baboon Mothers and Infants,* Harvard University Press, Cambridge.
Altmann, J., Altmann, S., and Hausfater, G., 1981, Physical maturation and age estimates of yellow baboons, *Papio cynocephalus*, in Amboseli National Park, Kenya, *Am. J. Primatol.* **1**:389–399.
Andelt, W. F., Althoff, D. P., and Gipson, P. S., 1979, Movements of breeding coyotes with emphasis on den site relationships, *J. Mammal.* **60**:568–575.
Anderson, D. C., Armitage, K. B., and Hoffmann, R. S., 1976, Socioecology of marmots: Female reproductive strategies, *Ecology* **57**:552–560.
Armitage, K. B., 1981, Sociality as a life-history tactic of ground squirrels, *Oecologia* **48**:36–49.
Arnold, S. J., 1981*a*, Behavioral variation in natural populations. I. Phenotypic, genetic, and environmental correlations between chemoreceptive responses to prey in the garter snake, *Thamnophis elegans, Evolution* **35**:489–509.
Arnold, S. J., 1981*b*, Behavioral variation in natural populations. II. The inheritance of a feeding response in crosses between geographic races of the garter snake, *Thamnophis elegans, Evolution* **35**:510–515.
Baerends, G. P., 1975, An evaluation of the conflict hypothesis as an explanatory principle for the evolution of displays, in: *Function and Evolution in Behavior: Essays in Honour of Professor Niko Tinbergen* (G. Baerends, C. Beer, and A. Manning, eds.), pp. 187–227, Oxford University Press, New York.
Baerends, G. P., 1976, The functional organization of behaviour, *Anim. Behav.* **24**,726–733.
Baerends-van Roon, J. M., and Baerends, G. P., 1979, *The Morphogenesis of the Behaviour of the Domestic Cat*, North-Holland, Amsterdam.
Baldwin, J. D., and Baldwin, J. I., 1973, The role of play in social organization: Comparative observations on squirrel monkeys (*Saimiri*), *Primates* **14**:369–381.
Baldwin, J. D., and Baldwin, J. F., 1976, Effects of food ecology on social play: A laboratory simulation, *Z. Tierpsychol.* **40**:1–14.
Barash, D. P., 1974, The evolution of marmot societies: A general theory, *Science* **185**:415–420.
Barash, D. P., 1976, Some evolutionary aspects of parental behavior in animals and man, *Am. J. Psychol.* **89**:195–217.
Barlow, G. W., 1977, Modal action patterns, in: *How Animals Communicate* (T. A. Sebeok, ed.), pp. 98–134, University of Indiana Press, Bloomington.
Barnett, S. A., 1958, An analysis of social behaviour in wild rats, *Proc. Zool. Soc. Lond.* **130**:107–152.
Bateson, P. P. G., 1976, Rules and reciprocity in behavioral development, in: *Growing Points in Ethology* (P. P. G. Bateson, and R. A. Hinde, eds.), pp. 401–421, Cambridge University Press, New York.
Bateson, P. P. G., 1978, How does behavior develop?, *Perspect. Ethol.* **3**:55–66.
Bateson, P., 1981, Discontinuities in development and changes in the organization of play in cats, in: *Behavioral Development: The Bielefeld Interdisciplinary Conference* (K. Immelmann, G. W. Barlow, L. Petrinovich, and M. Main, eds.), pp. 281–295, Cambridge University Press, New York.
Bateson, P., and Young, M., 1981, Separation from the mother and the development of play in young cats, *Anim. Behav.* **29**:173–180.

Bateson, P., Martin, P., and Young, M., 1981, Effects of interrupting cat mothers' lactation with bromocriptine on the subsequent play of their kittens, *Physiol. Behav.* **27**:841–845.

Beatty, W. W., Dodge, A. M., Traylor, K. L., and Meaney, M. J., 1981, Temporal boundary of the sensitive period for hormonal organization of social play in juvenile rats, *Physiol. Behav.* **26**:241–243.

Beatty, W. W., Dodge, A. M., Dodge, L. J., White, K., and Panksepp, J., 1982, Psychomotor stimulants, social deprivation and play in juvenile rats, *Pharmacol. Biochem. Behav.* **16**:417–422.

Beck, B. B., 1980, *Animal Tool Behavior,* Garland, New York.

Bekoff, A., 1976, Ontogeny of leg motor output in the chick embryo: A neural analysis, *Brain Res.* **106**:271–291.

Bekoff, A., 1978, A neuroethological approach to the study of the ontogeny of coordinated behavior, in: *The Development of Behavior: Comparative and Evolutionary Aspects* (G. Burghardt and M. Bekoff, eds.), pp. 19–41, Garland Press, New York.

Bekoff, A., 1981, Embryonic development of the neural circuitry underlying motor coordination, in: *Studies in Developmental Neurobiology: Essays in Honor of Viktor Hamburger* (W. M. Cowan, ed.), pp. 134–170, Oxford University Press, New York.

Bekoff, A., 1985, Development of locomotion in vertebrates: A comparative perspective, in: *The Comparative Development of Adaptive Skills: Evolutionary Implications* (E. Gollin, ed.), pp. 57–94, Erlbaum, Hillsdale, New Jersey.

Bekoff, M., 1976*a*, Animal play: Problems and perspectives, in: *Perspectives in Ethology* (P. Bateson and P. H. Klopfer, eds.), pp. 165–188, Plenum Press, New York.

Bekoff, M., 1976*b*, The social deprivation paradigm: Who's being deprived of what, *Dev. Psychobiol.* **9**:499–500.

Bekoff, M., 1977*a*, Socialization in mammals with an emphasis on nonprimates, in: *Primate Biosocial Development* (S. Chevalier-Skolnikoff and F. E. Poirier, eds.), pp. 603–636, Garland Press, New York.

Bekoff, M., 1977*b*, Quantitative studies of three areas of classical ethology: Social dominance, behavioral taxonomy, and behavioral variability, in: *Quantitative Methods in the Study of Animal Behavior* (B. A. Hazlett, ed.), pp. 1–46, Academic Press, New York.

Bekoff, M., 1977*c*, Mammalian dispersal and the ontogeny of individual behavioral phenotypes, *Am. Nat.* **111**:715–732.

Bekoff, M., 1977*d*, Social communication in canids: Evidence for the evolution of a stereotyped mammalian display, *Science* **197**:1097–1099.

Bekoff, M., 1978*a*, A field study of the development of behavior in Adelie penguins: Univariate and numerical taxonomic approaches, in: *The Development of Behavior: Comparative and Evolutionary Aspects* (G. Burghardt and M. Bekoff, eds.), pp. 177–202, Garland Press, New York.

Bekoff, M., 1978*b*, Social play: Structure, function, and the evolution of a cooperative social behavior, in: *The Development of Behavior: Comparative and Evolutionary Aspects* (G. Burghardt and M. Bekoff, eds.), pp. 367–383, Garland Press, New York.

Bekoff, M., 1978*c*, Behavioral development in coyotes and eastern coyotes, in: *Coyotes: Biology, Behavior, and Management* (M. Bekoff, ed.), pp. 97–126, Academic Press, New York.

Bekoff, M., 1981*a*, Development of agonistic behavior: Ethological and ecological aspects, in: *Multidisciplinary Approaches to Aggression Research* (P. F. Brain and D. Benton, eds.), pp. 161–178, Elsevier, Amsterdam.

Bekoff, M., 1981*b*, Mammalian sibling interactions: Genes, facilitative environments, and the coefficient of familiarity, in: *Parental Care in Mammals* (D. J. Gubernick and P. H. Klopfer, eds.), pp. 307–346, Plenum Press, New York.

Bekoff, M., 1984, Social play behavior, *BioScience* **34:**228–233.

Bekoff, M., 1985, Evolutionary perspectives of behavioral development, *Z. Tierpsychol.*, **69:**166–167.

Bekoff, M., and Byers, J. A., 1981, A critical reanalysis of the ontogeny and phylogeny of mammalian social and locomotor play: An ethological hornet's nest, in: *Behavioral Development: The Bielefeld Interdisciplinary Conference* (K. Immelmann, G. W. Barlow, L. Petrinovich, and M. Main, eds.), pp. 296–337, Cambridge University Press, New York.

Bekoff, M., and Fox, M. W., 1972, Postnatal neural ontogeny: Environment-dependent and/or environment expectant?, *Dev. Psychobiol.* **5:**323–341.

Bekoff, M., and Wells, M. C., 1980, The social ecology of coyotes, *Sci. Am.* **242:**112–120.

Bekoff, M., and Wells, M. C., 1981, Behavioural budgeting by wild coyotes: The influence of food resources and social organization, *Anim. Behav.* **29:**794–801.

Bekoff, M., and Wells, M. C., 1982, Behavioral ecology of coyotes: Social organization, rearing patterns, space use, and resource defense, *Z. Tierpsychol.* **60:**281–305.

Bekoff, M., and Wells, M., 1986, Social behavior and ecology of coyotes, *Adv. Study Behav.*, **16:**251–338.

Bekoff, M., Ainley, D. G., and Bekoff, A., 1979, The ontogeny and organization of comfort behavior in Adelie penguins, *Wilson Bull.* **91:**255–270.

Bekoff, M., Byers, J. A., and Bekoff, A., 1980, Prenatal motility and postnatal play: Functional continuity?, *Dev. Psychobiol.* **13:**225–228.

Bekoff, M., Diamond, J., and Mitton, J. B., 1981*a*, Life-history patterns and sociality in canids: Body size, reproduction, and behavior, *Oecologia* **50:**386–390.

Bekoff, M., Tyrrell, M., Lipetz, V., and Jamieson, R., 1981*b*, Fighting patterns in young coyotes: Initiation, escalation, and assessment, *Aggressive Behav.* **7:**225–244.

Bekoff, M., Daniels, T. J., and Gittleman, J. L., 1984, Life history patterns and the comparative social ecology of carnivores, *Annu. Rev. Ecol. Syst.* **15:**191–232.

Bengtsson, H., and Ryden, O., 1981, Development of parent–young interaction in asynchronously hatched broods of altricial birds, *Z. Tierpsychol.* **56:**255–272.

Berger, J., 1979, Social ontogeny and behavioural diversity: Consequences for Bighorn sheep *Ovis canadensis* inhabiting desert and mountain environments, *J. Zool. Lond.* **188:**251–266.

Berger, J., 1980, The ecology, structure and functions of social play in bighorn sheep (*Ovis canadensis*), *J. Zool. Lond.* **192:**531–542.

Berger, J., 1981, The role of risks in mammalian combat: Zebra and onager fights, *Z. Tierpsychol.* **56:**297–304.

Berman, C. M., 1982*a*, The ontogeny of social relationships with group companions among free-ranging rhesus monkeys I. Social networks and differentiation, *Anim. Behav.* **30:**149–162.

Berman, C. M., 1982*b*, The ontogeny of social relationships with group companions among free-ranging infant rhesus monkeys II. Differentiation and attractiveness, *Anim. Behav.* **30:**163–170.

Bernstein, I. S., 1967, Defining the natural habitat, in: *First Congress of the International Primatology Society*, pp. 177–179.

Bertram, B. C. R., 1976, Kin selection in lions and in evolution, in: *Growing Points in Ethology* (P. P. G. Bateson and R. A. Hinde, eds.), pp. 281–301, Cambridge University Press, New York.

Biben, M., 1983, Comparative ontogeny of social behavior in three south American canids, the maned wolf, crab-eating fox and bush dog: Implications for sociality, *Anim. Behav.* **31**:814–826.

Blueweiss, B., Fox, H., Kudzma, V., Nakashima, D., Peters, R., and Sams, S., 1978, Relationships between body size and some life history parameters, *Oecologia* **37**:257–272.

Boorman, S. A., and Levitt, P. R., 1980, *The Genetics of Altruism*, Academic Press, New York.

Borchelt, P. L., 1977, Development of dustbathing components in bobwhite and Japanese quail, *Dev. Psychobiol.* **10**:97–103.

Bowen, D., 1978, Prey size and coyote social organization, Ph.D. Thesis, University of British Columbia, Vancouver, B. C., Canada.

Breed, M. D., and Bekoff, M., 1981, Individual recognition and social relationships, *J. Theor. Biol.* **88**:589–593.

Brown, J. L., 1978, Avian communal breeding systems, *Annu. Rev. Ecol. Syst.* **9**:123–155.

Brown, J. L., 1980, Fitness in complex avian social systems, in: *Evolution of Social Behavior: Hypotheses and Empirical Tests* (H. Markl, ed.), pp. 115–128, Verlag Chemie, Weinheim.

Brown, J. L., and R. P. Balda, 1977, The relationship of habitat quality to group size in Hall's babbler (*Pomatostomus halli*), *Condor* **79**:312–320.

Bruner, J., 1974, Nature and uses of immaturity, in: *The Growth of Competence* (K. J. Connolly and J. S. Bruner, eds.), pp. 11–48, Academic Press, New York.

Burgess, J. W., and Shaw, E., 1979, Development and ecology of fish schooling, *Oceanus* **22**:11–17.

Burghardt, G. M., and Bekoff, M., eds., 1978, *The Development of Behavior: Comparative and Evolutionary Aspects,* Garland Press, New York.

Byers, J. A., 1977, Terrain preferences in the play of Siberian ibex kids (*Capra ibex sibirica*), *Z. Tierpsychol.* **45**:199–209.

Byers, J. A., 1980, Play partner preferences in Siberian ibex, *Capra ibex sibirica, Z. Tierpsychol.* **53**:23–30.

Byers, J. A., 1981, The significance of play, *Science* **212**:1493–1494.

Byers, J. A., 1983, Social interactions of juvenile collared peccaries, *Tayassu tajacu* (Mammalia: Artiodactyla), *J. Zool. Lond.* **201**:81–96.

Byers, J. A., 1984*a*, Behavioral development in pronghorn, submitted for publication.

Byers, J. A., 1984*b*, Natural variation in early experience and behavioral development of pronghorn, submitted for pulbication.

Byers, J. A., 1984*c*, Play in ungulates, in: *Play in Animals and Humans* (P. K. Smith, ed.), pp. 45–65, Blackwell, London.

Byers, J. A., 1985, Why is play confined to endotherms?, submitted for publication.

Byers, J. A., and Bekoff, M., 1981, Social, spacing, and cooperative behavior of the collared peccary, *Tayassu tajacu, J. Mammal.* **62**:767–785.

Cairns, R. B., 1973, Fighting and punishment from a developmental perspective, *Nebraska Symp. Motiv.* **1973**:59–124.

Cairns, R. B., 1979, *Social Development: The Origins and Plasticity of Interchanges,* Freeman, San Francisco.

Camenzind, F. J., 1978, Behavioral ecology of coyotes (*Canis latrans*) on the National Elk refuge, Jackson, Wyoming, Ph.D. Thesis, University of Wyoming, Laramie.

Caro, T. M., 1981, Sex differences in the termination of social play in cats, *Anim. Behav.* **29**:271–279.

Cavalli-Sforza, L. L., and Feldman, M. W., 1981, *Cultural Transmission and Evolution: A Quantitative Approach,* Princeton University Press, Princeton, New Jersey.

Chalmers, N. R. 1980*a*, Developmental relationships among social, manipulatory, postural and locomotor behaviours in olive baboons, *Papio anubis, Behaviour* **74:**22–37.

Chalmers, N. R., 1980*b*, The ontogeny of play in feral olive baboons (*Papio anubis*), *Anim. Behav.* **28:**570–585.

Chalmers, N. R., and Locke-Hayden, J., 1981, Temporal patterns of play bouts in captive common marmosets (*Callithrix jacchus*), *Anim. Behav.* **29:**1229–1238.

Chalmers, N. R., and Locke-Haydon, J., 1984, Correlations among measures of playfulness and skillfullness in captive common marmosets (*Callithrix acchus jacchus*), *Dev. Psychobiol.* **17:**191–208.

Charlesworth, B., 1980, *Evolution in Aged-Structured Populations,* Cambridge University Press, Cambridge.

Chauvin, R., and Muckensturm-Chauvin, B., 1980, *Behavioral Complexities,* New International Universities Press, New York.

Chevalier-Skolnikoff, S., 1974, The ontogeny of communication in the stumptailed macaque, *Macaca speciosa, Contrib. Primatol.* **2:**174.

Cheverud, J. M., Rutledge, J. J., and Atchley, W. R., 1983, Quantitative genetics of development: Genetic correlations among age-specific trait values and the evolution of ontogeny, *Evolution* **37:**895–905.

Clark, A. B., 1978, Sex ratio and local resource competition in a prosimian primate, *Science* **201:**163–165.

Clark, A. B., and Wilson, P. S., 1981, Avian hatching adaptations: Hatching asynchrony, brood reduction, and nest failure, *Q. Rev. Biol.* **56:**253–277.

Clark, A. B., and Wilson, D. S., 1985, The onset of incubation in birds, *Amer. Nat.* **125:**603–611.

Clutton-Brock, T. H., and Albon, S. D., 1979, The roaring of red deer and the evolution of honest advertisement, *Behaviour* **69:**144–169.

Cole, K. S., and Noakes, D. L. G., 1980, Development of early social behaviour of rainbow trout, *Salmo gairdneri* (Pisces, Salmonidae), *Behav. Proc.* **5:**97–112.

Cole, L. C., 1954, The population consequences of life history phenomena, *Q. Rev. Biol.* **29:**103–137.

Collias, N. E., 1943, Statistical analysis of factors which make for success in initial encounters between hens, *Am. Nat.* **77:**519–538.

Coss, R. G., 1978, Development of face aversion by the jewel fish (*Hemichromis bimaculatus*, Gill 1862), *Z. Tierpsychol.* **48:**28–46.

Coss, R. G., 1979, Delayed plasticity of an instinct: Recognition and avoidance of 2 facing eyes by the jewel fish, *Dev. Psychobiol.* **12:**335–345.

DeGhett, V., 1977, The social deprivation paradigm: A reply to Bekoff, *Dev. Psychobiol.* **10:**187–189.

Delcomyn, F., 1980, Neural basis of rhythmic behavior in animals, *Science* **210:**492–498.

Douglas-Hamilton, I., and Douglas-Hamilton, O., 1975, *Among the Elephants*, Viking, New York.

Downhower, J. F., and Armitage, K. B., 1981, Dispersal of yearling yellow-bellied marmots (*Marmota flaviventris*), *Anim. Behav.* **29:**1064–1069.

Dubost, G., and Feer, F., 1981, The behavior of the male *Antilope cervicapra* L., its development according to age and social rank, *Behaviour* **76:**62–127.

Dunbar, M. J., 1980, The blunting of Occam's razor, or to hell with parsimony, *Can. J. Zool.* **58:**123–128.

Edgerton, V. R., 1978, Mammalian muscle fiber types and their adaptability, *Am. Zool.* **18:**113–125.

Edgerton, V. R., Grillner, S., Sjostrom, A., and Zangger, P., 1976, in: *Neural Control of*

Locomotion (R. M. Herman, S. Grillner, P. S. G. Stein, and D. G. Stuart, eds.), Plenum Press, New York.

Eibl-Eibesfeldt, I., 1979, Human ethology—Concepts and implications for the study of man, *Behav. Brain Sci.* **2:**1–57.

Eisenberg, J. F., 1981, *The Mammalian Radiations: An Analysis of Trends in Evolution, Adaptation, and Behavior,* University of Chicago Press, Chicago.

Emlen, S. T., 1978, The evolution of cooperative breeding in birds, in: *Behavioural Ecology: An Evolutionary Approach* (J. R. Krebs and N. B. Davies, eds.), pp. 245–281, Sinauer, Sunderland, Massachusetts.

Emlen, S. T., 1980, Ecological determinism and sociobiology, in: *Sociobiology: Beyond Nature/Nurture* (G. W. Barlow and J. Silverberg, eds.), pp. 125–150, Westview Press, Boulder, Colorado.

Emlen, S. T., 1982*a*, The evolution of helping. I. An ecological constraints model, *Am. Nat.* **119:**29–39.

Emlen, S. T., 1982*b*, The evolution of helping. II. The role of behavioral conflict, *Am. Nat.* **119:**40–53.

Estep, D. Q., and Bruce, K. E., 1981, The concept of rape in non-humans: A critique, *Anim. Behav.* **29:**1272–1273.

Etienne, A. S., Emmanuelli, E., and Zinder, M., 1982, Ontogeny of hoarding in the golden hamster: The development of motor patterns and their sequential coordination, *Dev. Psychobiol.* **15:**33–45.

Ewer, R. F., 1968, *Ethology of Mammals*, Plenum Press, New York.

Fabri, K. E., 1983, Angeborenes und erworbenes im Spiel der Tiere (Zur Allgemeinen Theorie der Ontogenese von Verhalten und Psyche bei Tieren), *Z. Psychol.* **191:**50–64.

Fagen, R. M., 1977, Selection for optimal-age-dependent schedules of play behavior, *Am. Nat.* **111:**395–414.

Fagen, R. M., 1978, Evolutionary biological models of animal behavior, in: *The Development of Behavior: Comparative and Evolutionary Aspects* (G. Burghardt and M. Bekoff, eds.), pp. 385–404, Garland Press, New York.

Fagen, R., 1981, *Animal Play Behavior,* Oxford University Press, Oxford.

Farr, J. A., 1980, The effects of juvenile social interaction on growth rate, size and age at maturity, and adult social behavior in *Girardinus matellicus* Poey (Pisces: Poeciliidae), *Z. Tierpsychol.* **52:**247–268.

Farr, L., and Andrews, R. V., 1978*a*, Rank-associated differences in metabolic rates and locomotor activity of dominant and subordinate *Peromyscus maniculatus, Comp. Biochem. Physiol.* **61A:**401–406.

Farr, L., and Andrews, R. V., 1978*b*, Rank-associated desynchronization and activity rhythms of *Peromyscus maniculatus* in response to social pressure, *Comp. Biochem. Physiol.* **61A:**539–542.

Fentress, J. C., 1978, *Mus musicus:* The developmental orchestration of selected movement patterns in mice, in: *The Development of Behavior: Comparative and Evolutionary Aspects* (G. M. Burghardt and M. Bekoff, eds.), pp. 321–342, Garland Press, New York.

Fentress, J. C., 1981, Order in ontogeny: Relational dynamics, in: *Behavioral Development: The Bielefeld Interdisciplinary Conference* (K. Immelmann, G. W. Barlow, L. Petrinovich, and M. Main, eds.), pp. 338–371, Cambridge University Press, Cambridge.

Fernald, R. D., and Hirata, N. R., 1979, The ontogeny of social behavior and body coloration in the African cichlid fish *Haplochromus burtoni, Z. Tierpsychol. 50:*180–187.

Ferron, J., 1981, Comparative ontogeny of behaviour in four species of squirrels (Sciuridae), *Z. Tierpsychol.* **55:**193–216.

Ferron, J., and Lefebvre, L., 1982, Comparative organization of grooming sequences in adult and young Sciurid rodents, *Behaviour* **81:**110–127.

Festa-Bianchet, M., and King, W. J., 1984, Behavior and dispersal of yearling Columbian ground squirrels, *Can. J. Zool.* **62:**161–167.

Fleming, T. H., 1979, Life-history strategies, in: *Ecology of Small Mammals* (D. M. Stoddart, ed.), pp. 1–61, Chapman and Hall, London.

Fossey, D., 1979, Development of the mountain gorilla (*Gorilla gorilla beringei*): The first thirty-six months, in: *The Great Apes* (D. A. Hamburg and E. R. McCown, eds.), pp. 139–184, Benjamin/Cummings, Menlo Park, California.

Fox, S. F., Rose, E., and Myers, R., 1981, Dominance and the acquisition of superior home ranges in the lizard *Uta stansburiana, Ecology* **62:**888–893.

Fraser, D., Thompson, B. K., Ferguson, D. K., and Darroch, R. L., 1979, The 'teat order' of suckling pigs. 3. Relation to competition within litters, *J. Agric. Sci. Camb.* **92:**257–261.

Gadgil, M., 1982, Changes with age in the strategy of social behavior, *Perspect. Ethol.* **5:**489–501.

Gaines, M. S., and McClenaghan, L. R., 1980, Dispersal in small mammals, *Annu. Rev. Ecol. Syst.* **11:**163–196.

Galef, B. G., 1981, The ecology of weaning: Parasitism and the achievement of independence by altricial mammals, in: *Parental Care in Mammals* (D. J. Gubernick and P. H. Klopfer, eds.), pp. 211–241, Plenum Press, New York.

Geist, V., 1978, *Life Strategies, Human Evolution, Environmental Design: Toward a Biological Theory of Health,* Springer, Berlin.

Georgescu-Roegen, N., 1971, *The Entropy Law and the Economic Process,* Harvard University Press, Cambridge.

Golightly, R., 1981, The comparative energetics of two desert canids: The coyotes (*Canis latrans*) and the kit fox (*Vulpes macrotis*), Ph.D. Thesis, Arizona State University, Tempe, Arizona.

Gould, J., 1975, Honey bees recruitment: The dance-language controversy, *Science* **189:**685–693.

Gould, S. J., 1977, *Ontogeny and Phylogeny,* Harvard University Press, Cambridge, Massachusetts.

Gould, S. J., and Lewontin, R. C., 1979, The spandrels of San Marco and the Panglossian paradigm: A critique of the adaptionist programme, *Proc. Roy. Soc. Lond. B* **205:**581–598.

Gowaty, P. A., 1982, Sexual terms in sociobiology: Emotionally evocative and, paradoxically, jargon, *Anim. Behav.* **30:**630–631.

Greenwood, P. J., 1980, Mating systems, philopatry and dispersal in birds and mammals, *Anim. Behav.* **28:**1140–1162.

Grillner, S., 1975, Locomotion in vertebrates—Control mechanisms and reflex interaction, *Physiol. Rev.* **55:**247–304.

Gubernick, D. J., and Klopfer, P. H., eds., 1981, *Parental Care in Mammals,* Plenum Press, New York.

Guhl, A. M., 1958, The development of social organization in the domestic chick, *Anim. Behav.* **6:**92–111.

Hadidian, J., 1980, Yawning in an old world monkey, *Macaca nigra* (Primates: Cercopithecidae), *Behaviour* **75:**133–147.

Hailman, J. P., 1967, The ontogeny of an instinct, *Behaviour* (Suppl. 15).

Hamburger, V., 1963, Some aspects of the embryology of behavior, *Q. Rev. Biol.* **38:**342–365.

Hamburger, V. H., 1980, Embryology and the modern synthesis in evolutionary theory, in: *The Evolutionary Synthesis: Perspectives on the Unification of Biology* (E. Mayr and W. B. Provine, eds.), pp. 97–112, Harvard University Press, Cambridge.

Hamilton, W. D., 1964, The genetical evolution of social behaviour, *J. Theor. Biol.* **7**:1–52.

Happold, M., 1976, The ontogeny of social behaviour in four conilurine rodents (Muridae) of Australia, *Z. Tierpsychol.* **40**:265–278.

Harcourt, A. H., 1978, Strategies of emigration and transfer by primates, with particular reference to gorillas, *Z. Tierpsychol.* **48**:401–420.

Harcourt, A. H., and Stewart, K. J., 1981, Gorilla male relationships: Can differences during immaturity lead to contrasting reproductive tactics in adulthood? *Anim. Behav.* **29**:206–210.

Harcourt, A. H., Fossey, D., and Sabater-Pi, J., 1981, Demography of *Gorilla gorilla, J. Zool. Lond.* **195**:215–233.

Harrington, F. H., Mech, L. D., and Fritts, S. H., 1983, Pack size and wolf pup survival: Their relationship under varying ecological conditions, *Behav. Ecol. Sociobiol.* **13**:19–26.

Hartsock, T. G., and Graves, H. B., 1976, Neonatal behavior and nutrition-related mortality in domestic swine, *J. Anim. Sci.* **42**:235–241.

Hartsock, T. G., Graves, H. B., and Baumgardt, B. R., 1977, Agonistic behavior and the nursing order in suckling piglets: Relationships with survival, growth and body composition, *J. Anim. Sci.* **44**:320–330.

Hausfater, G., and Gilmore, H. S., 1979, Age-related changes in the neutral tail carriage of anubis baboons (*Papio anubis*), *Mammalia* **43**:465–472.

Henderson, N., 1981, A fit mouse is a hoppy mouse: Jumping behavior in 15-day-old *Mus musculus, Dev. Psychobiol.* **14**:459–472.

Henry, J. D., 1985, The little foxes, *Nat. Hist.* **94**:46–57.

Hinde, R. A., 1959, Some recent trends in ethology, in: *Psychology: A Study of a Science* (S. Koch, ed.), pp. 561–610, McGraw-Hill, New York.

Hinde, R. A., 1975, The concept of function, in: *Function and Evolution in Behaviour* (G. Baerends, C. G. Beer, and A. Manning, eds.), pp. 3–15, Oxford University Press, New York.

Hinde, R. A., 1982, *Ethology: Its Nature and Relations with Other Sciences,* Oxford, New York.

Hinde, R. A., ed., 1983, *Primate Social Relationships: An Integrated Approach,* Sinauer, Sunderland, Massachusetts.

Hofer, M. A., 1978, Hidden regulatory processes in early social relationships, *Perspect. Ethol.* **3**:135–166.

Holloszy, J. O., 1967, Biochemical adaptations in muscle, *J. Biol. Chem.* **242**:2278–2282.

Horn, H. S., 1978, Optimal tactics of reproduction and life-history, in: *Behavioural Ecology: An Evolutionary Approach* (J. R. Krebs and N. B. Davies, eds.), pp. 411–429, Sinauer, Sunderland, Massachusetts.

Horwich, R. H., 1972, The ontogeny of social behavior in the gray squirrel (*Sciurus carolinensis*), *Z. Tierpsychol.* (Suppl. 8).

Ikeda, H., 1983, Development of young and parental care of the Raccoon Dog *Nyctereutes procyonoides viverrinus* Temmick, in captivity, *J. Mammal. Japan* **9**:229–236.

Immelmann, K., 1972, Sexual and other long-term aspects of imprinting in birds and other species, *Adv. Study Behav.* **45**:147–174.

Immelmann, K., 1975*a*, The evolutionary significance of early experience, in: *Function and Evolution in Behaviour* (G. Baerends, C. Beer, and A. Manning, eds.), pp. 243–253, Oxford University Press, New York.

Immelmann, K., 1975*b*, Ecological significance of imprinting and early learning, *Annu. Rev. Ecol. Syst.* **6**:15–37.

Immelmann, K., Barlow, G. W., Petrinovich, L., and Main, M., eds., 1981, *Behavioral*

Development: The Bielefeld Interdisciplinary Project, Cambridge University Press, New York.

Jacobs, J., 1981, How heritable is innate behaviour?, *Z. Tierpsychol.* **55:**1–18.

Jacoby, C. A., 1980, Ontogeny of behavior in the dungeness crab, *Cancer magister* Dane 1952, Ph.D. Thesis, Stanford University, Palo Alto, California.

Jacoby, C. A., 1983, Ontogeny of behavior in the crab instars of the Dungeness Crab, *Cancer magister* Dana 1852, *Z. Tierpsychol.* **63:**1–16.

Jarvis, J. U. M., 1981, Eusociality in a mammal: Cooperative breeding in naked mole-rat colonies, *Science* **212:**571–573.

Jerison, H., 1973, *Evolution of the Brain and Intelligence,* Academic Press, New York.

Johns, W., and Armitage, K. B., 1979, Behavioral ecology of alpine yellow-bellied marmots, *Behav. Ecol. Sociobiol.* **5:**133–157.

Johnston, T. D., 1982, Selective costs and benefits in the evolution of learning, *Adv. Study Behav.* **12:**65–106.

Kikkawa, J., 1980, Winter survival in relation to dominance classes among silvereyes *Zosterops lateralis chlorocephala* of Heron Island, Great Barrier Reef, *Ibis* **122:**437–446.

King, J. A., 1973, The ecology of aggressive behavior, *Annu. Rev. Ecol. Syst.* **4:**117–138.

Kleiman, D. G., 1977, Monogamy in mammals, *Q. Rev. Biol.* **52:**39–69.

Klopfer, P. H., 1970, Sensory physiology and esthetics, *Am. Sci.* **58:**399–403.

Klopfer, P. H., 1981, The naked ape reclothed, *Am. J. Primatol.* **1:**301–305.

Knight, S. W., 1978, Dominance hierarchies of coyote litters, M. S. Thesis, Utah State University, Logan, Utah.

Kruijt, J. P., 1964, Ontogeny of social behaviour in Burmese red junglefowl, *Behaviour* (Suppl. 12).

Latour, P. B., 1981*a*, Interactions between free-ranging, adult male polar bears (*Ursus maritimus* Phipps): A case of adult social play, *Can. J. Zool.* **59:**1775–1783.

Latour, P. B., 1981*b*, Spatial relationships and behavior of polar bears (*Ursus maritimus* Phipps) concentrated on land during the ice-free season of Hudson Bay, *Can. J. Zool.* **59:**1763–1774.

Leach, E., 1981, Biology and social science; Wedding or rape?, *Nature* **291:**267–268.

Lehrman, D. S., 1953, A critique of Konrad Lorenz's theory of instinctive behavior, *Q. Rev. Biol.* **28:**337–363.

Ligon, J. D., 1981, Demographic patterns and communal breeding in the green woodhoopoe, *Phoeniculus purpureus,* in: *Natural Selection and Social Behavior: Recent Research and New Theory* (R. D. Alexander and D. W. Tinkle, eds.), pp. 231–243, Chiron Press, New York.

Lorenz, K. Z., 1981, *The Foundations of Ethology*, Springer, New York.

Lott, D. F., 1984, Intraspecific variation in the social systems of wild vertebrates, *Behaviour* **88:**266–325.

Mace, G. M., 1979, The evolutionary ecology of small mammals, Ph.D. Thesis, University of Sussex, Sussex, England.

MacArthur, R. H., and Wilson, E. O., 1967, *The Theory of Island Biogeography*, Princeton University Press, Princeton.

Macdonald, D. W., and Moelhman, P., 1982, Cooperation, altruism, and restraint in the reproduction of carnivores, *Perspect. Ethol.* **5:**433–467.

Magnuson, J. J., 1962, An analysis of aggressive behavior, growth, and competition for food and space in medaka (*Oryzias latipes* (Pisces, Cyprinodontidae)), *Can. J. Zool.* **40:**313–363.

Malcolm, J., and Marten, K., 1982, Natural selection and the communal rearing of pups in African wild dogs (*Lycaon pictus*), *Behav. Ecol. Sociobiol.* **10:**1–13.

Markstein, P. L., and Lehner, P. N., 1980, A comparison of predatory behavior between prey-naive and prey-experienced adult coyotes (*Canis latrans*), *Bull. Psychonom. Soc.* **15**:271–274.

Marler, P., 1981, Birdsong: The acquisition of a learned motor skill, *Trends NeuroSci.* **4**:88–94.

Marler, P., Dooling, R. J., and Zoloth, S., 1980, Comparative perspectives on ethology and behavioral development, in: *Comparative Methods in Psychology* (M. H. Bornstein, ed.), pp. 189–230, Erlbaum, Hillsdale, New Jersey.

Martin, P., 1982, The energy cost of play: Definition and estimation, *Anim. Behav.* **30**:294–295.

Martin, P., 1984, The time and energy costs of play behaviour in the cat, *Z. Tierpsychol.* **64**:298–312.

Martin, P., and Caro, T. M., 1985, On the functions of play and its role in behavioral development, *Adv. Study Behav.* **15**:59–103.

Mason, W., 1979, Ontogeny of social behavior, in: *Handbook of Behavioral Neurobiology*. Vol. 3, *Social Behavior and Communication* (P. Marler and J. G. Vandenbergh, eds.), pp. 1–28, Plenum, New York.

Maynard Smith, J., 1980, A new theory of sexual investment, *Behav. Ecol. Sociobiol.* **7**:247–251.

Maynard Smith, J., and Price, G. R., 1973, The logic of animal conflict, *Nature* **246**:15–18.

Mayr, E., 1974, Behavior programs and evolutionary strategies, *Am. Sci. 62*:650–659.

McClure, P. A., 1981, Sex-biased litter reduction in food-restricted wood rats (*Neotoma floridana*), *Science* **211**:1058–1060.

McKinney, F., 1965, The comfort movements of Anatidae, *Behaviour* **25**:120–220.

Meaney, M. J., Didge, A. M., and Beatty, W. W., 1981, Sex-dependent effects of amygdaloid lesions on the social play of prepubertal rats, *Physiol. Behav.* **26**:467–472.

Meaney, M. J., Stewart, J., and Beatty, W. W., 1985, Sex differences in social play: The socialization of sex roles, *Adv. Study Behav.* **15**:1–58.

Meikle, D. B., and Vessey, S. H., 1981, Nepotism among rhesus monkey brothers, *Nature* **294**:160–161.

Michener, G. R., 1979, Spatial relationships and social organization of adult Richardson's ground squirrels, *Can. J. Zool.* **57**:125–139.

Michener, G. R., 1980, Differential reproduction among female Richardson's ground squirrels and its relation to sex ratio, *Behav. Ecol. Sociobiol.* **7**:173–178.

Michener, G. R., 1981, Ontogeny of spatial relationships and social behaviour in juvenile Richardson's ground squirrels, *Can. J. Zool.* **59**:1666–1676.

Michener, G. R., 1983, Kin identification, matriarchies, and the evolution of sociality in ground-dwelling sciurids, in: *Recent Advances in the Study of Mammalian Behavior* (J. F. Eisenberg and D. Kleiman, eds.), pp. 528–572, Special Publication Number 7, American Society of Mammalogists.

Miller, D. B., 1981, Conceptual strategies in behavioral development: Normal development and plasticity, in: *Behavioral Development: The Bielefeld Interdisciplinary Conference* (K. Immelmann, G. W. Barlow, L. Petrinovich, and M. Main, eds.), pp. 58–82, Cambridge University Press, New York.

Miller, L. L., Whimsett, J. M., Vandenbergh, J. G., and Colby, D. R., 1977, Physical and behavioral aspects of sexual maturation in male golden hamsters, *J. Comp. Physiol. Psychol.* **91**:245–259.

Mills, M. G. L., 1982, The mating system of the brown hyaena, *Hyaena brunnea* in the southern Kalahari, *Behav. Ecol. Sociobiol.* **10**:131–136.

Mock, D. W., 1981, Brood reduction in herons: The siblicide threshold hypothesis, Paper presented at the Animal Behavior Society Meetings, Knoxville, Tennessee.

Moehlman, P. D., 1979, Jackal helpers and pup survival, *Nature,* **277**:382–383.
Moehlman, P. D., 1981, Reply, *Nature* **289**:825.
Moehlman, P. D., 1983, Socioecology of silverbacked and golden jackals (*Canis mesomelas, C. aureus*), in: *Recent Advances in the Study of Mammalian Behavior* (J. Eisenberg and D. G. Kleiman, eds.), pp. 423–453, Special Publication Number 7, American Society of Mammalogists.
Muller-Schwarze, D., Stagge, B., and Muller-Schwarze, C., 1982, Play behavior: Persistence, decrease, and energetic compensation during food shortage in deer fawns, *Science* **215**:85–87.
Neuchterlein, G. L., 1981, Asynchronous hatching and sibling competition in western grebes, *Can. J. Zool.* **59**:994–998.
Nowicki, S., and Armitage, K. B., 1979, Behavior of juvenile yellow-bellied marmots: Play and social integration, *Z. Tierpsychol.* **51**:85–105.
Oakley, F. B., and Reynolds, P. C., 1976, Differing responses to social play deprivation in two species of macaque, in: *The Anthropological Study of Play: Problems and Prospects* (D. F. Lancy and B. A. Tindall, eds.), pp. 179–188, Leisure Press, Cornwell, New York.
O'Connor, R. J., 1978, Brood reduction in birds: Selection for fratricide, infanticide and suicide?, *Anim. Behav.* **26**:79–96.
Oppenheim, R., 1974, The ontogeny of behavior in the chick embryo, *Adv. Study Behav.* **5**:133–172.
Owens, D. D., and Owens, M. J., 1979, Communal denning and clan associations in brown hyenas (*Hyaena brunnea*, Thunberg) of the central Kalahari Desert, *Afr. J. Ecol.* **17**:35–44.
Owens, D. D., and Owens, M. J., 1984, Helping behaviour in brown hyenas, *Nature* **308**:843–845.
Owens, N. W., 1975, A comparison of aggressive play and aggression in free-living baboons, *Papio anubis, Anim. Behav.* **23**:757–765.
Packard, J., 1980, Deferred reproduction in wolves (*Canis lupus*), Ph.D. Thesis, University of Minnesota, Minneapolis, Minnesota.
Pagés, E., 1982, Jeu et socialisation: Aspect descriptif et theorique de l'ontogenese chez un microcebe prosimien malgache, *J. Psychol.* **3**:241–261.
Pagés, E., 1983, Identification, caracterisation et role du jeu social chez un prosimien nocturne, *Microcebus coquereli, Biol. Behav.* **8**:319–343.
Panksepp, J., 1979, The regulation of play: Neurochemical controls, *Neurosci. Abstr.* **5**:172.
Panksepp, J., 1981, The ontogeny of play in rats, *Dev. Psychobiol.* **14**:327–332.
Parker, G. A., 1974, Assessment strategy and the evolution of fighting behaviour, *J. Theor. Biol.* **47**:223–243.
Parry, G. D., 1981, The meanings of *r*- and *K*-selection, *Oecologia* **48**:260–264.
Patenaude, F., 1984, The ontogeny of behavior of free-living beavers (*Castor canadensis*), *Z. Tierpsychol.* **66**:33–44.
Pellis, S. M., 1981, A description of social play by the Australian magpie *Gymnorhina tibicen* based on Eshkol–Wachman notation, *Bird Behav.* **3**:61–79.
Petrinovich, L., 1981, A method for the study of development, in: *Behavioral Development: The Bielefeld Interdisciplinary Project* (K. Immelmann, G. Barlowe, L. Petrinovich, and M. Main, eds.), pp. 90–130, Cambridge University Press, New York.
Poole, A., 1979, Sibling aggression among nestling ospreys in Florida Bay, *Auk* **96**:415–416.
Poole, T. B., 1966, Aggressive play in polecats, *Symp. Zool. Soc. Lond.* **18**:23–44.
Popp, J. L., and DeVore, I., 1979, Aggressive competition and social dominance theory: Synopsis, in: *The Great Apes* (D. Hamburg and E. R. McCown, eds.), 317–338, Benjamin/Cummings, Menlo Park, California.

Price, T. D., and Grant, P. R., 1985, The evolution of ontogeny in Darwin's finches: A quantitative genetic approach, *Amer. Nat.* **125:**169–188.
Rasa, O. A. E., 1979, The effects of crowding on the social relationships and behaviour of the dwarf mongoose (*Helogale undulata rufula*), *Z. Tierpsychol.* **49:**317–329.
Rasa, O. A. E., 1984, A motivational analysis of object play in juvenile dwarf mongooses (*Helogale undulata rufula*), *Anim. Behav.* **32:**579–589.
Reed, E. S., 1981, The lawfulness of natural selection, *Am. Nat.* **118:**61–71.
Reinhardt, V., and Reinhardt, A., 1982, Mock fighting in cattle, *Behaviour* **81:**1–13.
Richmond, G., and Sachs, B. D., 1980, Grooming in norway rats: The development and adult expression of a complex motor pattern, *Behaviour* **75:**82–96.
Richter, W., 1982, Hatching asynchrony: The nest failure hypothesis and brood reduction, *Am. Nat.* **120:**828–832.
Riedman, M. L., 1982, The evolution of alloparental care and adoption in mammals and birds, *Q. Rev. Biol.* **57:**405–435.
Rijksen, H. D., 1981, Infant killing: A possible consequence of a disputed leader role, *Behaviour* **78:**138–168.
Rohwer, S., and Ewald, P. W., 1978, The cost of dominance and advantage of subordination in a badge signaling system, *Evolution* **35:**441–454.
Rood, J. P., 1978, Dwarf mongoose helpers at the den, *Z. Tierpsychol.* **48:**277–287.
Rood, J. P., 1983, The social system of the dwarf mongoose, in: *Advances in the Study of Mammalian Behavior* (J. F. Eisenberg and D. G. Kleiman, eds.), pp. 454–488, Special Publication No. 7, American Society of Mammalogists.
Rowley, I., 1981, The communal way of life in the splendid wren, *Malurus splendens, Z. Tierpsychol.* **55:**228–267.
Rubenstein, D. I., 1982, Reproductive value and behavioral strategies: Coming of age in monkeys and horses, *Perspect. Ethol.* **5:**469–487.
Rushen, J., 1982, Development of social behaviour in chickens: A factor analysis, *Behav. Processes* **7:**319–333.
Ryden, H., 1975, *God's Dog*, Coward, McCann, and Geoghegan, New York.
Ryden, O., and Bengtsson, H., 1980, Differential begging and locomotory behaviour by early and late hatched nestlings affecting the distribution of food in asynchronously hatched broods of altricial birds, *Z. Tierpsychol.* **53:**209–224.
Sackett, G. P., Sameroff, A. J., Cairns, R. B., and Suomi, S. J., 1981, Continuity in behavioral development: Theoretical and empirical issues, in: *Behavioral Development: The Bielefeld Interdisciplinary Conference* (K. Immelmann, G. W. Barlow, L. Petrinovich, and M. Main, eds.), pp. 23–57, Cambridge University Press, New York.
Safriel, U. N., 1981, Social hierarchy among siblings in broods of the oystercatcher *Haematopus ostralegus, Behav. Ecol. Sociobiol.* **9:**59–63.
Saggese, E. P., and Tullar, B. F., 1974, Possible predation of a gray fox on a red fox pup, *N. Y. Fish Game J.* **21:**86.
Schaffer, W. M., 1974, Selection for optimal life histories: The effects of age structures, *Ecology* **55:**291–303.
Schenkel, R., 1967, Submission: Its features and function in the wolf and dog, *Am. Zool.* **7:**319–329.
Schleidt, W. M., and Shalter, M. D., 1973, Stereotypy of a fixed action pattern during ontogeny in *Coturnix coturnix coturnix, Z. Tierpsychol.* **33:**35–37.
Scott, J. P., 1962, Hostility and aggression in animals, in: *Roots of Behavior* (E. L. Bliss, ed.), pp. 167–178, Hafner, New York.
Scott, J. P., and Fredericson, E., 1951, The causes of fighting in mice and rats, *Physiol. Zool.* **24:**273–309.

Sharatchandra, H. C., and Gadgil, M., 1980, On the time-budget of different life-history stages of chital (*Axis axis*), *J. Bombay Nat. Hist. Soc.* **75:**949–960.

Simpson, M. J. A., 1976, The study of animal play, in: *Growing Points in Ethology* (P. P. G. Bateson and R. A. Hinde, eds.), pp. 385–400, Cambridge University Press, New York.

Simpson, M. J. A., Simpson, A. E., Hooley, J., and Zunz, M., 1981, Infant-related influences on birth intervals in rhesus monkeys, *Nature* **290:**49–51.

Smith, P. K., 1982, Does play matter: Functional and evolutionary aspects of animal and human play, *Behav. Brain Sci.* **5:**139–184.

Stacey, P. B., 1982, Female promiscuity and male reproductive success in social birds and mammals, *Am. Nat.* **120:**51–64.

Stallcup, J. A., and Woolfenden, G. E., 1978, Family status and contributions to breeding by Florida scrub jays, *Anim. Behav.* **26:**1144–1156.

Stamps, J., 1983, The relationship between ontogenetic habitat shifts, competition and predator avoidance in a juvenile lizard (*Anolis aeneus*), *Behav. Ecol. Sociobiol.* **12:**19–33.

Stearns, S. C., 1976, Life-history tactics: A review of the ideas, *Q. Rev. Biol.* **51:**3–47.

Stearns, S. C., 1977, The evolution of life history traits, *Annu. Rev. Ecol. Syst.* **8:**145–171.

Stearns, S. C., 1980, A new view of life-history evolution, *Oikos* **35:**266–281.

Stearns, S. C., and Crandall, R. E., 1981, Quantitative predictions of delayed maturity, *Evolution* **35:**455–463.

Stearns, S. C., and Sage, R. D., 1980, Maladaptation in a marginal population of the mosquito fish, *Gambusia affinis*, *Evolution* **34:**65–75.

Stein, D. M., and Stacey, P. B., 1981, A comparison of infant–adult male relations in a one-male group with those in a multi-male group for yellow baboons (*Papio cynocephalus*), *Folia Primatol.* **36:**264–276.

Stinson, C. H., 1979, On the selective advantage of fraticide in reptiles, *Evolution* **33:**1219–1225.

Stirling, I., 1970, Studies on the behaviour of the South Australian fur seal, *Arctocephalus forsteri* (Lesson), *Aust. J. Zool.* **19:**267–273.

Sundquist, M. E., 1981, The social organization of tigers (*Panthera tigris*) in Royal Chitawan National Park, Nepal, *Smithsonian Contrib. Zool.* **336:**1–98.

Sussman, R. W., 1977, Socialization, social structure, and ecology of two sympatric species of *Lemur*, in: *Primate Bio-Social Development* (S. Chevalier-Skolnikoff and F. E. Poirier, eds.), pp. 515–528, Garland Press, New York.

Symons, D., 1978*a*, *Play and Aggression: A Study of Rhesus Monkeys*, Columbia University Press, New York.

Symons, D., 1978*b*, The question of function: Dominance and play, in: *Social Play in Primates* (E. O. Smith, ed.), pp. 193–230, Academic Press, New York.

Taylor, G. T., 1979, Reinforcement and intraspecific aggressive behavior, *Behav. Neural Biol.* **27:**1–24.

Taylor, R. H., 1962, The Adelie penguin *Pygoscelis adelie* at Cape Royds, *Ibis* **104:**176–204.

Thelen, E., 1981, Rhythmical behavior in infancy: An ethological perspective, *Dev. Psychol.* **17:**237–257.

Thiessen, D., Pendergrass, M., and Young, R. K., 1983, Development and expression of autogrooming in the Mongolian gerbil, *Meriones unguiculatus*, *J. Comp. Psychol.* **97:**187–190.

Thomas, P. O., and Taber, S. M., 1984, Mother–infant interaction and behavioral development in southern right whales, *Eubalaena australis*, *Behaviour* **88:**42–60.

Thorne, B. L., 1981, Genetic consequences of variation in sib maturation schedules, *Acta Biotheoret.* **30:**219–227.

Tinbergen, N., 1951, *The Study of Instinct*, Oxford University Press, New York.
Tinbergen, N., 1963, On aims and methods of ethology, *Z. Tierpsychol.* **20**:410–433.
Tooker, C. P., and Miller, R. J., 1980, The ontogeny of agonistic behavior in the blue gourami *Trichogaster trichopterus* (Pisces, Anabintoidei), *Anim. Behav.* **38**:973–938.
Trivers, R. L., 1971, The evolution of reciprocal altruism, *Q. Rev. Biol.* **46**:35–57.
Trivers, R. L., 1972, Parental investment and sexual selection, in: *Sexual Selection and the Descent of Man* (B. Campbell, ed.), pp. 139–179, Aldine, Chicago.
Trivers, R. L., and Willard, D. E., 1973, Natural selection of parental ability to vary the sex ratio of offspring, *Science* **179**:90–92.
Tschanz, B., and Hirsbrunner-Scharf, M., 1975, Adaptations to colony life on cliff ledges: A comparative study of guillemot and razorbill chicks, in: *Function and Evolution in Behaviour: Essays in Honour of Professor Niko Tinbergen* (G. Baerends, C. Beer, and A. Manning, eds.), pp. 358–380, Oxford University Press, New York.
Vadasz, C., Kobor, G., and Lajtha, A., 1983, Genetic dissection of a mammalian behaviour pattern, *Anim. Behav.* **31**:1029–1036.
Van der Molen, P. P., 1984, Bi-stability of emotions and motivations: An evolutionary consequence of the open-ended capacity for learning, *Acta Biotheor.* **33**:227–251.
Van Honk, C., and Hogeweg, P., 1981, The ontogeny of the social structure in a captive *Bombus terrestris* colony, *Behav. Ecol. Sociobiol.* **9**:111–119.
Van Iersel, J. J. A., and Bol, A. C. A., 1958, Preening by two tern species: A study on displacement behaviour, *Behaviour* **13**:1–88.
Von Saal, F. S., 1981, Variation in phenotype due to random intrauterine positioning of male and female fetuses in rodents, *J. Reprod. Fertil.* **62**:633–650.
Von Schantz, T., 1981, Evolution of group living, and the importance of food and social organization in population regulation: A study on the red fox (*Vulpes vulpes*), Ph.D. Thesis, Lund University, Sweden.
Wachtel, M. A., Bekoff, M., and Fuenzalida, C. E., 1978, Sparring by mule deer during rutting: Class participation, seasonal changes, and the nature of asymmetric contests, *Biol. Behav.* **3**:319–330.
Walters, J., 1981, Inferring kinship from behaviour: Maternity determinations in yellow baboons, *Anim. Behav.* **29**:126–136.
Warner, R. R., 1980, The coevolution of behavioral and life-history characteristics, in: *Sociobiology: Beyond Nature/Nurture* (G. W. Barlow and J. Silverberg, eds.), pp. 151–188, Westview Press, Boulder, Colorado.
Welker, W. I., 1971, Ontogeny of play and exploratory behaviors: A definition of problems and a search for conceptual solutions, in: *The Ontogeny of Vertebrate Behavior* (H. Moltz, ed.), pp. 171–228, Academic Press, New York.
Western, D., 1979, Size, life history and ecology in mammals, *Afr. J. Ecol.* **17**:185–204.
Wells, M. C., and Bekoff, M., 1981, An observational study of scent marking in free-ranging coyotes, *Anim. Behav.* **29**:332–350.
Wells, M. C., and Bekoff, M., 1982, Predation by wild coyotes: Behavioral and ecological analyses, *J. Mammal.* **63**:118–127.
Wiley, R. H., 1973, The strut display of male sage grouse: A "fixed" action pattern, *Behaviour* **47**:129–152.
Wiley, R. H., 1974, Evolution of social organization and life history patterns among grouse, *Q. Rev. Biol.* **49**:201–227.
Wiley, R. H., 1981, Social structure and individual ontogenies: Problems of description, mechanism, and evolution, *Perspect. Ethol.* **4**:105–133.
Williams, G. B., 1966, *Adaptation and Natural Selection*, Princeton University Press, Princeton.

Wilson, E. O., 1975, *Sociobiology: The New Synthesis,* Harvard University Press, Cambridge.

Wilson, E. O., 1977, Animal and human sociobiology, in: *The Changing Scenes in the Natural Sciences, 1776–1976* (C. E. Goulden, ed.), pp. 273–281, Academy of Natural Sciences, Philadelphia.

Wilson, R. H., 1974, Agonistic postures and latency to the first interaction during initial pair encounters in the red jungle fowl, *Gallus gallus, Anim. Behav.* **22**:75–82.

Wilson, S. C., 1973, The development of social behaviour in the vole (*Microtus agrestis*), *Zool. J. Linn. Soc.* **52**:45–62.

Wilson, S. C., 1982, The development of social behaviour between siblings and non-siblings of the voles *Microtus ochrogaster* and *Microtus pennsylvanicus, Anim. Behav.* **30**:426–437.

Wilson, S. C., and Kleiman, D. G., 1974, Eliciting and soliciting play, *Am. Zool.* **14**:341–370.

Zimen, E., 1975, Social dynamics of the wolf pack, in: *The Wild Canids: Their Systematics, Behavioral Ecology and Evolution* (M. W. Fox, ed.), pp. 336–362, van Nostrand Reinhold, New York.

6

Mantle Degassing Unification of the Trans-K–T Geobiological Record

DEWEY M. McLEAN

INTRODUCTION

About 65 million years ago (Mya) some extinctions occurred on our planet that are so notable that they define the boundary between two of the three eras of the Phanerozoic Eon. These are the Mesozoic and Cenozoic eras and, on a finer scale, the Cretaceous and Tertiary periods. The cause of the extinctions has long been the subject of debate, with hypotheses spanning the gamut from sudden global catastrophe via exotic extraterrestrial event to gradual bioevolutionary turnover via natural terrestrial processes. My belief is that a lack of understanding of critical time-rock relationships at and about the Cretaceous–Tertiary (K–T) boundary has produced conceptual asymmetry and, in fact, the illusion of catastrophe (McLean, 1981a). After 10 years of integrating elements of about two dozen fields of science in a search for principles on the involvement of the carbon cycle in the evolution of the earth's climate and life, in which I have concentrated upon the K–T contact record because it is rich in information, I see no K–T contact catastrophe in the oceans or on the lands, and certainly no simultaneous global catastrophe. Instead, I see gradual turnover during a K–T transition interval that spans hundreds of thousand of years. In this chapter, I attempt to remove asym-

---

DEWEY M. MCLEAN • Department of Geological Sciences, Virginia Polytechnic Institute and State University, Blacksburg, Virginia 24061.

metry and to unify the K–T transition data base via natural mantle-degassing perturbation of the earth's carbon cycle.

BACKGROUND

Extinctions have long been used as a "natural" boundary to differentiate Cretaceous and Tertiary strata on the assumption that a worldwide change in the physical regime triggered globally simultaneous extinctions (Jeletzky, 1962). In the oceans, about 90% of the genera of Cretaceous coccolithophorids and planktonic foraminifera became extinct; their extinctions define the marine Cretaceous–Tertiary (K–T) contact. Stratotypes (type sections) for the latest Cretaceous and earliest Tertiary are marine sections in Europe. The latest Maastrichtian Stage stratotype is in the Netherlands, and that for the earliest Tertiary Danian Stage in Denmark. Paleontologically, the contact has long been thought to be sharply defined by the seemingly abrupt disappearance of nearly 90% of the genera of both coccolithophorids and planktonic foraminifera. However, it is now known from the study of DSDP Site 524 in the south Atlantic that the coccolithophorid extinctions occurred during Early Tertiary, about 50,000 years later than the foraminiferal extinctions (Perch-Nielsen *et al.,* 1982). As will be discussed here, there is no evidence in the marine extinctions of the global darkening, cooling, or catastrophe predicted by the Alvarez asteroid hypothesis (Alvarez *et al.,* 1980) which proposes that a giant asteroid slammed into the earth 65 Mya, blasting so much dust into the stratosphere that the earth turned dark and cold, analogous to the nuclear winter concept.

On the lands the dinosaurs became extinct (I use the term "dinosaurs" in a nonsystematic sense for archosaurs such as ceratopsians, hadrosaurs, theropods, etc.). It is claimed (Alvarez, 1983) that "the dinosaurs were suddenly wiped out as a direct result of the asteroid impact" and that the trauma was the "worst one that has ever been recorded on the earth." In fact, as I will discuss, time-rock relationships indicate that the dinosaurs disappeared in the Early Tertiary, later than the marine extinctions. That their extinctions were not part of a great catastrophe was indicated by Schopf (1982), who noted that at the time of the extinctions only about 20 species distributed among 15 genera and 10 families remained.

Whereas the marine K–T contact is relatively well defined, the terrestrial K–T contact is not. Over vast areas of the Rocky Mountain region, a hiatus separates Cretaceous strata from Tertiary that, like a temporal "black hole," has swallowed up the critical trans-K–T record. Continuous trans-K–T record is reported to exist in some areas; however, the Cretaceous and Tertiary strata are so lithologically similar that a K–T contact has been difficult to define. It is in this scenario that the last remains of the dinosaurs came into use for picking the K–T contact on the unverified assumption that they were chronostratigraphically useful fossils. Later, a thin coal just above the last dinosaur bones was chosen as the K–T contact on the assumption that it represents simultaneous sedimentation from northern Colorado to Canada (Brown, 1962). Recently, a palynological K–T contact picked on the basis of the disappearance of some pollen taxa has become the working K–T contact. Unfortunately, the chroneity of the terrestrial K–T contact with the marine K–T has not been established. Because the terrestrial K–T contact is the basis for knowledge of trans-K–T bioevolution in western North America, its chroneity relative to the marine K–T contact must be established before claims of trans-K–T climate change based on terrestrial floras, trans-K–T vertebrate evolution, etc., can be accepted. The legitimate starting point is to determine the cause of the marine extinctions (McLean, 1985*a,b*), which is the reason for our subdividing the Mesozoic and Cenozoic eras, and then to examine the terrestrial scenario relative to the marine. I will begin by integrating the Deccan Traps volcanism into the trans-K–T scenario.

For my discussion of the K–T transition scenario, I focus upon the time of the marine extinctions that define the K–T boundary [65 Mya, after Harland *et al.* (1982)], and that of the early Tertiary dinosaurian extinctions, relating both to the trans-K–T mantle-degassing perturbation of the carbon cycle associated with the Deccan Traps flood basalt volcanism in India.

DECCAN TRAPS VOLCANISM/MANTLE DEGASSING

Several years ago (McLean, 1981*b*), I first linked Deccan Traps flood basalt volcanism in India to the trans-K–T carbon cycle perturbation and extinctions. Since that time, I have quantified the rate of release of mantle CO_2 during the Deccan Traps volcanism to that of background, and suggested that the addition of the Deccan Traps mantle CO_2 release to all other sources (all other volcanism, fumaroles, hot springs, etc.) would have caused a buildup of CO_2 in the earth's atmosphere and in the marine mixed layer.

Starting at about the time of the marine extinctions that define the global K–T contact [65 Mya, following the geological time scale of Har-

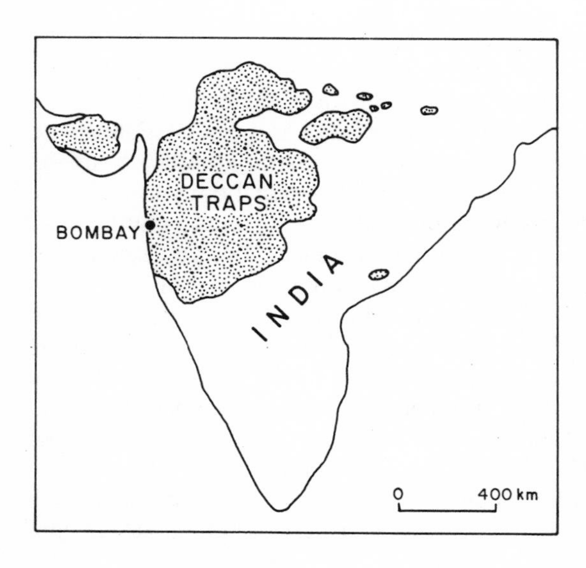

FIG. 1. India, showing the modern distribution of Deccan Traps basalts. Original lava coverage was much more extensive, as indicated by basaltic outliers in eastern India.

land *et al.* (1982)] was one of the greatest episodes of volcanism in the Phanerozoic Eon. This was the Deccan Traps flood basalt volcanism (Fig. 1), which flooded about 2.6 × 10^6 km^3 of India with basaltic lavas (Pascoe, 1964). According to Kaneoka (1980), the isotopic ages of the Deccan Traps concentrate around 66–60 Mya, with the main volcanic activity around 65 Mya. Wensink *et al.* (1979) note that most isotopic dates are early Tertiary, "around the Tertiary/Cretaceous boundary." Klootwijk (1979) cites a age range of 65–60 Mya, noting that the age of the earliest Deccan Traps activity is not much earlier than 65 Mya.

Recent work by Klootwijk (1979) and Wensink *et al.* (1979) provide new information on the duration of the main Deccan Traps volcanic activity (Fig. 2). The former notes that activity occurred within an older reversed and a younger normal polarity epoch. The latter note that most lavas were erupted during the older reversed interval, which they designate the Deccan reversed magnetic polarity epoch, with some volcanism continuing into the overlying Nipani normal epoch. According to the time scale of Harland *et al.* (1982), an isotopic age of 65 Mya places the bulk of Deccan Traps activity in reversed polarity chron R29. Lower and upper boundaries of R29 are 65.39 and 64.86 Mya; thus, the bulk of the Deccan Traps lavas may have been erupted over a short duration of about 0.53 million years. The duration of both R29 and N29 is 1.36 million years.

Extrusion of 2.6 × 10^6 km^2 of basalts onto the earth's surface during a short time of between 0.53 and 1.36 million years would have released

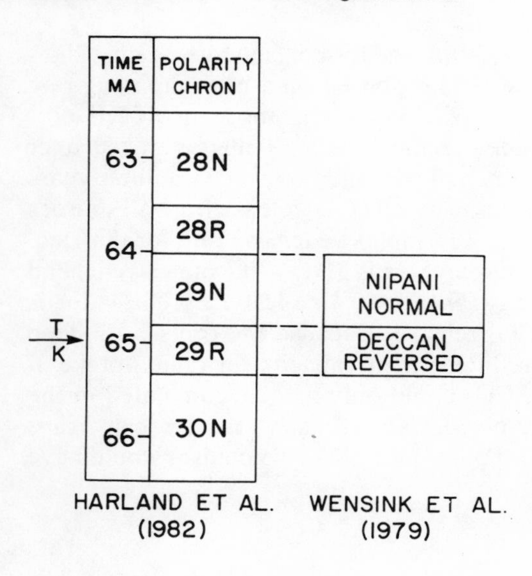

FIG. 2. Wensink *et al.* (1979) note that the bulk of the Deccan Traps basalts were extruded during the Deccan reversed interval, with the main volcanic activity around 65 Mya; according to the Harland *et al.* (1982) time scale, most Deccan Traps basalts were extruded during magnetic polarity chron R29.

vast amounts of mantle CO_2 into the atmosphere. An estimate of CO_2 release can be gained as follows [after Leavitt (1982)]:

$$m\mathrm{CO_2} = \sum_i \frac{s f_{\mathrm{dg}} v d}{\mathrm{mwCO_2}}$$

where mCO_2 is the number of moles of CO_2 released, s is the original weight fraction of CO_2 in the magma, f_{dg} is the degassing fraction during eruption, v is the volume of erupted material, d is the solid density of the erupted material, $mwCO_2$ is the molecular weight of CO_2 (=44), and i numbers the individual eruption. For Deccan Traps CO_2 release, I use $s = 0.005$, $f_{dg} = 0.6$, $v = 2.6 \times 10^6$ km^3, $d = 2.9 \times 10^{15}$ g km^{-3}, and $mwCO_2 = 44.0$. For s, 0.002 would be a "best estimate" and 0.005 a "liberal" one (Leavitt, 1982); I use the liberal value. For f_{dg}, 0.6 is a "best estimate" (Leavitt, 1982). For v, using Pascoe's (1964) estimate of 2.6×10^6 km^2 original coverage of the Deccan Traps with an average original thickness of 1 km provides an estimate of original volume of 2.6×10^6 km^3. Estimated mCO_2 release calculated on the entire Deccan Traps lava pile is 5×10^{17} (mCO_2 of the modern atmosphere is 5.6×10^{16}).

Knowledge of the volume and duration of eruption of the Deccan Traps lava pile allows annual rates of basalt and CO_2 production to be estimated relative to modern MORB (mid-ocean ridge basalt) and total mantle CO_2 release from all sources. For the former, modern ocean ridges have a total length of about 6×10^4 km; based on a half-spreading rate of 1 cm/year and basalt thickness of about 1 km, modern MORB production is 1.2 km^3/year. The rate for the Deccan Traps averaged over 1.36 million years is 1.9 km^3/year, and averaged over 0.53 million years is 4.9 km^3/year. For the latter, average mantle CO_2 release from all sources [data from Leavitt (1982)] is 4.1×10^{12} moles/year. The rate for the Deccan Traps averaged over 1.36 million years is 3.9×10^{11} moles/year, and averaged over 0.53 million years is 9.6×10^{11} moles/year. The latter is nearly 25% of annual mantle CO_2 release. Because the trans-K–T deep oceans were warm (14–15°C), reducing their capacity for rapid uptake of vast amounts of volcanic CO_2, CO_2 could only have accumulated in the atmosphere and in the marine mixed layer. Clearly, the Deccan Traps CO_2 contribution in addition to all other trans-K–T volcanism would have perturbed the earth's carbon cycle.

DECCAN TRAPS: K–T ISOCHRON TIMING EVENT

The trans-K–T marine biological and chemical records have all the earmarks of a major mantle degassing event; temporal relationships point to the Deccan Traps as the perturbing influence. The age of the K–T contact isochron is 65 Mya (Harland *et al.*, 1982); it falls within polarity chron R29. The Deccan Traps earliest, and major, activity is dated isotopically at about 65 Mya, which also falls within polarity chron R29. Thus, within the resolution of dating techniques, the marine extinctions that define the marine (and global) K–T contact isochron were simultaneous with the Deccan Traps volcanism. A Deccan Traps mantle CO_2 release of nearly 25% that of all sources, added to the CO_2 contribution by all other sources, would have triggered ecological instability, causing the marine extinctions and thus serving as the global K–T contact timing event.

Evidence of vigorous K–T contact mantle degassing abounds. Prior to the terminal Cretaceous marine extinctions, global ecological stability prevailed and marine productivity was high (Thierstein, 1981). During and after the marine extinction event, instability prevailed and marine productivity was reduced; instability was most severe during the interval of deposition of the K–T contact record. At the K–T contact, dead ocean conditions [severely reduced mixed-layer photosynthesis and $CaCO_3$ production; concept after Baes (1982)] seem to have prevailed (McLean, 1985*a*,*b*); this condition, preserved in the record as the K–T contact, was simultaneous with early Deccan Traps activity. Low productivity continued into the early Tertiary, coevally with the main volcanic activity.

As CO_2 builds up in the atmosphere, it also builds up in the mixed layer of the oceans (top 50–100 m). The highly selective nature of the terminal Cretaceous marine extinctions with regard to both the habitat and organisms most severely affected point to a CO_2 buildup in the mixed layer of the oceans. The extinctions took place in the mixed layer, and involved mostly planktonic foraminifera and coccolithophorids (calcareous microplanktonic protozoans and algae, respectively). Planktonic foraminifera begin dissolving at pH 7.8 (Boltovsky and Wright, 1976); coccolithophorids begin losing their coccoliths at pH 7.3–7.0. The extinctions were not an instantaneous event as often cited. In fact, as has recently been determined (Perch-Nielsen *et al.,* 1982), Cretaceous taxa found in Tertiary sediments, long thought to have been reworked, have the same isotopic values as the Tertiary taxa, showing that both lived in the same water mass. Thus, gradual bioevolutionary turnover is seen even in the terminal Cretaceous marine extinctions; the marine extinction event spanned at least 50,000 years (Hsu *et al.,* 1982*a*). Thus, Kent's (1977) estimate of the 10^4-year duration of the marine extinctions based on studies of the Gubbio, Italy, section cannot be correct. The concept of catastrophe seems to have originated from studies of sections with hiatuses at the K–T contact; secondary truncation of ranges along an illusory seemingly isochronous datum has created illusions of catastrophe (McLean, 1981*a*).

Recently, an iridium enrichment "spike" (often in a thin clay layer at or about the cutoff of the geological ranges of Cretaceous taxa that has been used as the K–T contact) has come into use as the new K–T contact. At or about the Ir/clay contact shifts occur in ^{13}C and ^{18}O isotopic values (Figs. 3 and 4), coeval with drops in mixed-layer photosynthesis and $CaCO_3$ production and with iridium enrichment. The isotopic shifts, iridium enrichment, and drop in $CaCO_3$ were not instantaneous events; all began before the deposition of the Ir/clay itself and persisted into the early Tertiary; the perturbation spanned at least tens of thousands to hundreds of thousands of years; the duration of the iridium enrichment event spanned 10^4–10^5 years (Officer and Drake, 1985). Pre-Ir/clay ^{13}C depletion is recorded at El Kef, Caravaca, Biarritz, DSDP Site 524, and DSDP Site 465A (Fig. 3). Because mantle CO_2 is depleted in ^{13}C (about 5 per mil) and surface water ^{13}C is controlled by exchange with the atmosphere and by the amount of carbon removed versus that replenished (Kroopnick *et al.,* 1977), and because the ^{13}C record is preserved in biogenic $CaCO_3$ produced in the mixed layer coevally with Deccan Traps mantle degassing, I propose that ^{13}C depeltion is reflective, at least in

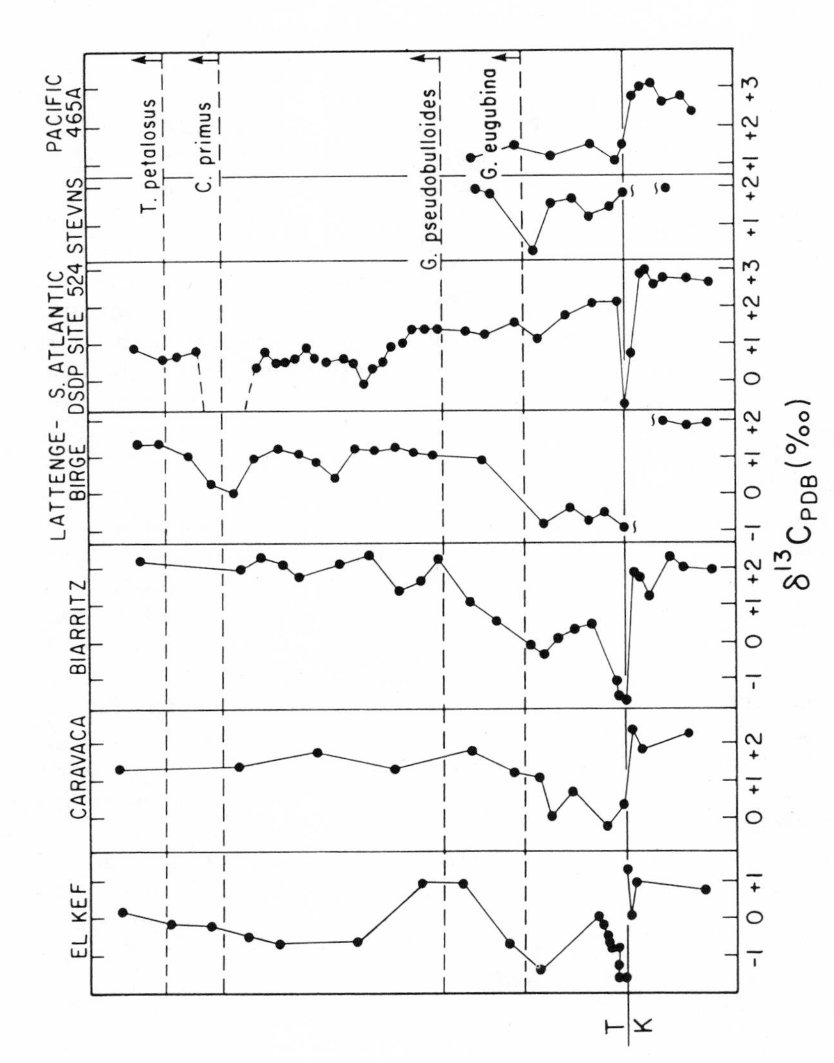

FIG. 3. Carbon-13 records at nearly complete trans-K–T marine sections, showing trans-K–T ^{13}C depletion coeval with Deccan Traps volcanic activity. [After Perch-Nielsen *et al.* (1982).]

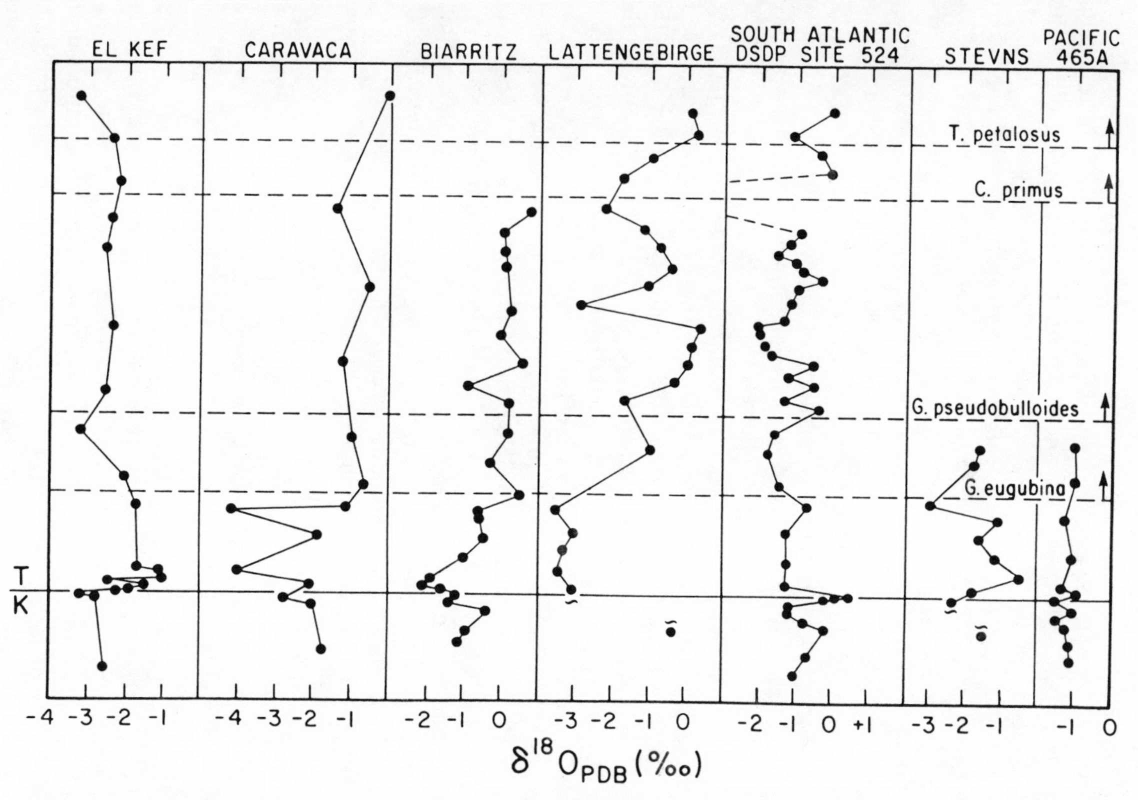

FIG. 4. Oxygen-18 records at nearly complete trans-K–T marine sections, showing trans-K–T ^{18}O depletion coeval with Deccan Traps volcanic activity; the ^{18}O record seems to be latitude dependent. [After Perch-Nielsen *et al.* (1982).

part, of mantle degassing. I also propose that the K–T boundary iridium enrichment represents mantle release (discussed below, p. 305).

Algae remove CO_2 from both the atmosphere and the mixed layer, converting it to particulates that settle into and store CO_2 in the deep oceans, maintaining low pCO_2 in the atmosphere/mixed layer; this is the Williams–Riley pump (Dyson, 1982); its failure today would raise atmospheric CO_2 levels severalfold (National Research Council Geophysics Study Committee, 1977). Modern total failure of the pump with no CO_2 uptake by other sinks, based on a modern mixed layer net primary productivity of 2.1×10^{15} moles carbon/year, would release 5.3×10^{16} moles of carbon in 25 years, doubling atmospheric CO_2. Trans-K–T dead-ocean conditions can be linked to a failure of the Williams–Riley pump. Evidence of pump failure is seen in the failure of the coccolithophorids. Hsu *et al.* (1982b) note the nearly total suppression of photosynthetic activities by planktonic organisms. Perch-Nielsen *et al.* (1982) suggest that a ^{13}C decrease of about 3 per mil in Tertiary sediments at the K–T contact (Fig. 3) is the result of greatly reduced photosynthesis. For reduced mixed-layer $CaCO_3$ production, Thierstein (1981) noted that the collapse of the coccolithophorids resulted in a sharp decrease of carbonate supply to the ocean floor. Trans-K–T cutoff of $CaCO_3$ is seen in sections with nearly complete trans-K–T stratigraphic sections. At El Kef, $CaCO_3$ content decreases sharply at the K–T boundary, from 37 to 0% in the boundary clay (Perch-Nielsen, 1982). At DSDP Site 524, $CaCO_3$ decreases to 2% in the boundary clay (Hsu *et al.*, 1982*b*). Summarizing to this point, one can say that dead-ocean conditions reflected in the K–T contact bioevolutionary, sedimentological, and chemical records, all controlled by Deccan Traps mantle degassing, are preserved in the record as the K–T contact. In short, the Deccan Traps volcanism is the K–T isochron timing event.

"STRANGELOVE" OCEAN

In the Early Tertiary, beginning at the time of the marine extinctions, and coeval with CO_2-induced chemical change in the mixed layer, thoracosphaerids, braarudosphaerids, and small planktonic foraminifera dominated the mixed layer plankton. In reference to this unusual assemblage, Hsu and McKenzie (1985) have coined the term "Strangelove" oceans. The duration of Strangelove conditions ranges between the Hsu and McKenzie estimate of repopulation of the oceans 100,000 years after Ir/clay deposition, and the Haq *et al.* (1977) estimate of dominance of thoracospherids, etc., for about 1–2 million years. At the same time one has low surface-water productivity for 1–2 million years following the marine crisis (Arthur and Dean, 1982). Ecological instability persisted for hundreds of thousands of years following the marine extinction event (Thierstein, 1981). Hsu *et al.* (1982*b*) suggested a return to normal surface water enriched in ^{13}C after 500,000 years with complete recovery of mixed layer fertility. On the lands, the dinosaurs became extinct during late Strangelove time. The record indicates gradual extinctions of the dinosaurs (Van Valen and Sloan, 1975; Clemens *et al.* 1981; Archibald and Clemens, 1982).

TRANS-K–T TERRESTRIAL SETTING

During the Late Cretaceous, the North American western interior was covered by a north–south-trending epeiric sea 200–1000 miles wide known as the Western Interior Cretaceous Seaway, stretching from the Gulf of Mexico northward to Alaska (Fig. 5). Strata deposited in the Seaway record several marine transgressions and regressions. Near the end of the Cretaceous, strata deposited along the western margin of the seaway (the Rocky Mountain region) during the Maastrichtian Bearpaw R4 regression incorporated the final record of the dinosaurs.

During the K–T transition time, Laramide mountain-building occurred in the Rocky Mountain region. Rivers flowing eastward from the uplifted areas carried sediments to the Cretaceous Seaway, producing eastward-prograding delta systems that forced the shoreline eastward (Fig. 6). In the Seaway, the finest grained sediments settled farthest from shore as the prodelta shales of the Pierre (~Bearpaw) Formation; these shales intertongued with delta front marine sands of the Fox Hills Formation, which migrated seaward over the Pierre Formation. Migrating seaward over and intertonguing with the Fox Hills were continental Cretaceous-age delta plain sediments of the Lance and Hell Creek formations and, over them, the continental delta plain sediments of the Tertiary-age Fort Union Formation (Fig. 7). Freshwater swamps and marshes that developed along the regressing R4 strandline produced the lignites of the Lance–Hell Creek and Fort Union formations. A lignite (or lignite zone) known as the "Z" coal (just above the highest dinosaur remains) has long been used as the K–T contact.

In the Western Interior Cretaceous Seaway nearly continuous sedimentation incorporated excellent ammonite faunas interbedded with volcanic ash beds (bentonites); sanidine and biotite from the bentonites

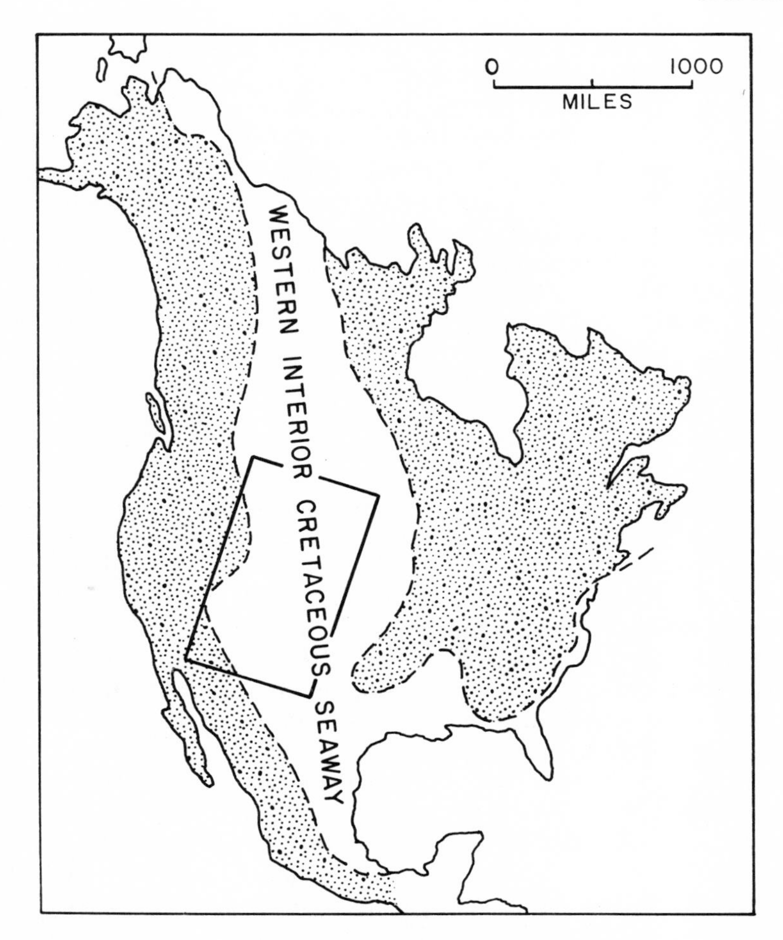

FIG. 5. North America during late Cretaceous time divided into eastern and western subcontinents by the Western Interior Cretaceous Seaway; the Rocky Mountain region is outlined by the rectangle.

have been dated isotopically and integrated with the ammonite zonation, allowing Obradovich and Cobban (1975) to develop a time scale for the Cretaceous of the western interior of North America. They estimated an age of 64–65 Mya for the K–T contact. On the basis of more recent K–Ar constants, Lanphere and Jones (1978) estimate the age of the K–T boundary in North America at 65–66 Mya. The most recent time scale by Harland *et al.* (1982) indicates an age of 65 Mya for the marine, and thus global, K–T contact.

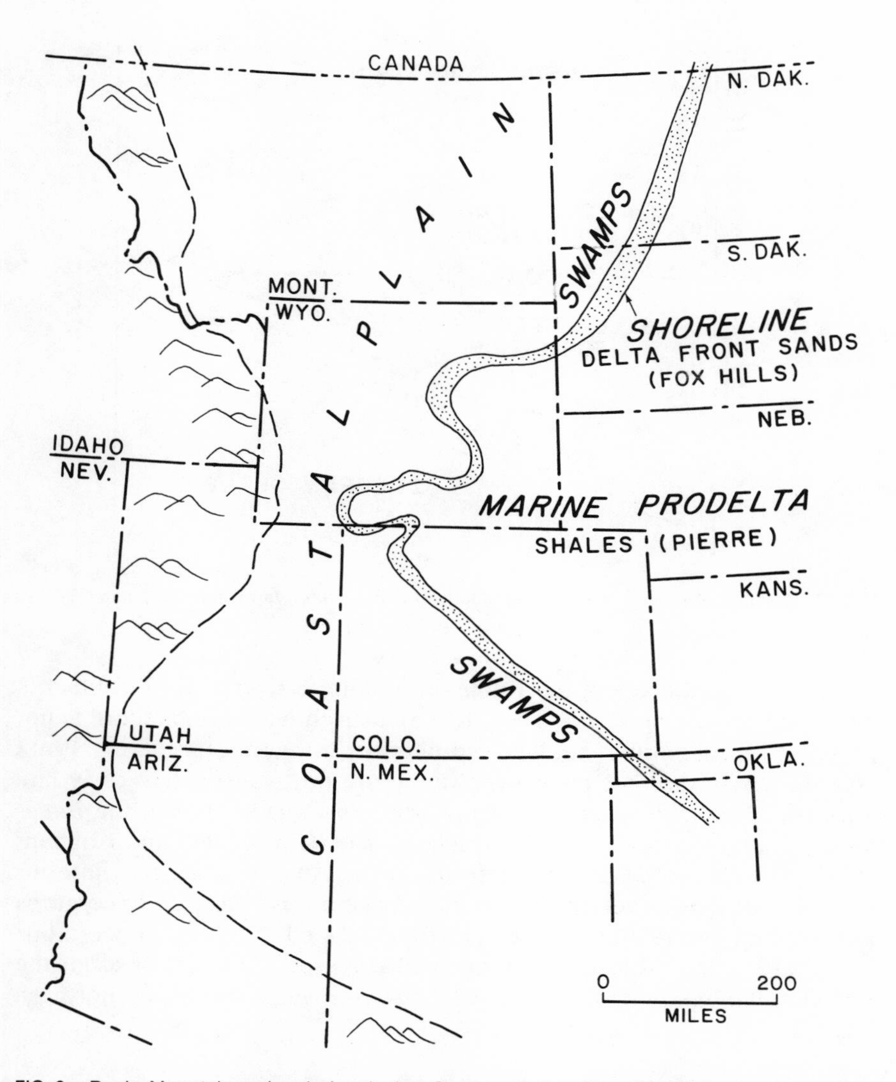

FIG. 6. Rocky Mountain region during the late Cretaceous *Baculites clinolobatus* range zone time 67–68 Mya, showing a broad coastal plain between mountains on the west and the western shoreline of the Western Interior Cretaceous Seaway; the shoreline is marked by Fox Hills delta front sands. [After Weimer (1977).]

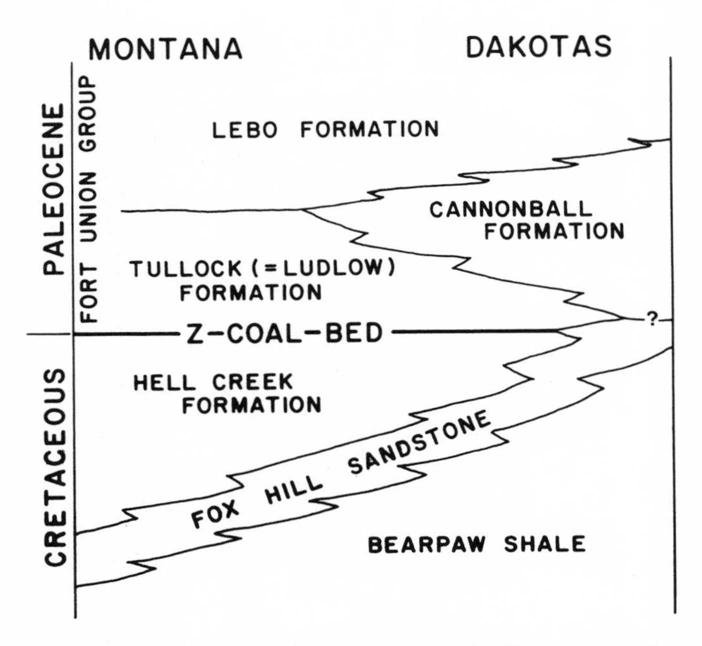

FIG. 7. Geological section of the Rocky Mountain region. [After Van Valen and Sloan (1977).]

However, the age of the Rocky Mountain terrestrial K–T contact is not based on marine strata. The latest Maastrichtian marine record is not preserved in the American western interior; the youngest marine strata contain *Discoscaphites cheyennesis,* and are no younger than early late Maastrichtian. The youngest marine strata in Canada contain the *Baculites grandis* zone of early Maastrichtian age (Obradovich and Cobban, 1975). The latest Maastrichtian strata in the American western interior are nonmarine beds referred to as the "Triceratops beds" or Triceratops Zone (Jeletzky, 1960). Jeletzky (1960) suggested that the "Triceratops beds" are of late Maastrichtian age and correspond to part or all of the *Belemnitella junior* and *Belemnella casimirovensis* zones of northern Europe.

CHRONEITY OF MARINE VERSUS TERRESTRIAL K–T CONTACTS

Recently, a pollen composition change involving the disappearance of some pollen taxa several meters above the highest dinosaur bones has become the "working" K–T contact. In the Raton Basin of New Mexico and Colorado, and at some sites in Montana, the pollen-based K–T contact coincides with iridium enrichment (both are discussed below), lending support to the terminal Cretaceous asteroid catastrophe theory (Alvarez *et al.,* 1980). By the new theory, a giant asteroid 6–10 km in diameter struck the earth at the end of the Cretaceous, plunging the earth into dust-veiled darkness, with abrupt global extinctions synchronous in the seas and on the land. The basis for the asteroid hypothesis is an iridium enrichment at the marine Cretaceous–Tertiary boundary. Iridium, a siderophile ("iron-lover"), was swept to the earth's iron core early in the earth's history, leaving crustal rocks depleted. Because some asteroids are enriched in iridium, K–T contact iridium enrichment is thought to represent iridium-rich dust settling to the earth after impact. By this mechanism, the iridium/dust layer is a chronohorizon (a stratigraphic surface, or thin bed, that is everywhere isochronous) that marks the global K–T contact. Because of a seeming close association of iridium enrichment and mass extinctions both on land and in the sea, Smit and Van Der Kaars (1984) and other catastrophists state that it seems logical to assume that both were caused by the same event—a large impact.

Finding an easily identifiable global K–T isochron marked by Ir/clay and extinctions would be an extraordinary stroke of good fortune for the science of geology. So valuable would such an isochron be that we must in good spirit, be we catastrophists or gradualists, integrate all available data to test for the possibility of a global K–T isochron. However, such a test, based on a sensible reading of the data, indicates that such simplicity does not exist. In fact, the trans-K–T record is abstruse and fraught with subtle time-rock illusions of isochroneity. Diachroneity seems the rule.

The seemingly apparent extinction datums of the terrestrial dinosaurs and vascular plants are themselves the upper boundaries of biostratigraphic units. The North American Commission on Stratigraphic Nomenclature (1983, Article 48) notes that "boundaries of most biostratigraphic units, unlike the boundaries of chronostratigraphic units, are both characteristically and conceptually diachronous. An exception is an abundance biozone boundary that reflects a mass mortality." At question here is: Do the terrestrial extinctions really reflect mass mortality, or an overly simplistic interpretation? On an equally important point, reference with regard to the Rocky Mountain region to "Cretaceous" and "Tertiary" pollen taxa is by definition only, and not by proof of temporal equivalence to marine strata of the same ages.

Recall that the global K–T isochron is defined in marine strata of magnetic polarity chron R29 at about 65 Mya (Harland *et al.,* 1982). This is the isochron to which the terrestrial K–T contact must be correlated. I will now attempt to correlate temporally K–T contacts from the Raton Basin (New Mexico and Colorado), eastern Montana, and the Red Deer Valley, Alberta, Canada, with the marine K–T contact.

In the Raton Basin, the stratigraphic position of the K–T boundary has long been controversial. Placed by Lee (1917) at the base of the Raton Formation, the paleobotanist Brown (1943, 1962) placed the contact in the lower portion of the Raton Formation about 15 m above the base of the formation. Ash and Tidwell (1976) note that the Raton Formation contains a lower, relatively barren flora of Cretaceous age consisting of *Palmocarpon* (possible palm fruit), *Sabalites* (palm leaf), and *Paleoaster,* a whorl of "leaves" of unknown taxonomy and function (Ash and Tidwell, 1976), and an upper flora of Paleocene age consisting of about 50 species (after Brown, 1962); no gymnosperms have been reported from the upper flora, which is mostly made up of angiosperms. The presence of *Paleoaster* in the lower flora led Brown to conclude it was of Late Cretaceous age; the species is known only from Cretaceous-age strata.

Pillmore *et al.* (1984) note that the Raton Basin K–T contact was later identified by Tschudy (1973) 80 m above the base of the Raton Formation on the basis of an abrupt disappearance of several pollen taxa: *Proteacidites, Tilia wodehousei sensu* Anderson, *Trisectoris,* and *Trichopeltinites,* referred to collectively as the "*Proteacidites* assemblage." The "extinction" datum is at the top of a 1- to 2-cm-thick clay that contains iridium (Orth *et al.,* 1981, 1982; Pillmore *et al.,* 1984). Tschudy (1973) shows the pollen genus *Aquillapollenites,* whose disappearance along with some other taxa is used to define the K–T contact in Montana and Alberta, disappearing about 30 m below the K–T contact defined on the basis of the disappearance of *Proteacidites* species. Tschudy does not correlate his pollen-based K–T contact temporally with the marine K–T contact, which occurs in polarity chron R29.

For Montana and surrounding areas, the stratigraphically highest dinosaur remains have long been instrumental in picking the K–T contact. Use of the dinosaurs in separating the Lance and Fort Union formations (Maastrichtian and Paleocene, respectively) dates back to Stanton and Knowlton's (1897) equating the top of the highest dinosaur-bearing strata to the top of the Lance. Later, Brown (1952), following Calvert's (1912) formula for picking the K–T contact at the lowest persistent bed of lignite, advocated searching an area for remains of dinosaurs as high as they can be found, and then looking for the first good coal zone no matter how thin. He states that "the base of this zone is the Cretaceous–Paleocene contact." Brown suggested that the lowest lignite zone indicates renewal of coal formation after a long, coal-barren interval, and represents a wide-

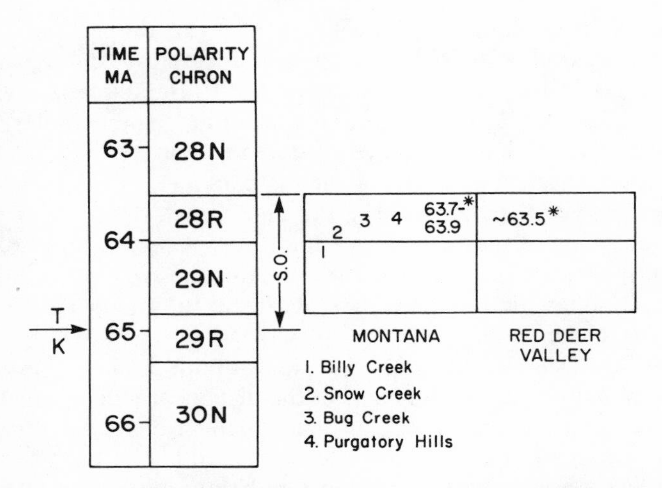

FIG. 8. The K–T contact at several sites in eastern Montana and the Red Deer Valley, Alberta, Canada, relative to the time scale of Harland *et al.* (1982); in both regions, the K–T contact occurs in a magnetic reversed interval. Isotopic dates at the K–T contact of 63.7–63.9 Mya in Montana (Baadsgaard and Lerbekmo, 1983) and about 63.5 Mya in the Red Deer Valley (Lerbekmo *et al.*, 1979) suggest the reversed interval to be R28. S.O., "Strangelove" oceans.

spread, simultaneous phenomenon throughout eastern Montana, Wyoming, and the Dakotas. Thus it was the "first good coal zone" following a long interval of coal-barren sedimentation that Brown (1962) designated as the K–T contact on the apparent assumption that "Calvert's basal coal zone, as proved at all localities east of the present Continental Divide and north of the Colorado–Wyoming line to the Canadian border" reflected simultaneous deposition of a coal bed, or zone, over the entire region. In short, the coal has been routinely treated as a chronohorizon that itself is isochronous with the marine K–T contact. This philosophy, which is routinely followed in palaeontological studies in the American western interior, is shown in Fig. 7 (after Van Valen and Sloan, 1977), where the "Z" coal is represented as an isochronous unit within time transgressive units. That the "Z" coal is not isochronous over its extent is demonstrated by Archibald *et al.* (1982), who showed that in eastern Montana the "Z" coal occurs in both normal and reversed zones. At the Billy Creek (Garfield County) section the "Z" coal occurs in Ft. Peck A + normal polarity zone, whereas at the Snow Creek (Garfield County), Bug Creek (McCone County), and Purgatory Hills (McCone County) sections the "Z" coal is in the Ft. Peck B − reversed polarity zone (Fig. 8).

The pollen-based K–T contact picked on disappearance of the genera *Aquillapollenites, Proteacidites,* etc., coincides with the "Z" coal.

Chroneity of the pollen-based K–T contact, the dinosaurian "extinctions," and the Ir/clay in the Montana region can now be compared with that of the marine K–T contact. Based on the Harland time scale, the 65-Mya date of the marine K–T contact locates it in magnetic polarity chron R29 (Fig. 8). As noted above, the K–T contact at several sites in Garfield and McCone Counties of eastern Montana is in a reversed polarity zone; the isotopic age of the "Z" coal in the Hell Creek drainage several kilometers southeast of the Billy Creek and Snow Creek sites of 63.7–63.9 Mya (Baadsgaard and Lerbekmo, 1983) locates the "Z" coal, and the K–T contact, in magnetic polarity chron R28; the Billy Creek K–T contact is in polarity chron N29.

In the Red Deer Valley, Alberta (Fig. 8), both the last dinosaurs and the K–T contact picked on the basis of the disappearance of most species of *Aquillapollenites,* etc., occur in a magnetic reversed interval (Lerbekmo *et al.*, 1979). The K–T contact is picked at the top of the Nevis coal. The last dinosaur remains occur 6 m below the K–T contact and 5 m below a bentonite with transparent, unaltered sanidine, which yielded radiometric ages of 63.0, 63.3, 63.5, and 63.5 Mya (Lerbekmo *et al.*, 1979). Based on sedimentation rates, the estimate of the duration of time between the last dinosaurs and the K–T contact is 90,000 years (Lerbekmo *et al.*, 1979). Thus, the dinosaurs disappeared about 63.5 Mya in the Red Deer Valley, and are younger than the marine extinctions that define the global K–T isochron. By definition, the dinosaurs became extinct during the Early Tertiary, during the late stages of Strangelove ocean conditions.

Summarizing to this point, both the pollen-based K–T contact and the dinosaurian extinctions are younger than the marine extinctions that define the global K–T isochron. The marine K–T contact occurs in magnetic polarity chron R29. However, on the lands, the Raton Basin pollen-based K–T contact occurs in polarity chron N29, whereas both the Montana and Red Deer Valley pollen-based K–T contacts occur in polarity chron R28. The terrestrial and marine extinctions were thus not synchronous, and do not reflect global mass mortality at the end of the Cretaceous via a geologically instantaneous catastrophic event. The trans-K–T extinctions were gradualistic in nature. Mass mortality cannot be demonstrated even within the Rocky Mountain region; diachroneity exists between the Raton Basin, Montana and Red Deer Valley sites, indicating that plants whose pollen is used to pick the K–T contact disappeared at different times in different places in the Rocky Mountain region. The Montana and Red Deer Valley palynological K–T contacts both occur within polarity chron R28; however, precise chroneity between the two localities cannot now be established.

CATASTROPHISM VERSUS GRADUALISM

The concept of terminal-Cretaceous, geologically sudden catastrophic extinctions on a global scale is not supported by the geobiological record. At DSDP Site 524, one of the few relatively complete trans-K–T sections known, Hsu *et al.* (1982*a*) showed that Cretaceous-age calcareous microplankton did not become extinct at the Ir/clay K–T boundary, but, in fact, persisted into the Early Tertiary before becoming extinct. Long thought to be reworked fossils, the Cretaceous taxa in Tertiary strata have the same isotopic values as the Tertiary taxa, showing that they coexisted in the same Tertiary water mass (Perch-Nielsen *et al.,* 1982). Thus, the marine record does not support a terminal Cretaceus catastrophe. In the last section it was shown that the terrestrial palynological K–T contact is not isochronous with the marine, and that the dinosaurs became extinct not at the end of the Cretaceous, but during the Early Tertiary. The trans-K–T record is one of gradual bioevolutionary turnover.

Catastrophism in the dinosaurian extinctions has received new life via the Alvarez *et al.* (1980) asteroid hypothesis, based on iridium enrichment at the K–T boundary [for an excellent review of the trans-K–T scenario, see Van Valen (1984)]; some extraterrestrial objects are enriched in iridium relative to the earth's crustal rocks. Supposedly, the impact of a 10-km-diameter asteroid injected so much dust into the stratosphere that the earth turned dark for several months, cutting off photosynthesis and simultaneously killing marine and terrestrial organisms, including the dinosaurs. By theory, the impact-related dust spread around the earth, settling out as a universal iridium-rich clay at the K–T contact.

The occurrence of iridium in the marine K–T boundary clay is interesting. Prior to developing my carbon cycle perturbation model in the late 1970s, I had examined and rejected the old extraterrestrial impact hypothesis as being incompatible with what I knew on K–T transition time-rock relationships. However, the new asteroid theory was not necessarily at odds with my own (McLean, 1978) model on CO_2-induced global warming, itself triggered by trans-K–T collapse of marine algae (the coccolithophorids). I had not been able to account for a reason for the collapse of the marine algae: maybe asteroid impact was the cause. I began to test the possibility that the Ir/clay might be a universal chronohorizon reflecting an extraterrestrial cause of the extinctions. I soon concluded that the Ir/clay was the result of natural terrestrial processes. At the 1981 Toronto AAAS meeting (McLean, 1981*b*), I proposed the Deccan Traps volcanism in India as the cause of both the trans-K–T extinctions and the Ir/clay, proposing mantle release of the iridium. I have continued my research on mantle degassing influence in the trans-K–T extinctions, and maintain that the iridium at the marine K–T contact represents mantle release onto the earth's surface.

Iridium has long been thought to be refractory, and virtually immobile upon the earth's surface. However, Zoller *et al.* (1983) report iridium in the gas phase of the modern volcano Kilauea at concentrations up to 2 $\times$ 10^4 times that in the basalt. This discovery is important to the trans-K–T scenario, and to origin of the K–T Ir/clay. Iridium can be released from the mantle via mantle plume volcanism and moved about upon the earth's surface. Fissure flow volcanism produces thermal plumes that can inject particulates into the stratosphere. During Deccan Traps volcanism, India was east of Africa and south of the equator. Prevailing winds in both hemispheres would have carried iridium-attached particulates to western Europe and the eastern South Atlantic, precisely where the marine Ir/clay is best developed (Officer and Drake, 1983). The smectite-rich character of the Ir/clay shows its volcanic origin (Rampino and Reynolds, 1983).

In the American Rocky Mountain region, iridium spikes at the palynological "K–T contact" are not isochronous with the marine, and thus not supportive of an abrupt global holocaust via asteroid impact. As noted in the preceding section, the marine Ir/clay clay is in magnetic polarity chron R29, whereas the Raton Basin and Montana Ir/clays are in N29 and R28, respectively. That the marine and terrestrial Ir/clays have different $^{187}Os/^{186}Os$ ratios and thus separate origins was shown by Luck and Turekian (1983).

The distribution of iridium in the Raton Basin and Montana Ir/clays argues against an origin via asteroid dust fallout. Iridium in dust settling out from impact would be distributed uniformly throughout the clay; it is not. In the Raton Basin sites (Pillmore *et al.*, 1984), the iridium occurs at the top of the clay layer. At the York Canyon site the iridium does not occur in the clay, but in the overlying coal. For the Ir/clay spike of the Hell Creek area, Montana, Smit and Van Der Kaars (1984) state that iridium "occurs in the top of a thin clay layer."

According to the asteroid theory, global darkness devastated the plant kingdom; sense would suggest that the plant kingdom would have had to reestablish itself and that during recovery, plants would have been relatively sparse. Yet, coals—indicative of luxuriant plant growth—rest immediately upon the Ir/clay at all localities. Except for a few taxa that disappear at the palynological K–T contact, the plant kingdom shows no signs of devastation. Most plausibly, climate change killed off relict plant communities on the vast coastal plain left behind as the epeiric sea regressed from North America; climate change may also have been instrumental in concentrating iridium via chemical change associated with swamp and marsh development. Finally, according to the asteroid theory the stratigraphically highest dinosaur bones should be encased within the Ir/clay layer; in fact, the highest dinosaur bones are always several meters *below* the palynological K–T boundary and the Ir/clay.

Krassilov's (1978) suggestion of an abrupt floral change at the K–T contact is not supported by the stratigraphic record. As I have argued previously (McLean, 1981a), Krassilov's stratigraphic information indicates that his four successive floral horizons only *correspond* to the faunal zones of *Inoceramus uwajimensis* (Coniacian), *Anapachydiscus naumanni* (Santonian or early Campanian), *I. orientalis* (early Campanian), and *Pachydiscus subcompressus* (late Maastrichtian). Overlying the Maastrichtian are tuffaceous strata of the Boshniakovian Formation, which grades laterally into marine beds with an impoverished fauna of small bivalves of "Danian aspect." Krassilov's stratigraphic foundation is thus too imprecise to verify the idea of catastrophe among terrestrial floras at the K–T contact.

As noted in previous sections, Tschudy's palynological composition change, which is being widely accepted both as the K–T contact in the Rocky Mountain region and as evidence that the plant kingdom was severely perturbed at the K–T contact, occurs in the Early Tertiary magnetic polarity chron R28. The marine extinctions that define the global K–T contact isochron occur in R29.

Other evidence cited as proof of a sudden global catastrophe via asteroid impact—diaplectic quartz found at the palynological K–T contact in the Rocky Mountain region (Bohor *et al.*, 1984) and sanidine spherules in the marine K–T boundary clay (Smit and Klaver, 1981)—can be of either terrestrial or extraterrestrial origin.

Summarizing to this point, one can say that there is no convincing evidence whatsoever of an asteroid impact-induced catastrophe at the end of the Cretaceous Period. The record is one of gradual bioevolutionary turnover during the K–T transition.

GLOBAL WARMING IN THE DINOSAURIAN EXTINCTIONS

Several years ago I proposed CO_2-induced "greenhouse" conditions as a factor in the dinosaurian extinctions, evoking coupling between climatic warming and reproductive dysfunction. High environmental temperatures are known to be detrimental to reproduction among many types of animals, including reptiles, birds, and mammals. In fact, several studies have indicated trans-K–T global warming. Margolis *et al.* (1977), interpreting the isotope record from several DSDP sites, state that the K–T boundary occurs during a time of significant global warming of bottom and surface waters. The Boersma *et al.* (1979) study of over 200 samples from Atlantic Ocean DSDP Sites 384, 86, 95, 152, 144, 20C, 21, 356, 357, and 329 shows a significant trans-K–T temperature rise at the surface and in deep waters of the Atlantic Ocean; their analysis of Douglas and Savin's (1973) data also indicates trans-K–T warming for the Pacific. Ekdale and Bromley (1984) indicate that the K–T boundary strata in Denmark "contain evidence to support a model of a terminal Mesozoic greenhouse"; the stable isotope record suggests a warming of 12°C in the Danish Seaway at the end of the Maastrichtian.

Oxygen-18 data from several of the most complete trans-K–T sections known indicate CO_2-induced global warming beginning about the K–T contact and peaking in the Early Tertiary. At DSDP Site 524, Hsu *et al.* (1982*b*) note a post-Cretaceous bottom water temperature increase of 5°C (from 16 to 21°C); the maximum temperature, reached in the Early Tertiary about 30,000 years after deposition of the K–T boundary Ir/clay, was about 10°C above background. Surface water warming of 2–10°C was attributed by Hsu *et al.* (1982*a*) to CO_2-induced greenhouse conditions.

Smit (1982) notes that ^{18}O data from the Gredero and Biarritz sections support a "catastrophic temperature rise" after the K–T boundary event. Perch-Nielsen *et al.* (1982) show variable results (Fig. 4). They note that geographic position plus changes in paleocirculation produced the oxygen isotope stratigraphy at the various locations, noting also that changes in global temperatures affect the higher latitude waters more than the lower ones.

Older interpretations of K–T contact cooling based on terrestrial paleofloras in the Rocky Mountain region are equivocal. An example is Van Valen and Sloan's (1977) suggestion that a decrease in the proportion of entire-margined leaves across the K–T contact indicates a decrease in temperature. Dolph and Dilcher (1979) indicate from studies of foliar physiognomy–climate relationships of modern floras in the Carolinas that leaf physiognomy and climate are not closely related, and that paleoclimate interpretations can be made only in light of certain precautions.

The CO_2-induced global warming in the Early Tertiary was simultaneous with the Early Tertiary disappearance of the dinosaurs during magnetic polarity chron R28. Interestingly, Erben *et al.* (1983) reported pathological dinosaurian egg-shell thinning in Early Tertiary strata from France and Spain, whereas Maastrichtian shells were normal. Fassett (1982) reported dinosaur bones above the level of the pollen-based K–T contact in the San Juan Basin of New Mexico.

CONCLUSIONS

Marine extinctions of calcareous microplankton 65 Mya that define the K–T contact were synchronous with the early main pulse of Deccan Traps flood basalt volcanism in India, which, in terms of the great volume of basalts deposited and its brevity of duration, was a unique Phanerozoic event. Prior to the Deccan Traps volcanism, steady state between the release of mantle CO_2 and uptake by surficial sinks seems to have prevailed. Beginning at the time of the marine extinctions and continuing into the Early Tertiary, ecological instability prevailed (Thierstein, 1981); unusual conditions in the oceans, indicated by the dominance in the marine plankton of thoracosphaerids, braarudosphaerids, and small foraminifera, are termed "Strangelove" conditions (Hsu and McKenzie, 1985), and were coeval with Early Tertiary Deccan Traps volcanism and with global warming as indicated by ^{18}O studies. On the lands, the dinosaurs became extinct during the Early Tertiary late "Strangelove" time perhaps 1.5 million years after the marine extinctions. The trans-K–T record is one of gradual bioevolutionary turnover for both the marine calcareous microplankton and the dinosaurs. Deccan Traps mantle-degassing perturbation of the carbon cycle accounts for the extinctions and for trans-K–T iridium enrichment in the marine record.

REFERENCES

Alvarez, L. W., 1983, Experimental evidence that an asteroid impact led to the extinction of many species 65 million years ago, *Proc. Natl. Acad. Sci. USA* **80**:627–642.

Alvarez, L. W., Alvarez, W., Asaro, F., and Michel, H. V., 1980, Extraterrestrial cause for the Cretaceous–Tertiary extinction, *Science* **208**:1095–1108.

Archibald, J. D., and Clemens, W. A., 1982, Late Cretaceous extinctions, *Am. Sci.* **70**:377–385.

Archibald, J. D., Butler, R. F., Lindsay, E. H., Clemens, W. A., and Dingus, L., 1982, Upper Cretaceous–Paleocene biostratigraphy, Hell Creek and Tullock Formations, northeastern Montana, *Geology* **10**:153–159.

Arthur, M. A., Dean, W. E., 1982, Changes in deep ocean circulation and carbon cycling at the Cretaceous–Tertiary boundary (abstract), in: *American Association for the Advancement of Science, 148th National Meeting*, p. 48.

Ash, S. R., and Tidwell, W. D., 1976, Upper Cretaceous and Paleocene floras of the Raton

Basin, Colorado and New Mexico, in *New Mexico Geological Society Guidebook 27th Field Conference, Vermejo Park* (R. C. Ewing and B. S. Kues, eds.), pp. 197–203.
Baadsgaard, H., and Lerbekmo, J. F., 1983, Rb–Sr and U–Pb dating of bentonites, *Can. J. Earth Sci.* **20:**1282–1290.
Baes, C. F., Jr., 1982, Effects of ocean chemistry and biology on atmospheric carbon dioxide, in: *The Carbon Review: 1982* (W. C. Clark, ed.), pp. 189–204, Clarendon Press, Oxford.
Boersma, A., Shackleton, N., Hall, M., and Given, Q., 1979, Carbon and oxygen isotope records at DSDP site 384 (North Atlantic) and some Paleocene paleotemperatures and carbon isotope variations in the Atlantic Ocean, *Initial Rep. Deep Sea Drilling Project* **43:**695–717.
Bohor, B. F., Foord, E. E., Modreski, P. J., and Triplehorn, D. M., 1984, Mineralogical evidence for an impact event at the Cretaceous–Tertiary boundary, *Science* **224:**867–869.
Boltovskoy, E., and Wright, R., 1976, *Recent Foraminifera,* Junk, The Hague.
Brown, R. W., 1943, Cretaceous–Tertiary boundary in the Denver Basin, Colorado, *Geol. Soc. Am. Bull.* **54:**65–86.
Brown, R. W., 1952, Tertiary strata in eastern Montana and western North and South Dakota, *Billings Geol. Soc. Guidebook,* 3rd Ann. Field Conf: 89–92.
Brown, R. W., 1962, Paleocene Flora of the Rocky Mountains and Great Plains, Geological Survey Professional Paper 375, U. S. Government Printing Office, Washington, D. C.
Calvert, W. R., 1912, Geology of certain lignite fields in eastern Montana, *U.S. Geol. Survey Bull.* **471:**187–201.
Clemens, W. A., Archibald, J. D., and Hickey, L. J., 1981, Out with a whimper not a bang, *Paleobiology* **7:**293–298.
Dolph, G. E., and Dilcher, D. L., 1979, Foliar physiognomy as an aid in determining paleoclimate, *Palaeontographica* **170B:**151–172.
Douglas, R. G., and Savin, S. M., 1973, Oxygen and carbon isotope analyses of Cretaceous and Tertiary foraminifera from the central North Pacific, *Initial Rep. Deep Sea Drilling Project* **17:**591–605.
Dyson, F. J., 1982, Balance sheets of the carbon and oxygen cycles, in: *The Long-Term Impacts of Increasing Atmospheric Carbon Dioxide Levels* (G. J. MacDonald, ed.), pp. 43–56, Ballinger, Cambridge.
Ekdale, A. A., and Bromley, R. G., 1984, Sedimentology and ichnology of the Cretaceous–Tertiary boundary in Denmark: Implications for the causes of the terminal Cretaceous extinction, *J. Sedimentary Petrol.* **54:**681–703.
Erben, H. K., Ashraf, A. R., Krumsiek, K., and Thein, J., 1983, Some dinosaurs survived the Cretaceous "final event" (abstract), *Terra Cognita* **3:**KA 6.
Fassett, J. E., 1982, Dinosaurs in the San Juan Basin, New Mexico, may have survived the event that resulted in creation of an iridium-enriched zone near the Cretaceous/Tertiary boundary, in: *Geological Implications of Impacts of Large Asteroids and Comets on the Earth* (L. T. Silver and P. H. Schultz, eds.), pp. 435–447, Geological Society of America Special Paper 190.
Haq, B. U., Perch-Nielsen, K., and Lohman, G. P., 1977, Contribution to the Paleocene nannofossil biogeography of the central and southwest Atlantic Ocean (Ceara Rise and Sao Paulo Plateau, DSDP Leg 39), *Initial Rep. Deep Sea Drilling Proj* **39:**841–848.
Harland, W. B., Cox, A. V., Llewllyn, P. G., Pickton, C. A. G., Smith, A. G., and Walters, R., 1982, *A Geologic Time Scale,* Cambridge University Press, Cambridge.
Hsu, K. J., and McKenzie, J. A., 1985, A Strangelove ocean in the earliest Tertiary, in: *The Carbon Cycle and Atmospheric CO_2: Natural Variation Archean to Present* (E. T. Sundquist and W. Broecker, eds.), Geophysical Monograph Series, Vol. 32, pp. 487–492.

Hsu, K. J., He, Q., McKenzie, J. A., Weissert, H., Perch-Nielsen, K., Oberhansli, H., Kelts, K., LaBrecque, J., Tauxe, L., Krahenbuhl, U., Percival, S. F., Wright, R., Karpoff, A. M., Petersen, N., Tucker, P., Poore, R. Z., Gombos, A. M., Pisciotto, K., Carman, M. F., and Schreiber, E., 1982*a,* Mass mortality and its environmental and evolutionary consequences, *Science* **216:**249–256.

Hsu, K. J., McKenzie, J. A., and He, Q. X., 1982*b,* Terminal Cretaceous environmental and evolutionary changes, in *Geological Implications of Impacts of Large Asteroids and Comets on the Earth,* (L. T. Silver and P. H. Schultz, eds.), pp. 317–328, Geological Society of America Special Paper 190.

Jeletzky, J. A., 1960, Youngest marine rock in the western interior of North America and the age of the Triceratops-beds; with remarks on comparable dinosaur-bearing beds outside North America, in: *International Geological Congress Report 21st Session, Norden, Part V, The Cretaceous–Tertiary Boundary* (T. Sorgenfrei, ed.), pp. 25–40, Copenhagen.

Jeletzky, K. J., 1962, The allegedly Danian dinosaur-bearing rocks of the globe and the problem of the Mesozoic–Cenozoic boundary, *J. Paleontol.* **36:**1005–1018.

Kaneoka, I., 1980, Ar^{40}/Ar^{39} dating on volcanic rocks of the Deccan Traps, India, *Earth Planetary Sci. Lett.* **46:**233–243.

Kent, D. V., 1977, An estimate of the duration of the faunal change at the Cretaceous–Tertiary boundary, *Geology* **5:**769–771.

Klootwijk, C. T., 1979, A review of paleomagnetic data from the Indo-Pakistani fragments of Gondwanaland, in *Geodynamics of Pakistan* (A. Farah and K. A. DeJong, eds.), Geological Survey of Pakistan, Quetta, pp. 41–80.

Krassilov, V. A., 1978, Late Cretaceous gymnosperms from Sakhalin and the terminal Cretaceous event, *Paleontology* **21:**893–905.

Kroopnick, P. M., Margolis, S. V., and Wong, C. S., 1977, Carbon-13 variations in marine carbonate sediments as indicators of the CO_2 balance between the atmosphere and oceans, in: *The Fate of Fossil Fuel CO_2 in the Oceans* (N. R. Andersen and A. Malahoff, eds.), pp. 295–321, Plenum Press, New York.

Lanphere, M. A., and Jones, D. L., 1978, Cretaceous time scale for North America, in: *Contributions to the Geological Time Scale* (G. V. Cohee, M. F. Glaessner, and H. D. Hedberg, eds.), pp. 259–268, American Association of Petroleum Geologists, Tulsa.

Leavitt, S. W., 1982, Annual volcanic carbon dioxide emission: An estimate from eruption chronologies, *Environ. Geol.* **4:**15–21.

Lee, W. T., 1917, Geology of the Raton Mesa and Other Regions in Colorado and New Mexico, U. S. Geological Survey Professional Paper 101, pp. 9–221.

Lerbekmo, J. F., Evans, M. E., and Baadsgaard, H., 1979, Magnetostratigraphy, biostratigraphy and geochronology of Cretaceous–Tertiary boundary sediments, Red Deer Valley, *Nature* **279:**26–30.

Luck, J. M., and Turekian, K. K., 1983, Osmium-187/osmium-186 in manganese nodules and the Cretaceous–Tertiary boundary, *Science* **222:**613–615.

Margolis, S. V., Kroopnick, P. M., and Goodney, D. E., 1977, Cenozoic and late Mesozoic paleooceanographic and paleoglacial history recorded in circum-Antartic deep sea sediments, *Marine Geol.* **25:** 131–147.

McLean, D. M., 1978, A terminal Mesozoic "greenhouse": Lessons from the past, *Science* **201:**401–406.

McLean, D. M., 1981*a,* A test of terminal Mesozoic "catastrophe", *Earth Planetary Sci. Lett.* **53:**103–108.

McLean, D. M., 1981*b,* Terminal Cretaceous extinctions and volcanism: A link (abstract), *American Association for the Advancement of Science, 147th National Meeting,* p. 128.

McLean, D. M., 1985*a,* Mantle degassing induced dead ocean in the Cretaceous–Tertiary transition, in: *The Carbon Cycle and Atmospheric CO_2: Natural Variation Archean to Present* (E. T. Sundquist and W. Broecker, eds.), Geophysical Monograph Series, Vol. 32, pp. 493–503.

McLean, D. M., 1985*b,* Deccan Traps mantle degassing in the terminal Cretaceous marine extinctions, *Cretaceous Res.,* in press.

National Research Council Geophysics Study Committee, 1977, in: *Energy and Climate,* p. 17, National Academy of Sciences, Washington, D. C.

North American Commission on Stratigraphic Nomenclature, 1983, North American Stratigraphic Code, *Am. Assoc. Petroleum Geol. Bull.* **67:**841–875.

Obradovich, J. D., and Cobban, W. A., 1975, A time-scale for the late Cretaceous of the western interior of North America, *Geol. Soc. Can. Special Paper* **13:**31–54.

Officer, C. B., and Drake, C. L., 1983, The Cretaceous–Tertiary transition, *Science* **219:**1383–1390.

Officer, C. B., and Drake, D. L., 1985, Terminal Cretaceous environmental events, *Science* **227:**1161–1167.

Orth, C. J., Gilmore, J. S., Knight, J. D., Pillmore, C. L., Tschudy, R. H., and Fassett, J. E., 1981, An iridium abundance anomaly at the palynological Cretaceous–Tertiary boundary in northern New Mexico, *Science* **214:** 1341–1343.

Orth, C. J., Gilmore, J. S., Knight, J. D., Pillmore, C. L., Tschudy, R. H., and Fassett, J. E., 1982, Iridium abundance measurement across the Cretaceous/Tertiary boundary in the San Juan and Raton Basins of northern New Mexico, in: *Geological Implications of Impacts of Large Asteroids and Comets on the Earth* (L. T. Silver and P. H. Schultz, eds.), pp. 423–433, Geological Society of America Special Paper 190.

Pascoe, E. H., 1964, *A Manual of the Geology of India and Burma,* Vol. III, Publ. Govt. India, Calcutta.

Perch-Nielsen, K., McKenzie, J., and He, Q., 1982, Biostratigraphy and isotope stratigraphgy and the "catastrophic" extinction of calcareous nannoplankton at the Cretaceous/Tertiary boundary, in: *Geological Implications of Impacts of Large Asteroids and Comets on the Earth* (L. H. Silver and P. H. Schultz, eds.), pp. 353–371, Geological Society of America Special Paper 190.

Pillmore, C. L., Tschudy, R. H., Orth, C. J., Gilmore, J. S., and Knight, J. D., 1984, Geologic framework of nonmarine Cretaceous–Tertiary boundary sites, Raton Basin, New Mexico, *Science* **223:**1180–1183.

Rampino, M. R., and Reynolds, R. C., Clay mineralogy of the Cretaceous–Tertiary boundary clay, *Science* **219:**495–498.

Schopf, T. J. M, 1982, Extinction of the dinosaurs: A 1982 understanding, in: *Geological Implications of Impacts of Large Asteroids and Comets on the Earth* (L. H. Silver and P. H. Schultz, eds.), pp. 415–422, Geological Society of America Special Paper 190.

Smit, J., 1982, Extinction and evolution of planktonic foraminifera after a major impact at the Cretaceous/Tertiary boundary, in: *Geological Implications of Impacts of Large Asteroids and Comets on the Earth* (L. T. Silver and P. H. Schultz, eds.), pp. 329–352, Geological Society of America Special Paper 190.

Smit, J., and Klaver, G., 1981, Sanidine spherules at the Cretaceous–Tertiary boundary indicate a large impact event, *Nature* **292:**47–49.

Smit, J., and Van Der Kaars, S., 1984, Terminal Cretaceous extinctions in the Hell Creek area: Compatible with catastrophic extinctions, *Science* **223:**1177–1179.

Stanton, T. W., and Knowlton, F. H., 1897, Stratigraphy and paleontology of the Laramie and related formations in Wyoming, *Geol. Soc. America Bull.* **8:**127–156.

Thierstein, H. R., 1981, Late Cretaceous nannoplankton and the change at the Cretaceous–Tertiary boundary, *Soc. Econ. Paleontol. Mineral. Spec. Pub.* **32:**355–394.

Tschudy, R. H., 1973, The Gasbuggy Core—A palynological appraisal, in: *Cretaceous and Tertiary Rocks of the Southern Colorado Plateau* (A Memoir of the Four Corners Geological Society) (J. E. Fassett, ed.), pp. 131–143.

Van Valen, L. M., 1984, Catastrophes, expectations, and the evidence, *Paleobiology* **10:**121–137.

Van Valen, L., and Sloan, R. E., 1977, Ecology and extinction of the dinosaurs, *Evol. Theory* **2:**37–64.

Weimer, R. J., 1977, Stratigraphy and tectonics of western coals, in: *Geology of Rocky Mountain Coal* (D. K. Murray, ed.), pp. 9–27, Colorado Geological Survey, Denver.

Wensink, H., Boelrijk, N. A. I. M., Hebeda, E. H., Priem, H. H. A., Verdurmen, E. A. T., and Verschure, R. H., 1979, Paleomagnetism and radiometric age determinations of the Deccan Traps, India, in: *Fourth International Gondwana Symposium: Papers* (B. Laskar and C. S. Raja Rao, eds.), Vol. II, pp. 832–849, Hindustan Publishing Corporation, Delhi.

Zoller, W. H., Parrington, J. R., and Kotra, J. M., 1983, Iridium enrichment in airborne particles from Kilauea Volcano: January 1983, *Science* **222:**1118–1120.

Index

Garbage As You Like It

by

Jerome Goldstein

Illustrations by Virginia Howie

RODALE BOOKS, INC.
Emmaus, Penna. 18049

Standard Book Number 87596-040-5
Library of Congress Card Number 70-83761

printed in U.S.A.
E-634
First Printing—June, 1969

Dedicated to the people
everywhere who believe we can solve
(or at least do better with)
today's "unsolvable" problems.

Introduction

How does your city dispose of its garbage? Does it burn it . . . bury it . . . or just dump it? This book hopes to make you care about what happens to garbage *after* it leaves your home, and *before* we're all buried in it.

A very large percentage of your tax dollar goes for disposing of the wastes we all generate. It costs us $3 billion each year. For our money, we really should not have to breathe it in the air, or drink it in our water, or smell it. Yet that is what is happening all over the United States. It's up to you, if anything worthwhile is going to be done.

I remember reading a great, but frightening, story by Shirley Jackson called *The Lottery*. In it, the village residents hold a town meeting of sorts which climaxes in the stoning of the person who drew the "winning" lottery ticket.

At the rate we are going on now, Miss Jackson's grim fantasy might be played out with garbage instead of stones. In an effort to find a place to dump the village trash, an annual lottery just might be needed.

The "winner" could find his homestead turned into a dump.

The longer we wait, the tougher and more critical the problem of waste disposal becomes. We can't afford to delay any longer.

Public money spent for the collection, treatment and disposal of wastes is so great that the cost is exceeded by only education and road-building. But in spite of the money we spend, the job is being done poorly. We're making a dump of our entire nation!

If we are going to solve the national crisis which exists today, we must be willing to experiment . . . to think big and new. Even the words we use must change. We must stop trying to "throw the garbage away." There is no longer any "away."

The "new" word (and concept) you are about to discover is *compost*. Though it goes back to antiquity, compost never has been really discussed in terms of Twentieth Century garbage. I hope to convince you that composting is the revolutionary idea that can save us all from being buried in our own wastes.

This then is a call to each and every one of us to "Rally 'Round the Heap!'" But the heap in this case is not a compost pile in the backyard garden. No, indeed. The heap is the millions and millions of tons of our cities' wastes—the garbage, the paper, the sewage sludge—that can be treated so as not to pollute our air, water and soil. The treatment is composting.

Contents

CHAPTER I

"Behold This Compost! Behold It Well!"

—Walt Whitman

In Spring, 1960, Rodale Press—a publishing company in Emmaus, Pennsylvania—brought out the first issue of *Compost Science*. I was and still am the editor of this journal that has article after article on the garbage and other wastes you help to produce. If we had an editorial motto, it would read: ALL THE TRASH THAT'S FIT TO PRINT.

From that very first issue, we have tried to get a message across to government officials in cities and states as well as the federal government . . . to mayors, councilmen and public works directors . . . to researchers in industries and universities. Our message is simply this:

> Getting rid of wastes by burning them in an incinerator or dumping them or trying to "lose" them in rivers is wrong. It's wrong from a national viewpoint, because of pollution and destruction of our nation's resources. It's wrong from a personal health standpoint. It's wrong—and here is where the action is—it's wrong from a dollars and cents basis. It costs more money to burn wastes than it does to SAVE them.

Yes, that's what we've been writing about for many

years. It's cheaper, healthier, and wiser to treat wastes and build mountains out of them, if necessary, than to burn them or bury them.

You as an individual citizen produce the garbage and wastes—more than a ton a year; you pay for whatever method the people you elect decide to use; you should know what happens to the garbage and other trash that leaves your home. You might as well face up to it. Simply trying to forget about your trash just doesn't work. It comes back to haunt you—in the air you breathe, in the water you drink, in the mess you see.

* * * * * *

Perhaps I've been closer to garbage than most people, but like I say, you haven't been too far away yourself whether or not you've realized it. As editor of *Compost Science,* I've visited garbage dumps and treatment plants and research laboratories in the United States and Europe. (Other world travelers may rave about Paris restaurants. I boast of lunch in the neat alcove next to the dumping hopper of a Dutch compost plant.)

* * * * * *

The crisis we face today has been amply stressed by special panels of experts brought together by the President of the United States, Congress, governors, scientific associations and private foundations. Books, magazine articles, editorials, TV programs have made it abundantly clear that—

There are more of us,
living in cities,

producing more garbage.

There are no more,
"easy" spots to dump garbage
and forget about it.

In this book, I hope to make you realize that original thinking and bold action are vital to prevent our nation from becoming more of a dump than it already is. Again and again, I want to make it clear that we have the technical know-how to treat our wastes without polluting the environment. We are already spending the money in attempts to burn or bury the wastes. So let us use that same money—in many cases *less* money—to make a potential resource of those wastes.

To solve the waste crisis, we do not have to spend more money or invent new machines. What we must do is accept a new philosophy about garbage, a philosophy that makes us realize once and for all that we don't have the means to destroy our wastes without having them come back to plague us.

Once you accept the fact that the waste crisis is actually a social disease, it follows then that society must apply the cure. That is why this book is written not for the engineer or technician, but for you as members of society. You and I, as individual citizens, are the ones who will truly decide when and how our nation's pollution problems will be solved.

* * * * * *

Pollution today is more of a public policy question than a technological problem. We must realize that fact. Because you and I may be professionally untrained does not mean we have an excuse or a handicap against becoming directly involved in the business of improving our environment. Indeed, a case can be made that if the pollution problem is really going to be solved, it must be done by the socially-aware, albeit technically-deficient.

* * * * * *

A slogan goes a long way toward dramatizing a solution to a critical problem. We remember "A Chicken in Every Pot," "A Car in Every Garage," and a "Five-Cent Cigar." I've got a slogan: "Compost in Every Dump."

Gardeners will most likely recognize the term "compost," since many home vegetable and flower growers (especially organic gardeners) make a soil conditioner known as compost from leaves, weeds, manure, grass clippings, etc. Compost is a practice recommended for gardeners by everyone from the fellow down the street who grows those great tomatoes to the Washington office of the U. S. Department of Agriculture. Compost is a biblical word, and Shakespeare has Hamlet advise: "Do not spread the compost on the weeds, to make them ranker."

In his poem, *This Compost,* Walt Whitman gave this quality to the process:

> "Behold this compost! behold it well!
> Perhaps every mite has once form'd part of a sick
> person—yet behold!
> The grass of spring covers the prairies,
> The bean bursts noiselessly through the mould in
> the garden,
> The delicate spear of the onion pierces upward,

The apple-buds cluster together on the apple-branches,
The resurrection of the wheat appears with pale
visage out of its graves.
What chemistry!
That the winds are not really infectious,
That all is clean forever and forever,
That the cool drink from the well tastes so good,
That blackberries are so flavorous and juicy,
That the fruits of the apple-orchard and the orange-
orchard, that melons, grapes, peaches, plums,
will none of them poison me,
That when I recline on the grass I do not catch any
disease
Now I am terrified at the Earth, it is that calm and
patient,
It grows such sweet things out of such corruptions,
It turns harmless and stainless on its axis, with such
endless succession of diseased corpses,
It distils such exquisite winds out of such infused
fetor,
It gives such divine materials to men, and accepts
such leavings from them at last."

But don't confuse the "Compost" term for cities with the poetical, magical, symbolic product of the backyard heap. I've been put on the defensive too long by public health officials who are ardent backyard composters but who believe incineration is the only method for city wastes. They confuse the methods and economics of a garden hobby with the methods and economics of a city responsibility.

The term "Compost" as it applies to city wastes, first and foremost, is a treatment process not an end product. As a treatment process, it should be judged with alternative methods as to costs, pollution and long-term national policy. By definition, the compost process destroys the disease germs in raw wastes and yields a stabilized residue that doesn't smell, smolder or in any way offend the eye, ear, nose or throat. At this point, I make no claims that it is good for anything. However, I do maintain that it is not bad for anything. And to those of you who have been around city garbage dumps, that is not an insignificant quality.

Of all alternative methods for treating municipal wastes, compost is the only process that creates a potentially-useful material. It *conserves* as it treats, as opposed to a process like incineration that *destroys* as it treats. Compost is an optimistic practice, conserving wastes in the belief that future researchers will develop uses for this waste. And, compared to other treatment methods available to cities, composting wastes is economical.

At this moment, *nearly* 20 *per cent* of all the city refuse in the Netherlands is converted into compost, and this quantity is increasing every year. Why can the Dutch make use of compost and we can't? The answer is simple—The Dutch Government strongly promotes composting, while the United States of America does not.

Back in 1952, a special body was founded by the Minister of Agriculture of the Netherlands, named "Compost Foundation." It has these tasks:

1. To make investigations and to advise the municipalities in the field of refuse disposal and composting;
2. To stimulate the use of refuse compost for land improvement; and
3. To promote the erection of compost plants.

Municipalities who want to construct a compost plant can make use of finances without interest charges.

Under this plan, 14 new compost plants have been built in the past decade.

The United States needs a national policy to create compost from garbage. We need a "Compost Foundation U.S.A." If this were done, much of the waste which now destroys our waterways would instead build our lands.

There is no reason why we should not learn from the experience of a country like the Netherlands. Do we have such unlimited resources that we should dissipate what we have been blessed with? We cannot afford to do so, and we must act while there is still time. The Dutch faced an immediate problem and reacted positively. We face a perpetual problem increasing in intensity, and act we must!

CHAPTER II

Making A Resource Out Of Wastes

THANKSGIVING . . . and America sits down to eat. The turkey and the trimmings, the cranberries and the pies. You can tell how good a life we lead by the stuff in the garbage can on Sunday.

Things have changed quite a bit since those first Thanksgivings. The turkeys of the Pilgrims weren't wrapped in paper; the cranberries never reached the can; the pies didn't come in aluminum plates; and they didn't "enjoy" the luxury of no-deposit bottles.

But far more than food preparation and packaging have changed. Just think of the numbers of pilgrim descendants now concentrated in urban areas.

You and the wastes you produce are causing a fantastic problem. The contents of your garbage can are making a dump of our land. And all the experts can predict is that the dump is getting worse.

* * * * * *

"Garbage" has a wide range of meanings. To many, it's the left-overs from meals, paper and plastic cartons, etc.—from orange peels and lettuce leaves to the morning newspaper. To others, garbage is a "bunch of junk," anything you don't want. Garbage is bottles and cans, plastic toys, old tires, sofas, rubber tires—even junked automobiles. Professionals in the field call it "solid waste" to differentiate it from liquid waste.

Garbage comes into existence when its owner no longer wants it. If I buy a carton of eggs, the carton soon becomes garbage to me. However, if I have a neighbor who pays 5 cents for each egg carton, that carton is of value to me. So garbage often shifts around depending upon the demand.

While persons have different viewpoints on what garbage is, cities and public health officials have tried to standardize the definition of garbage. To them, garbage is a fraction of the solid wastes that must eventually be treated by the city. It has many different sources and qualities, and varies by season and area.

As defined by the American Public Works Association, garbage "is the animal and vegetable waste resulting from the handling, preparation, cooking and serving of foods." It originates primarily in home kitchens, stores, markets, restaurants, hotels, and other places where food is stored, prepared, or served.

Refuse comprises all of the solid wastes of the community, coming from homes, institutions, industry and agriculture. It includes garbage, rubbish, ashes, dead animals, bulky wastes, abandoned cars, demolition wastes and sewage teatment residues. It includes the organic—paper, garbage, rags, grass, leaves, wood and yard trimmings. And refuse also includes the inorganic—metals, tin cans, stones, glass, bottles and other mineral refuse.

In the United States, the per capita production of solid wastes (garbage, paper, cans, mentioned above) increased from 2.75 pounds per day in 1920 to 4.5 pounds per day in 1965. Studies in the San Francisco area showed that the rate of production comes close to 15 to 17 lbs./person/day, if *all* solid wastes are included. The present total for the nation is more than 165 million tons, going up to 260 million tons by 1980.

The annual cost to our nation's economy for collecting and disposing of these wastes is astronomical—topped only by expenditures for schools and roads. The figure is between $3 and $4 billion each year.

By the year 2,000, municipal sewage returns are expected to double. By that time, 95 per cent of our population will be living in urban areas. Waste disposal through city sewer systems will average some 132 gallons per day per person.

(Solid wastes also include abandoned cars, agricultural and industrial wastes, but for purposes of this book, the main wastes referred to are those produced by city residents and treated by city agencies. The same attitude regarding utilization, however, applies to all wastes, if solutions are to be achieved.)

Not until the last few years have most city governments even admitted to having a problem with solid waste disposal. The policy has been to spread refuse around and hope it won't be noticed. This policy, of course, never was right, but at least, for the most part, succeeded — few noticed or complained about the spreading dumps.

Fortunately or unfortunately, we have outgrown that luxury.

A 1968 report on Waste Management by New York's Regional Plan Association recalled how "the New York region developed in a magnificent natural environment which was not easy to spoil. Close to 3,000 miles of tidal shoreline break up the land mass with numerous bodies of water, backed up by the Atlantic Ocean. Several mountain chains . . . a rich forest cover . . . prevailing winds provide natural ventilation superior to many large metropolitan areas.

"But a concentration of 19 million people on 2,300 square miles of built-up land has led to increasing concentrations of wastes in water, on land, and in the air.

The prospect of 30 million people living here by the year 2,000 threatens much more serious deterioration of the environment unless waste management is substantially reorganized."

The Regional Plan Association of New York report says that 200,000 tons of particulate matter (air pollutants) are discharged into metropolitan New York's air from the burning of refuse, either in apartment house and city incinerators or open dumps.

Other solid wastes in the New York metropolitan area (half of which are paper) add up to 17 million tons. About 30 per cent is incinerated, while most of the balance is disposed of in sanitary landfill and open dumps. *Only some 4 per cent is salvaged.*

The solid wastes story in and around New York City is repeated again and again all over the country. Wastes are either burned, buried or flushed—moved by either trucks, trains or barges—BUT RARELY TREATED IN SUCH A WAY THAT THEIR RAW MATERIALS COULD EVER BE USED AGAIN!

We must make a resource out of our wastes. Each year, more than a half billion dollars worth of packaging materials wind up as refuse. More accurately, they wind up as litter. Here's a breakdown of the annual U. S. production of packaging materials as reported by the President's Science Advisory Committee on solid wastes:

36,000,000,000 lbs. of paper (188 pounds per person).

48,000,000,000 metal cans (250 per person).

26,000,000,000 bottles and jars (135 per person).

65,000,000,000 metal and plastic caps and crowns (338 per person).

State Senator Whitney North Seymour, Jr., of New York has made a study of the "no deposit, no return" trend. The first "n-d,n-r" soft drink bottle was introduced in 1948. By 1950 total shipments amounted to only 168,000 gross. By 1960 this had increased ten times to 1,730,000 gross bottles. By 1965 the figures had skyrocketed to 6,989,000 gross soft drink bottles. Current predictions indicate a further jump to 28,200,000 gross soft drink "n-d,n-r" bottles by the year 1970. In that same period it is expected that annual sales of "n-d,n-r" beer bottles will be 55,800,000 gross.

You can see how quickly the solid waste problem can be made infinitely greater by a change in packaging methods. It's obvious how population concentrations in cities will overwhelm current facilities even more. Discarded materials in such astronomical amounts pose a nightmare of uncertainty for those who must solve the problem and we who must live with it.

Government control over packaging materials offers one possible approach. Another can be to enforce households to sort out each type of trash. Charges can be levied upon the producer; Sen. Seymour suggests charging container manufacturers a "disposal fee" for each bottle or can shipped into a state. But such charges would quickly be passed on to the consumer. And there would still be the problem of what to do with the bottles and cans anyway.

The only solution is developing a built-in system of re-use. Such a system cannot be governed by present standards enforced by the salvage industry which is made up of the junk and scrap dealers. They buy only a tiny fraction of the total junk our society produces. *We cannot let those firms create our nation's waste treatment policy.*

* * * * * *

Our cities have been basing decisions on waste management upon whatever procedure is cheapest for the disposer (the Public Works Department, for example). But the decision may prove to be costliest for the city dwellers 25 years hence. A new brand of economics is vital to waste management. We can no longer afford to leave the decisions on waste management up to leaders in our public works department . . . to sanitary engineers . . . or to conventional-minded public health officials. *They need our help to change!*

Warnings of eventual calamity come from our most respected scientists. What are we doing about it? Very, very little. We've had presidential commissions; we've had foundation studies; we've had special panels of city, industrial and university authorities. Each publication has been more depressing than its predecessor. But we fail to act in a way that recognizes the problem as being less technical than social.

To prepare an eloquent analysis of waste management and then expect the city's garbage agencies to do something effective is like giving the Kerner Report on Civil Disorders to the city police departments to administer. Just as the city police department doesn't have the power or the qualifications to eliminate urban conditions in ghettos that breed riots, so city public works departments alone can't bring about effective solutions to waste disposal. This is not to say that a better job cannot be done by both police and public works departments. But it must be recognized that for the best job, the conditions creating the problem must be corrected.

Wastes which you and I produce are making our cities and countryside unfit. It's time you and I let the right people in government know we want the right solution!

We must stop thinking that the No. 1 and No. 2 methods are bury or burn. If we are to use any B treatment for wastes, let us think of BUILD.

CHAPTER III

Composting -- Waste Treatment for the Future

Compost is one of those words that, to be understood, needs a frame of reference. Definitions range from the coldly analytical to the hotly emotional. Here are some examples of what I mean:

Compost Definitions

To engineers:

A controlled "selective incineration" conducted at temperatures as high as 160° F. to 170° F. resulting in a usable, stable safe "ash" of high organic content.

An aerobic, thermophilic (high temperature) degradation of putrescible material in refuse by microorganisms.

A waste management "factory."

To gardeners, conservationists and the "hopeful" in all fields:

A mixture for fertilizing or renovating the land.

A process for making humus out of a heap of decomposing wastes.

A practice every home gardener should use to convert weeds, leaves, grass clippings, kitchen and garden wastes into a useful soil conditioner.

The very symbol of life in the soil and the cycle of life.

The most effective way to "recycle" wastes.

To city officials:

A highly controversial method for treating garbage that should be "looked into" when the dump is forced to close and incinerator bids come in $5 million higher than anticipated.

To skeptics in all fields:

An unproven, over-ballyhooed process with no

application to U. S. conditions since the finished compost can never be sold. A method that is okay for Europe and other parts of the world but "it can't happen here."

To the millions who produce wastes and help create the present crisis:

You, the reader of this book, are in this category. You may never have heard the word "compost" before, let alone have any firsthand knowledge of soil improvement. By the time you have finished this book, I hope your definition of the word "compost" will be "hope for a new, cleaner, healthier, pleasanter kind of world."

* * * * * *

For the purposes of this book and its application to our nation's solid waste problems, composting means the controlled treatment of garbage and other common wastes so that a hygienically-safe end product is the result. The process ordinarily involves materials-handling, sorting, shredding, heat generation, temperature and air controls, and stock-piling. When composting is also considered as a "manufacturing" process yielding a marketable product, packaging and shipping would be included. Depending upon salvage markets for paper, metals, plastics, etc., the thoroughness of sorting will vary.

When evaluated as a waste treatment process, large-scale composting is technically sound, hygienically proven and economically feasible.

The composting method holds the key to whether

or not we will be able to save our nation from becoming one large dump. The return of organic wastes to the soil in a manner which does not violate environmental laws is the most sensible, the most economical, the most constructive treatment process. Instead of having an uncontrolled flood of wastes to contend with, we find ourselves blessed with a continuous source of soil-building material. Certainly humanity should not ignore such a potential that could offer a tremendous boost to the world's ability to feed its inhabitants.

To be valuable as a soil-builder or as raw material for developing other products, compost need not be bagged and marketed by conventional means. Nor, in fact, need the compost method be justified now by claiming that all of the end product will be used in a short period of time.

Composting as a process should be judged successful simply because it treats wastes in a manner that destroys the dangerous germs and polluting agents in raw wastes, and conserves what is potentially useful. As one authority from the University of California puts it, a simple composting process is a way of returning wastes to the soil with minimum insult to nature.

We now have sufficient technology—the machines and the know-how—to treat many wastes so that the end product can be used as a soil conditioner. Or the composted wastes can merely be stock-piled—with no odor, at minimum expense, and can offer a natural resource to be tapped any time our economy is ready to make use of it.

Just as this country needs a new attitude toward

waste utilization, so must it adopt a more realistic approach to the concept of composting. No longer can we afford the luxury of the pure scientist saying about composting: "This method holds potential for the future." We must wake up to the fact that this method holds the key to *today's* waste treatment problems.

* * * * * *

Composting has gotten off to a slow start in the United States. In many countries in Europe, the idea behind composting—putting wastes to use—fitted in well with agricultural conditions, location of cities near farm areas, etc. As a result, Europe has had a number of composting plants and the method is recognized as an effective waste treatment alternative to landfill and incineration.

In the United States, however, composting has always been considered something of a gimmick. In the beginning, before pollution became so obvious a threat to our society, too much stress was given to the magical

qualities of composting—the "garbage to gold" idea. As a result, the idea attracted speculative investors ready to make their fortune in the garbage composting business. Excessive promises were made to cities; mayors and councilmen were told that composting could be done for practically nothing since sale of the end product would pay for the entire treatment costs.

In practice, things didn't work out that way. Only a fraction of the compost was sold and that fraction did not always cover expenses. Sometimes the private companies who contracted with the city went into bankruptcy, and the garbage went back to the dump.

Actually, the main loser in the composting trials to date has been the composting idea itself. The city signed a contract—often for 20 years—to deliver its garbage to a plant built by the composting company, agreeing to pay a set price. When the plant closed down, the company didn't seem to suffer too much from the bankruptcy proceedings in many cases. But once again the scoffers could say: "You see, I told you composting doesn't work in the United States."

Like the skeptics who hung around Kitty Hawk waiting for the Wright Brothers to fail, composting has had more than its share of sideline critics. Unfortunately, they've had more than their share of opportunities to say "I told you so" when another plant closed down. But if these same people stood around the incinerators or landfill dumps in this country, they would at least have a better notion of the nature of the problem. Garbage, sewage and other organic wastes are one big mess to tackle. Getting off the ground with an effective waste treatment method has given us more of a problem than the one which faced the Wright Brothers.

The criticism of composting's lack of success in the U. S. must be equated with the lack of success of incinerators. Not one incinerator presently operating in the U.S.A. satisfies air pollution quality standards. Concerning landfills, there's a dump within a few minutes of our nation's Capitol that would make any of the compost plants forced to close down look good by comparison.

* * * * * *

Few technical publications in the waste treatment field feature articles on composting. At conferences concerning solid waste treatment, composting rarely gets a major spot on the agenda. Because there never has been a trade industry group, such as that representing incineration or landfill, there have been no major workshops comparable to those developed by the companies actively engaged in those fields. Nor has there been much expertise in composting developed by the consulting engineers who advise municipalities.

To overcome this lack in the communications field, Rodale Press began publishing Compost Science in 1960. This quarterly magazine has reported technical and practical developments in the field of large-scale composting. The statement of purpose in our first issue read as follows:

> "At the outset, we want to make it clear that we have no process to sell or product to promote. We are publishers and editors thoroughly convinced that there is a need to conserve this coun-

try's as well as the world's natural resources. We believe that converting municipal and industrial organic wastes into useful products would be an effective step forward in a long-range conservation program. And along with the conservation benefits is an aspect equally as important—that of developing a treatment process that does not create subsequent problems in water or air pollution."

Actually, the composting process does nothing more than set up a systematic materials-handling procedure for garbage and other wastes. Garbage is the raw material, and the manufacturing operation involves receiving, processing and turning out "finished products." Dr. Harold B. Gotaas, Dean of Northwestern University's Technological Institute describes the operations in a large-scale composting plant this way in his textbook, *Composting: Sanitary Disposal and Reclamation of Organic Wastes:*

1. Reception of the refuse—hopper with conveyors for trucks to unload.

2. Sorting—removing materials that people throw away for which there is a commercial demand. Such materials may be rags, metal cans, hard plastics, paper, etc. These are segregated mechanically, magnetically or manually, then taken to scrap dealers.

3. Preparation of the compost—shredding, pulverizing or grinding to make decomposition easier and faster.

4. Stabilization area—open or enclosed location where garbage actually "purifies" itself of any disease

germs, harmful parasites, etc., through creation of high temperatures and microorganisms. In this stage, controls measure temperature and regulate air and moisture in the refuse.

5. Removal of material to "holding" site. Treated compost can be stockpiled on small acreage and used as needed. Depending upon regional conditions, compost can be used to reclaim public or private lands, as an ingredient in fertilizer, or as raw material by industry.

Over the years, Compost Science has published reports from experts all over the world. These men are engineers, chemists, microbiologists, biologists, agricultural scientists as well as ecologists, conservationists and aware citizens. They are knowledgeable in their fields. I believe the following excerpts from their reports will make it clear to you just how much testing has already been done on composting and why it is ready to be adopted now.

Feasibility of Composting

> Any waste treatment process should be feasible; safe from public-health, accident, and nuisance hazards; and economically competitive with other satisfactory treatment methods. Research has shown the technical feasibility of composting U. S. municipal wastes. Numerous plant installations abroad have shown the practicality of composting refuse or refuse-sludge. Most of the European plants are publicly owned and operated and, while generally eco-

nomically competitive with other means of refuse disposal, are nonprofit-making.

—John S. Wiley
Sanitary Engineering Director (Retired)
Public Health Service
Dept. of Health, Education & Welfare

Every Disposal Method Costs Money

Nowadays every method of refuse disposal costs money. Good, aesthetic and hygienic methods cost more than bad, unaesthetic and unhygienic methods. We are of the opinion that composting is one of the most attractive methods because part of the refuse is employed for maintaining the fertility level of the soil. As far as costs are concerned, at this moment composting can compete with other justified methods of refuse disposal in the Netherlands.

—B. Teensma
Stichting "Compost"
Amsterdam, Netherlands

Preventing Domestic Fly Problem

In the Bureau of Vector Control, we entertain great ambition for the composting process since we sincerely believe that this practice, conducted extensively, represents the most feasible basis for *preventing* the domestic fly problem.

—Richard F. Peters, Chief
Bureau of Vector Control
Department of Public Health
Berkeley, California

Principle for Success

Without doubt the composting of refuse can be a most successful form of disposal in many cases, providing the following principle is understood at the outset:

> It is not prudent to expect to show a profit as a result of the operation. Refuse disposal is a public service which in any event must be carried out and there is no justification for expecting compost production to be a profit-making industry any more than say, land reclamation or incineration.
>
> —A. G. Davies
> Chief Public Health Inspector
> Woking Urban District Council
> Woking, Surrey, England

Composting and Sewage Sludge

One of the major advantages of composting refuse is that many other troublesome solid organic wastes also can be handled in the process. For example, John Wiley of the Public Health Service found that a number of composting plants in Europe were successfully utilizing either raw or digested sewage sludge. Research and experience with pilot plants indicates that sewage sludge could also be processed successfully along with refuse under United States conditions. C. Golueke and H. B. Gotaas state . . . "It would appear that any amount of sludge could be added to refuse for composting provided the sludge was dewatered so that the moisture content of the mixture would be satisfactory."

—Ralph J. Black
Office of Solid Wastes
Public Health Service
Cincinnati, Ohio

Public Health and Refuse Disposal

In my opinion, solid and liquid refuse material should be considered as a unit, since they pose similar disposal problems which must be solved. And I believe that composting refuse and sludge offers a method that is able to coordinate the interests of engineers and agriculturists as well as hygienists. In composting, the emphasis is on the use of hygienically-acceptable urban refuse, which carries no danger to humans or animals.

Through composting, the biologically-directed, self-purification process of waste products can be supported and their pollution dangers eliminated. These wastes can then be returned to the soil as an ideal medium which improves the soil.

—Dr. K. H. Knoll
Medical Faculty
Justus-Von-Liebig University
Giessen, Germany

Composting Municipal Garbage in Israel

Interest in composting municipal garbage has developed rapidly in the past five years in Israel due to the growing awareness that modern engineering has put composting on a rational production basis.

The sanitary engineer has seen in composting a practical and sanitary way of disposing of urban garbage, while the agriculturalist sees garbage composting as an important additional source of organic fertilizer much needed to replenish the tired and heavily worked soils of Israel, which have been tilled since Biblical times and before. This community of interests has resulted in a trend toward composting which may put Israel ahead of many other countries in this respect.

Agricultural authorities in Israel estimate that the country badly needs some one million cubic meters of additional organic fertilizer per year.

Composting all of the garbage from Israel's urban centers would make available about 260,000 cubic meters of compost annually which would supply about 25 per cent of the country's needs. From these figures it can be seen that the objective need for conversion of urban wastes to compost is a real one and that agriculture can absorb as much organic fertilizer as can be produced from garbage.

—Hillel Shuval
Chief Sanitary Engineer
Director, Division of Sanitation
Ministry of Health, Jerusalem
Israel

Composting Vegetable and Fruit Wastes

The results of preliminary investigations indicate that composting offers a feasible and esthetically acceptable method for disposal of high-moisture wastes, such as fruit and vegetable solids.

Dry materials, such as municipal compost or rice hulls, can be used to absorb the moisture from fruit waste solids and to adjust the moisture concentration of the mixture within the range optimum for high rate composting.

Offensive odors did not develop in the fruit waste solids. During the initial stage of the compost process slight fruit fermentation orders were noticeable. The odor of ammonia could be detected in the final stage of the process. Fly breeding in compost was not a problem. Fruit pieces exposed in the compost pile attracted flies but egg deposition was not observed.

—Walter A. Mercer, Walter W. Rose
National Canners Association
Research Foundation
Berkeley, California

What It Is and Why

The high rate composting of domestic refuse may be described as a process in which materials such as carbohydrates, fats, and proteins are oxidized and stabilized biologically by thermophilic (heat-loving) organisms in a few days. The end product is a stable organic material ordinarily referred to as humus.

Almost any organic material—feathers, bones, leaves, grass clippings, wood, garbage, rubbish, and paper may be composted. Composting disposes of most all organic refuse from the home without odor, smoke, flies or bugs, and the pathogenic (harmful) organisms are destroyed. The addition of compost humus to the soil helps to dilute and reduce toxic substances already in the soil (pesticides, herbicides, etc.). The compost humus acts as a soil conditioner as it helps to retain water. Compost from domestic refuse including garbage usually contains significant amounts of nitrogen, phosphorous, and potash.

—J. W. Clark
Professor of Civil Engineering
New Mexico State University

CHAPTER IV

How One U.S. City Got A Compost Plant

Houston, Texas, is one of the few cities in the United States to have a composting plant. Located on Lawndale Road not far from the Houston Ship Channel and a sewage treatment facility, the plant offers a good example of what is involved in composting refuse. The story of how composting came to Houston is also typical of the trials and tribulations of waste management.

* * * * * *

"Like breathing air at the bottom of an 8,000-foot dump." That's how life for Houston residents was described several years ago when the *Houston Post* ran a series of articles on local pollution. It went on to point out:

"Air pollution will stop whenever enough people rise up and demand that all individuals and interests stop using the public air as a private dumping grounds for disposal of their waste materials. It will cost the offenders money, of course, if they are compelled to give up their free dumping ground. But it is an expense that they now are passing on to the public at large. These costs already are high in terms of health as well as dollars and are increasing all the time. The question is at what point enough people will become sufficiently aroused to demand effective action."

Lee McLemore, Houston City Councilman and chairman of the city's Garbage Committee, knew that the Houston incinerators were beyond repair and that the old garbage dump was simply running out. The city had its back against the garbage wall when Councilman McLemore decided that compost could play a definite part in his city's waste treatment problems.

On November 30, 1964, his Garbage Committee issued a "Report on Disposal of Residential, Commercial and Industrial Waste of the City of Houston." The report, which took two years to prepare, included these comments:

The problem we of the Committee faced from the beginning was the fact that in Houston there is no depressed land—such as canyons, dry river beds, or gravel pits—which would be suitable for sanitary fills and that, in any event, sanitary fills depress the surrounding real estate, causing slum areas and undesirable development, thereby causing a loss in tax revenue.

The answer seemed to lie within the three most popular methods known in the field today: (a) incinerators; (b) landfill; (c) compost, or other mechanical means, in disposing of garbage.

INCINERATION has been greatly improved in the past few years for the control of fly ash, rodents, and odors, subject to air and water pollution control, so that today modern incineration can take care of the average garbage needs in most localities. There still remains the ashes, which must be landfilled.

LANDFILL, as we have previously stated, is not desirable in low, flat areas where the water table is so close to the ground, such as in the Houston area. It is our suggestion that landfill be used only for solid materials, such as building materials, brick, rocks, and ashes from our incinerators. Such landfill sites will be abandoned sand pits, ravines, and other areas where reclaiming such areas could and would restore land values suitable for residential or commercial development.

COMPOST, or other mechanical methods. The majority of the Committee believe these methods, although new in America, merit a research and development program. Compost does eliminate odors, rodents, flies, and provides a useful soil-building product that is marketable. Private enterprise could invest money in this program for the building of such plants and the City could guarantee so many tons of garbage per day at a stated amount bid on by such companies to amortize their investment over a twenty year period.

* * * * * *

The councilman ran into a lot of opposition from the sanitary engineering experts on the city staff, but several advantages to composting made him stay with the idea. First, when it came to incinerators, the city would have to pay for the cost of building the plant—more than several million dollars worth. With composting, the cost of plant construction would be paid by the private company who got the contract for composting refuse. Second, Houston prides itself in its dramatic growth and civic accomplishments. Here was a chance to show the nation how to solve its waste problems. Third, the incinerator had not been built yet which didn't present an air pollution problem of some kind and cost at least one dollar more per ton than composting to operate. As stated above, landfills simply didn't work with the lack of suitable land areas.

After advertising for bids, the city signed contracts for 3 compost plants and 1 incinerator in March, 1965. One firm, as part of its agreement with the city, located its compost plant adjacent to a residential district and never really got underway because of complaints from neighbors. One compost plant ran into financial difficulties and the plant itself was never built. The one incinerator plant, paid for by the city, is still having major pollution problems. (You can see why McLemore says that garbage defeats more city councilmen than any other single issue.)

And that leaves one composting plant—Lone Star. I visited the plant in April, 1968, when the plant was just beginning its second year of operation.

* * * * * *

A city garbage truck had just driven up to the scales alongside the front entrance of Houston's composting plant, and a sanitation worker brought the "delivery ticket" to the weigh-in checker, as I got out of the car.

It was a rainy Thursday morning in early April, and a steady flow of trucks ran the same pattern—driving up to the scales, the counter in the front office automatically recording the weight, and then unloading its refuse onto the ramp at the back of the plant. During the day, some 100 deliveries would be made, carrying about 350 tons of refuse—about one-sixth of the trash collected in Houston—to the compost plant which cost about $2 million.

In a series of efficient steps, the refuse is sorted . . . ground and reground . . . digested for 4 days where it is stabilized enough to be stockpiled outdoors without creating offensive odors or attracting hungry rodents. Refuse is processed like the raw material it is—the speed of conveyor belts is varied as necessary to move refuse from one processing area to the next.

The 20-year contract between Houston and Lone Star Organics, Inc., operators of the composting plant, provided that the city would pay the plant $3.53 for each ton of garbage handled, with the price to be readjusted yearly on the basis of the federal cost-of-living index. A readjustment in April, 1967 set the fee to the city at $3.78 a ton for the first 300 tons and $3.40 for any excess tonnage received in a day.

City Councilman McLemore, the man who was determined to get a compost plant in Houston, believes the arrangement has been most successful. Comparable costs for an incinerator, which the city would have had to finance, amount to over $6 per ton so the city is saving a substantial amount of money since the plant began in early 1967.

Observes another councilman: "If we had 4 plants like this one operating around the city, our garbage problem would be solved."

Approximately 2,000 tons of refuse per week—100,-000 tons per year—are received. About two-thirds of this tonnage winds up as compost.

The plant operates 6 days per week, two shifts with approximately 25 men working each 8-hour shift. Heaviest collections are on Monday and Tuesday, with most trucks arriving at the plant between 9:30 a.m. and 2 p.m.

A typical sample of the material received at the refuse hopper goes like this: 75 per cent paper, cardboard and rags; 13 per cent garbage and garden trash; 7 per cent tin cans; 15 per cent large pieces of iron and steel; 2 per cent rubber, heavy plastics, etc.; 1 per cent glass and 0.5 per cent aluminum and non-ferrous metals.

The plant is located alongside a railroad siding so some salvage material—such as scrap metal—is loaded directly into freight cars. This material is delivered directly off a sorting belt. Much of the paper is salvaged, going through a baling machine and a contract has been signed so that all paper is sold to one paper salvage dealer. A similar arrangement has been signed for the salvage of tin cans.

Aluminum, brass, tires and heavy plastics and other nonferrous materials must be removed manually, since they are not picked up later by the magnetic separator. Likewise with large hunks of iron or steel which can damage the grinding mills. Noncompostables (2 to 3%) are no problem as clean landfill.

The dumped refuse is pushed onto a conveyor belt, leading to a sorting area. Sorting is done by as many as 12 men at the "picnic" table—6 on a side. Materials for salvage are picked out by hand and dropped onto cross-moving belts.

Grinding is an essential step in the composting process, since the uniformly-small sized material permits quick, controlled decomposition of organic wastes. At the Houston plant, a 500-horsepower hammermill accepts the material from the conveyor.

Still another separation is done magnetically—and that's how 25 tons of tin cans per day are removed. The cans are later transported by freight trains for cementation use in the mining and recovery of copper. The moving conveyor passes the compostable refuse on to a second hammermill which reduces particle size further.

The ground refuse is conveyed into one of four digestion tanks or bays—each 360-feet long, 20-feet wide and 8-feet deep. The material usually stays in the digester for four days.

The digesting unit is equipped with two blowers which can provide air to a plenum chamber underneath the compost bed. The air is directed into the composting material through the hollow floor with slotted panels at various points by dampers, and the object is to raise the temperature to 135° F. in the first 24 hours. At the end of 48 hours, the temperature should exceed 145° F. The final range is between 165° to 170° F.—considered ideal for thermophilic composting. Excess moisture can be drained off the material in the tanks.

The composting bed is turned over by a 15-ton agitator traveling on rails, which also empties the bed onto conveyor belts. Records are kept on the material for temperatures reached, air flow, moisture and nutrients.

After leaving the digestion tanks, the material is stockpiled outdoors, much of it is hauled to a site at Hastings about 20 miles away.

The company has formed a marketing division to distribute the finished compost under its brand name of *Alive*. A handsomely-printed folder proclaims the many virtues of the "superior organic compost" with its humus-like characteristics, high water-retention capability, etc., and what a job it can do on golf courses, lawns, nurseries, farms and gardens.

Metro Waste of Wheaton, Illinois—the parent company of Lone Star Organics—has been exploring other potential markets and uses. These include: Working with fertilizer companies to market the product as a blend of organics and chemicals; experimenting with the use of compost instead of wood fibers for hydromulching; working with a paper company on the use of compost for reforestation programs; working with people involved in reclamation and conservation work such as strip mining companies, land developers, and government agencies.

An article in the *Houston Post,* noting the opposition to other compost plants, included this observation: "Mrs J. O. Maddox, a veteran pollution fighter who lives about 8 blocks from the plant, went before the City Council and complained violently about the plant in its early days. Today she says: 'I think they're doing all right. We don't even seem to have as many flies out here as we used to have. I think they deserve a pat on the back.' " As part of its campaign to improve community relations, plant officials open the plant for inspection visits by area residents.

Well into its second year of operation, the composting plant is performing its job well and is showing how city wastes can be converted efficiently and economically into a useful material.

CHAPTER V

Why The Old Methods Don't Work

Ever since the first apple core hit the Garden of Eden, man (and woman) has dumped trash. But until recently, not as much stuff was allowed to become trash. Built-in obsolescence was unknown. The value of goods was measured by longevity. Cast-offs were remade with the garbage going to the hogs and chickens.

Dumps used to create relatively little hazard or unpleasantness. Plenty of room existed on the outskirts of any village, town or city to locate a dump, and besides, rats and flies were the accepted enemies of people on farms where most people lived.

But the times have changed entirely. Today millions upon millions of people crowd into city apartments with practically nowhere to keep anything. Closets, garages, attics and cellars (when those exist) must be cleared out regularly.

Space is at a premium outside as well as inside. As land values increase around large cities, space for such an unproductive and unsightly area as a public dump is out of the question. Sloppy procedures at dumps have rightfully earned their rejection. (A cubic foot of raw garbage can create 75,000 flies.)

Yet there's a good chance the contents of your garbage cans may wind up in an open dump—one that's getting further and further away and costing you more and more tax money for trucking. And the way people, apartments and housing developments are increasing, that dump may soon be closed up by the people living nearby . . . so another dump will be located further away.

When managed properly, a dump is not a dump but becomes a sanitary landfill. The sanitary landfill is defined by the American Society of Civil Engineers as: "A method of disposing of refuse on land without creating nuisances or hazards to public health or safety, by utilizing the principles of engineering to confine the refuse to the smallest practical area, to reduce it to the smallest practical volume, and to cover it with a layer of earth at the conclusion of each day's operation, or at more frequent intervals as may be necessary."

But sanitary landfill is the elegant-sounding name for what the National Academy of Sciences (NAS) committee calls "probably the most widely suspect" method of waste disposal.

There are three general kinds of landfill. First, that which is unrestricted in regard to material. This must be located in some spot where there is no possible chance of ground water pollution. Such sites are difficult to find and depend on thorough geological investigation. A bit more readily discovered spot is the site that can take ordinary municipal waste refuse, excluding the liquid wastes from industries of most kinds. Here, too, great care must be taken to select a spot where ground water will not be contaminated.

Finally, there is the kind of landfill which is used only for materials that will not decompose or leach. These can, supposedly, be placed almost anywhere that fill is needed, without any chance of later problems with pollution.

Sanitary landfill is essentially just dumping, except that a tractor first moves the earth away, then pushes it back over the rubbish and tamps it down. There are, supposedly, no odor, no flies, no vermin. So most communities do not object to sanitary landfill as vehemently as they do to open dumps.

But the problem of space still remains. Where will you find the spot meeting the specifications above where the people of outlying towns and suburbs will not object to having the city's wastes dumped in a landfill? The final product—a flat piece of well-tamped soil—can be made into a park or a golf course, a playground or a picnic ground, although it is not suitable for building homes for many years, and then, only with highly expensive foundations. This is because the earth must settle, and certain degradable materials in the garbage must decay, with some ill-smelling and possibly explosive gases rising to the surface of the ground. The tendency of the refuse to subside can cause broken water and sewer pipes.

The NAS book points out that there is a high probability of water pollution with sanitary landfills. And the ecological danger to marshlands and swamp areas is very real. Filling such spots with *anything* disrupts completely the chain of life that binds seaweed, algae, the millions of growing things in the water, fish, water birds and land birds, and thus, eventually man who also depends on the success of every link in this chain to assure him of enough food and the continued services of his best insect-eating friends, the birds. To say nothing, of course, of the educational and esthetic value of such lands and the pure and simple joy to be had of listening to the peepers in the spring, watching the flight of herons and bitterns, gathering the swamp grasses and reeds that grow along the edges.

Waste Management and Control, issued by a special government panel, from the National Academy of

Sciences, tells about the landfill practices of New York City which dumps about one-third of its solid wastes in a tidal marshland area on Staten Island. And "during the past 16 years New York City has filled 176 acres of low land in the East River and Westchester Creek with six million tons of refuse at a cost of $31.5 million. The cost was less than it would have cost to burn it and the 176 acres of land were 'reclaimed' for useful purposes."

The book continues, "Concern has been expressed over the effects of sanitary landfilling on the ecology of tidal and swamp lands. Undoubtedly the answer is one of degree but there is need for study of the broad, long-term requirements of our population on a regional or national basis. There are, for example, millions of acres of tidal and swamp lands on the east coast. The ecology of most of them should not be disturbed. However, the economics of waste disposal indicate that some of these regions—small in terms of total acreage though large in terms of solid waste disposal needs—should be committed to development by sanitary landfilling."

In cities bordering oceans, it had been customary to haul the solid wastes out to sea and dump them, or dump them close to land, thus filling in the coastline. The NAS book states, "no criteria or standards have been developed in this connection. Economic interest in solid-waste disposal complicates the present situation: in many instances land speculators achieve considerable profit, not only in the initial landfill of marshland or low areas but also through eventual subdivision and sale for residential development. This is not always the best use to which areas could have been put, in terms of the public interest."

One would have to judge that this last sentence is the understatement of the year, if one reads an article on San Francisco Bay which appeared in the March-April, 1968 issue of *Audubon* magazine. Says Harold Gilliam, the author, "To the four million (people who live around San Francisco Bay) this 50-mile long body of water is space and light and color, the central feature of the region and its greatest natural asset. It is a colossal theater for the drama of the weather—for gray clouds of winter trailing long banners of rain, for floods of summer fog that flow through the Golden Gate and over the hills to the bay in slow-motion cascades, for the changing play of light on a surface that reflects patterns of illumination from low-passing cumulus clouds or wisps of fog, that mirrors a spectrum of colors at sunset, then turns to a black gulf surrounded by glowing constellations of cities along the shores for 100 miles."

And what is becoming of San Francisco Bay? It has become the precinct of what Mr. Gilliam calls the "Filler Barons" who have managed to fill a third of the original bay with landfills and artificial islands and "threaten to shrink what is left to a thin, befouled channel." "In place of its vast expanses of marshland and open water will be mile after mile of subdivisions, tract housing, stores and factories."

He calls this exploitation "greed and avarice masquerading as progress."

Some of the figures Mr. Gilliam gives of the situation around San Francisco Bay are horrifying in their portent. The bay's original 680 square mile area has been diminished by filling and diking to 400 square miles. Of 300 square miles of natural marshland there are only 75 left. And almost everything else that remains in the way of shoreline around the bay is already marked for filling or diking by some city, county or private developer.

"One of the major incentives to filling is the opportunity it offers to dispose of domestic garbage," says Mr. Gilliam. "There are some 32 disposal sites around the shores of the bay, where dirt is spread over the refuse in 'sanitary landfills'. Water pollution officials frequently warn the garbage dumpers about fouling the waters by inadequate coverage."

He goes on to say, "There is urgent need for research on the economic feasibility of such rational methods as composting—reusing the refuse for soil enrichment and other constructive purposes."

Liquid wastes around the bay have been largely confined to sewage plants, within recent years, but the plants cannot keep up with the increasing amounts of water and large quantities spill into the bay. Mr. Gilliam also points out the grave danger of building homes and other buildings on filled land in an area so subject to earthquakes as is the southern California area. Some geologists declare all further building on filled land should be prohibited.

The complexities of the situation are manifest in the story of two planned "fills." A private corporation planned a fill of 1,000 feet into the bay and 2,000 feet along the waterfront at Sausalito. "Where there is now open water and a small sandbar frequented by shorebirds, there would be twenty buildings, including 130 apartment units, 160 hotel rooms, two restaurants, a coffee shop, boat facilities and a shopping area. About half of the fill would be covered by parking lots."

The other was a city plan to enlarge Berkeley to twice its present size by filling in the 4,000 acres of the bay which are owned by the city. Out of the immensely complex political and economic maneuvering around these two issues came a state-sponsored San Francisco Bay Conservation and Development Commission charged with submitting a plan to the 1969 California legislature, to provide for all future development of the shoreline and preservation of its scenic beauty. Since it began its work, the Commission has limited filling of the bay to about 110 acres, but, says Mr. Gilliam, pressure on the legislature by economic interests will be immense and can only be offset by activity of conservation groups of all kinds.

An even more frightening picture of what we may be doing to our natural environment by dumping wastes indiscriminately into the ocean is described by LaMont C. Cole, Professor of Ecology at Cornell, writing in the *New York Times Magazine,* March 31, 1968. He says that we are dumping vast quantities of pollutants into the oceans. These include, according to the Food and Drug Administration, perhaps a half million different chemical substances. Many of them are recently developed. We know little about the potential for harm of such products as pesticides, radioactive substances, detergents. No one has had time to test more than a minute fraction of these chemical substances on even a tiny part of the vast population of microscopic ocean plants which, Dr. Cole tells us, produce most of the world's oxygen, to say nothing of the ocean bacteria and microorganisms that produce its nitrogen.

If the Torrey Canyon had been carrying a cargo of herbicides rather than oil, the effect might have been to kill all these growing things in the North Sea—the growing things that are important for the process of photosynthesis, and what Dr. Cole calls "the ultimate disaster" might well have ensued. By this he means the disruption of the nitrogen cycle on which all living things on earth depend. "Depending upon which step broke down," he says, "the nitrogen in the atmosphere might disappear, it might be replaced by poisonous ammonia or it might remain unused in the atmosphere because the plants could not absorb it in gaseous form."

Wouldn't it take many years for such a catastrophe to occur? It might and it might not. As with so many things in our environment with which we meddle carelessly and recklessly, we just don't know. Dr. Cole gives the impression that such a catastrophe might happen almost overnight if we continue to pollute the oceans with wastes of completely unknown toxicity.

* * * * * *

Other alternatives for waste disposal are being tried in some areas. Philadelphia recently decided to ship its wastes to the northeastern Pennsylvania anthracite areas for disposal in abandoned strip mines. City council approved an ordinance authorizing a four-year experiment by a land reclamation company which will haul the trash to the strip mines in sealed railroad cars and dump it there. According to Philadelphia spokesmen, the program will help greatly in abating Philadelphia's air pollution problem, since it will cut down on the amount of wastes to be burned.

And, according to the folks in Philadelphia, the unsightly pits left by strip mines will be filled. For those readers who do not know what strip mines are, they are great gashes in the magnificently beautiful Appalachian mountains, dug by bulldozers, so that coal could be mined at the surface rather than through underground mines. In most places no effort has been made to refill the pits or reforest the vast areas left bare and eroded by this practice. To Philadelphians, it may appear that they are contributing a great blessing to the coal regions by filling in their strip mines. It may occur to more than a few inhabitants of the mountains that there may be better ways to fill strip mines than with Philadelphia garbage.

A similar "trash train" proposal has been made for San Francisco. Here's what the University of California's Food Protection and Toxicology Center has to say about that idea:

> "Out of the San Francisco garbage dilemma has come the *thoughtless suggestion* that Bay Area communities send their 5,000 tons of daily garbage some 350 miles to the Nevada desert by special 90-car trains on the assumption that the garbage would fertilize the desert. Such a suggestion again

emphasizes the widely prevailing ignorance on matters of conservation.

"The ecological role of the desert is as much to be maintained and preserved as are the other ecological factors which contribute to the total quality of the environment. The primary objective of the Brisbane ban on garbage is to preserve the environmental qualities of San Francisco Bay. Why, then, should the inherent values of desert lands be jeopardized by woeful ignorance?

"These lands have their useful role in Nature. They were not designed to become depositories for man-made wastes. *Nor yet can raw garbage alone ever become fertilizer.*"

* * * * * *

An alternative to dumping wastes is to burn them in an incinerator. This used to mean in the backyard, home or apartment house incinerator. But the big trouble with private units is the large amount of fly-ash—the tiny particles of unburned and partially-burned material—that blow away from the fire. Faulty incinerators have contributed heavily to the air pollution load, and have been banned by many communities. Unfortunately a ban on private incinerators doesn't improve the air per se. You've got to have a desirable alternative.

New York Mayor John Lindsay had to ask, "What are we going to do about all that garbage?", after New York City enforced new rules on apartment house incinerators. New York City's incinerators have their own

troubles, spewing out more than 17,000 tons of soot annually into the air.

The fact is that city incinerators are always air polluters. Studies carried out by the Environmental Pollution Committee of the San Francisco Chamber of Commerce showed that *no incineration plant in the United States is now meeting California air pollution standards.* Unbiased experts in the solid waste field have said that "all incinerators are sick, some being more chronically ill than others."

The comprehensive report issued by the National Academy of Sciences on *Waste Management and Control* observes: "While more efficient incinerators will provide a short-term solution to one facet of the land pollution problem, the magnitude of the problem by the year 2,000 will demand new answers based on systems which have not yet been designed. These solutions will probably take the form of systems which salvage and re-use all components of what is now waste, putting back into the soil those components that benefit the soil, and returning to industry and commerce those components that are industrially useful."

Even though incineration is costly and often a serious pollution offender, many cities still think of building an incinerator. Despite its troubles, an incinerator industry has developed, and bond issues and other public financing can be arranged without a mayor or a city council feeling its collective neck is sticking way out. There must be something about a stack that is comforting.

But what comes out of that stack can be most discomforting and what goes in to it can be most uneconomical. A large percentage of solid waste is noncombustible so must be diverted before it ever gets to the incinerator. And then a sizable amount of what does get through winds up as ash that must be disposed of.

* * * * * *

Why don't we still feed garbage to hogs? George R. Stewart in his book, *Not So Rich As You Think,* covers this disposal means. Trichinosis used to be the hazard associated with eating pork that had been fed on garbage. The public was well warned to cook pork thoroughly and garbage feeding went on.

Then the hogs became susceptible to a fatal disease which was spread by the garbage feeding. As Mr. Stewart says, "An industry can take chances at killing a few customers, but it cannot take chances with killing very many of its prime suppliers." So laws were passed making it mandatory for garbage to be sterilized before it became hog-food. This gets to be expensive and eventually hog feeding of garbage vanished from commercial stock farms.

Of course, hog feeding had another drawback to those wanting a disposal solution. The hog ate the garbage and, for every 6 pounds of garbage put on one pound of weight. Most of the remaining 5 pounds piled up as manure—so another disposal problem arose.

CHAPTER VI

Collect-and-Destroy -- The "Impossible" Job

Destroying wastes, at least attempting to destroy them, is the accepted goal of most professionals in sanitary engineering. All their training points in this direction. But with today's pollution standards (or at least awareness), destruction presents Public Health Service engineers, public works directors, etc., with an unfair, as well as impossible, task.

The ideas of *waste destruction,* either by burning or

dumping, *without pollution* have turned out to be irreconcilable bedfellows. Their inability to get together gives us all sorts of problems.

Rarely do engineers recommend that a city save its wastes and forget about concentrating on ways to destroy them. It's not that engineers can't perfect the technology and safeguards to conserve wastes. It's that most of today's engineers think they have not "finished the job" if they've merely converted the wastes into a hygienically-safe material that can remain in our environment without polluting it. They believe, from past training and standards, that wastes must be gotten rid of—if there is a residue, it must be buried or sold.

Who says that the engineer must make the wastes disappear? You and I along with millions of Americans do not want smoke . . . or smells . . . or rats . . . or dumps, but who said the wastes must be DESTROYED?

Destruction is not *an* end; it is *the* end, the end of any potential use to which resources in wastes could be put. To make a policy of destruction is to make a virtue of nothingness.

One astute observer of the American garbage scene, a researcher for the Public Health Service, puts it this way:

> "We act now like we had no faith in the future. If a man has cancer, we try to keep him alive as long as possible, hoping someone will develop a cure. Yet when it comes to potentially-rich wastes like city refuse, we only think of destruction. If we had faith in our system, we'd conserve these wastes

so that future discoverers would show how to use this material."

* * * * * *

Recycling, rather than destroying, has been recommended as the only real solution to waste treatment. The only trouble is that the recommendation is confined to the pages of a special Presidential commission's report, but is not included in a recommendation for a city about to build a new incinerator. There have been truly beautiful discussions of the philosophy of waste management. But the philosophy is restricted to generalities. As soon as a city is ready to replace its dump, we are told that recycling is an extravagance. If the local junk dealers don't want to "recycle," that is, buy any part of the wastes, then it is taken for granted that destruction is the only alternative.

The waste problem is much too critical to allow junk dealers to determine national policy on whether or not wastes can be utilized.

* * * * * *

Composting wastes is closer to recycling than any alternative method under consideration. Yet, here again, the philosophy in favor of recycling breaks down as soon as a market study shows that the finished compost can't be sold successfully as a soil conditioner. We must stop using a false "double standard" when it comes to evaluating composting as a treatment method.

There's little point in compost research and espousing a philosophy of recycling wastes, if everything depends upon sales of the end product. Remember, no one is saying:

"Hey, I got a great idea! Let's all start making wastes and then sell the stuff!"

We've been making wastes since the day we were born. We're in the business already—whether we like it or not. We have no choice about *having* the wastes. We do have a choice about whether to do a good or bad job when it comes to *treating* them.

* * * * * *

The Law of Conservation of Matter cannot be flaunted. While it is possible to change the nature of a substance, it is not possible to make it disappear entirely. Thus, the ultimate solution for garbage would be to change it into something useful.

* * * * * *

The reason the waste problem is so overwhelming is that while we know what is wrong with the old methods, we also know what is wrong with the new methods. Optimism is notably deficient.

What is really needed to solve the waste crisis is a willingness to take the kinds of risks that we would take during war-time. I think that Dr. Athelstan Spilhaus, who headed the National Academy of Science's committee reporting on Waste Management and Control, did a magnificent job of revealing the faulty logic used so far. In an article in the November-December, 1967 issue of *Scientist and Citizen,* Dr. Spilhaus, now president of Philadelphia's Franklin Institute, wrote:

I think that the first thing we have to face is that

everything that has been done so far simply whittles at the problem but does not really cut into it. We need some completely new thinking and some step jumps; we need to face this problem and take the kinds of risks that we would take in the crisis of war. Why is it that governments will take risks in war, and won't take corresponding risks for their people living in a crisis situation? . . .

We need new thinking and daring experimentation, and we need even a change in our vocabulary. What do we mean by waste disposal? What do we do when we dispose of waste? Do we throw it away? There is no "away" any more. It's like the lady who was asked what she did with her garbage. She was rather a lazy housekeeper. She said she just kicked it around until it got lost. But that is exactly what sanitary engineering people have been doing for centuries. The trouble is there is no place for it to get lost any more. There are too many of us.

We need to approach two very simple ideas. If there is no place to put it—if we cannot dispose of it and if we cannot use it—we had better not just spread it around like the lazy housewife. The first interim step and one that we must move to rather immediately is control of pollutants, whatever they may be; whether they be lethal chemicals or whether they be noise which may cause mental illness—the control of these pollutants at the source. That is the first simple thing.

How do you start controlling things at the

source? Many things have been suggested. What about glass bottles? You read on certain glass bottles that Federal law forbids their reuse. I think that this should be changed so that Federal law forbids their *not* being returned and reused. Maybe we should tax them by their half life—the time it takes a tin can to half rust away. But in any case, we must find some controls at the source, and control at the source need not merely be these ways of discouraging wastes that are strewn around, but control at the source means collecting wastes at the source before it is strewn around.

Regardless of what any economist tells me, I'm convinced by the second law of thermodynamics that it must be cheaper to collect something at the source than to scrape it off the buildings, wash it out of the clothes, and so forth.

The second law of thermodynamics tells us—among other things—that as a product moves through a series of changes, it loses some energy in each change. In order to reconstitute the product in a more complex, more useful form, new energy must be introduced—as heat, as man-hours of work, or in some other form. Even if what we want to do is simply regain parts of the original product that have been dispersed, rather than actually to reconstitute it, this is true. The earlier in the series of changes the new energy is introduced, the less energy will have been lost and the less new energy is required. Therefore, it must be cheaper in energy (and energy costs money) to collect waste at the source before it has undergone a long series of changes and dispersals. This is a kind of complete rethinking in economics.

The second thing that we must move towards, if we are going to do anything about this crisis situation, is to aim at complete recycling of the things we use. This means an even greater change in our thinking. We now have a great system that is based on a consumer economy. We think our civilization improves when the consumption increases. But we forget, most of us, that there is no such thing as a consumer. We are pleased when our standard of living goes up because it means that more poundage or tonnage of stuff is put into every consumer's hands every year. But we forget that he really consumes nothing. He uses it and then he discards exactly the same mass of material that we put into his hands.

We must change our conception of the consumer and talk about him as a user. Then we can think of recycling the material he uses. The most conspicuous thing to think of recycling is the automobile. We have a magnificent system to dig stuff out of the ground, refine iron and other metals, make it into a complex piece of machinery, and produce an automobile that costs only as much per pound as coffee. Then we have a magnificent marketing system, a tremendous mass distribution system, to put this car into the hands of the consumer. Well, he uses it a little bit, fouls the atmosphere with it a little bit, and then he and our whole system forget about it. Can we have a mass collection system, a mass disassembly plant, and massively reuse used automobiles? Can we adopt this as a basic concept of our economy? This would take a little reorganization of American industry, but it is the only ultimate solution. That is why this is a crisis situation because people must become aware that this is the only solution, and that whittling will do no good.

We can reuse some of our waste heat, and at the same time control some of our pollution by burning refuse in modern incinerators which do not permit the smoke or effluents to go into the air, and at the same time use the heat for power production. In this way we would be controlling the wastes at the source and going the further way of using the heat for a useful purpose.

The way toward recycling is a long one, and there will be many things that we will not be able to reuse. There are many things that it is obvious we can reuse now, but many that we don't know yet how to reuse. In the meantime, we should still collect these at the source, segregate them and store them where we can get at them, for two reasons: if we store them where we can get at them, then we know where they are so we know they are not doing us any harm; secondly, as things progress, we may find a way to use them, and then we can mine them. Before we have worked out a way to recycle the automobiles we ought to make an Alfred P. Sloan Memorial Mountain of junked automobiles somewhere in the flat land which we could use for skiing on in the meantime and mine when our high grade iron ores give out. And this is not a joke. In Johannesburg, South Africa, when they first mined gold, they dumped the waste right next to the mine shaft, building great slag dumps. Then when the cyanide process came along they used them. There was enough gold left in those dumps so it was worth reprocessing them. Years and years later they found there was some uranium in the dumps and they processed them a third time. Of course, it was not by design that they concentrated these things; it was just plain laziness, but because they were concentrated, they could be reused.

So the final thing is to design for recycling or design things to be degradable in a way which will be useful and not harmful to the soil.

These are the simple aims that we need to work toward in this waste management and control. These are the only ways I know of controlling waste. Difficult as it may be to introduce control at the source and recycle, it would be much more difficult to persuade people that they must use less and less of everything, lowering their standard of living. That is the only other way I know of meeting this crisis situation.

CHAPTER VII

National Program for Using Wastes

Compost is a "safe material" that doesn't look, smell or act like the raw stuff it came from. This material is no better or worse than, for example, the "dirt" in your backyard or in the local park.

Our country could use a steady supply of that kind of "dirt." Nobody talks about burning dirt. Sometimes, in fact, dirt has turned out to be quite valuable. Dirt has built our nation and its industries—coal, oil, pharmaceuticals, chemicals. So why not adopt a national policy of making "dirt" out of solid wastes?

Actually, I prefer the name *compost* to dirt despite the many problems in semantics.

As mentioned earlier, the word compost has been in use for a great many centuries. But it has always been used in a gardening and farming sense, that is, as a material to apply to fields to improve crop growth.

Periodically, the idea of creating something valuable from a substance as worthless as garbage has been publicized in features on "Garbage to Gold!" Each time, the reader discovered how hidden riches could be gained by composting garbage and then selling the soil conditioner to an eager suburban garden market. Always the prices of comparable soil conditioners would be quoted; in some cases, the value would be in the hundreds of dollars per ton. And think how many tons of compost could be produced in a single city every day of the year!

But that's where the gold mine vanished. Merely because a number of stores may be able to sell a limited amount of soil conditioner for a few weeks out of the year does not mean that 1,000 times that amount can be sold for the same price . . . or indeed for any price.

In the U.S., compost plants have been operated in Oakland, San Fernando, Sacramento, Phoenix, Houston, St. Petersburg, Mobile and Altoona. But in just about every case, the continuation of the plant was or is dependent upon sale of the end product. In just about every case, the compost plant was run by a private company which contracted with the city to accept its refuse at so much per ton. The company, following the accepted tradition, theorized that it would get its profit from sale of the end product. It rarely works out that way, but what is most important, it should not *have* to work out that way.

Here are some suggestions for changing the rules that have governed the use of composting by cities in the past:

1. The companies now in the "Composting Business" should offer equipment for sale to cities that would allow them to compost their wastes. Equipment would include shredders and grinders, conveyors, digestors, windrow formers, special aerating devices, temperature controls, etc. The machinery would be available for sale in the same way that incineration equipment can be purchased.

Currently, compost companies have approached cities on the basis of accepting refuse on a flat charge of about $3.50 per ton. The companies have, in most cases, deluded both the cities and themselves with the idea that most, if not all, of the finished compost could be successfully marketed as a commercial soil conditioner. As a result, what starts off as an engineering method for treating wastes quickly changes to a marketing challenge for treated wastes. The transition should not have to be made.

If composting companies must persist in the immediate future in operating the equipment in the compost plant site, then the contract with the city should provide for a *disposal site* (or variety of sites) as well as a flat treatment cost. In other words, the composted residue is the property of the city, not a company.

2. Cities, or better yet *regional disposal authorities,* might look at the composting method as a way to "save" their present landfill site, prolonging its use for many more years. As practiced now, landfill requires

a soil cover and new areas for burial. Shredding and simple low-cost composting treatment could provide for continued use of many landfill sites that would otherwise be forced to close down if conventional practices were followed. (A study under way at a Madison, Wisconsin, landfill is demonstrating the possibilities of this technique. The practice of grinding prior to landfill and composting is discussed in the Appendix.)

3. Many qualified agronomists and engineers are convinced that if a stockpile of finished compost were available, a gradual demand for its use as a soil conditioner would develop. But don't misunderstand me. I do not mean that a public clamor will arise to buy up all the compost around. I mean that a "use philosophy" will eventually emerge that will see compost used to stop erosion on river and channel banks, to improve soil structure on commercial farms and orchards, to landscape new highways and reclaim public lands.

When it comes to uses for raw materials (like former wastes) you either have faith in the future and the ability of researchers . . . or you bury your head in the illusory solution offered by burned or buried wastes.

Federal agencies, state and city officials must reasses their approach to the solid waste problem. Recycling wastes back into our environment should be given top priority.

Limiting research grants to studies of pathogens in garbage and how composting eliminates them is not enough. Such research is important, but by and large, most authorities seem to be convinced that composting either already eliminates pathogens—that is, harmful germs—or can be modified to eliminate them.

Research grants should be given to economists and social scientists as well as engineers and regional disposal authorities to see how treated wastes could be used, how salvage markets could be expanded, how the life of a landfill could be extended if some form of composting were used, how the cost of treating wastes by composting and stockpiling compares to incineration and stockpiling. (There seems to be a strong tendency among incinerator-advocates to gloss over the disposal of non-combustibles and incinerator residue, while stressing the problems of disposing of composted refuse. Also, air pollution controls have added tremendously to the costs for constructing and operating incinerators.)

Millions of gardeners in the United States, led by organic gardeners, are familiar with the idea of compost that is not for sale. Home gardeners readily accept the idea that organic wastes of all kinds—grass clippings, garbage, weeds, manure, leaves—are not just stuff to be joked about, fretted about or destroyed as quickly as possible. Gardeners are used to "saving" wastes and using them.

The simple, direct philosophy of the home gardener must be adopted by our city, state and federal governments. The idea is so obvious—to treat wastes by composting, to stockpile this "safe" material, to use as much of it as we can. As time goes on, more and more uses for this potential resource will emerge; more efficient equipment will become available to spread compost over the soil, to transport it to areas where its raw materials can be economically and efficiently reused. In the meantime, we are protecting our environment and saving our tax money.

Research has already indicated the following possible uses for organic wastes: vitamins, antibiotics, animal feed, protein, adhesives, alcohol, charcoal, paper pulp, yeast, mushroom-growing, building panels, packaging materials, hardboard, synthetic cloth, detergents, fiber, plastics, mineral retrieval, and of course, soil reclamation.

* * * * * *

The time has come to stop whipping the "dead horse" of composting—and start racing with a new one. To talk of composting as a waste treatment process and then reject that process because the end product can't be sold is a luxury our nation can no longer afford.

Composting is an intermediate process for treating

refuse *prior* to its ultimate disposal. As a treatment process, it should be compared to alternative methods as to pollution, health hazards, economy, engineering, soundness and applicability to local conditions.

If city officials are interested in seeing pollution from wastes prevented, then they must understand the distinction between *compost as a process* and *compost as a product*. Until this difference is obvious, no serious discussion of composting as a solution to our pollution woes can be carried out.

Since the Congress passed the Solid Waste Disposal Act of 1965, more official research than ever before has been done on composting. Temperature studies, microbiological studies, pathogen studies, methodology studies, sampling studies, and literature studies. From time to time, I have an opportunity to talk with some of the men who are making these studies. And I get the impression that these researchers are doing their studies in the shadow of the knowledge that the finished compost cannot be sold. And thus, since they accept the old guidelines for evaluating the compost method, the process is really not to be recommended to U. S. cities *despite* the outcome of their research.

The reason government funds are available in the first place is not because the compost lobby is so strong. It's because the current treatment methods are so sick! The most objective authorities point out that all incinerators are "sick to some degree, with some being chronically ill." Sanitary landfills are running out of space, or have been operated very much like open dumps. It's not that more people want to buy compost than before; it's that composting can be cheaper in every way than incineration and more sanitary than landfill. Incineration merely transfers much of the waste into the air where it continues to be one of our most urgent problems. Sanitary landfills create future problems of gases, odors, and settling land.

* * * * * *

Garbage (whether composted or not) disposal is a *municipal* responsibility, plain and simple.

Trucks owned or contracted by a city pick up and haul the garbage. For many reasons, like shortage of disposal sites and pollution hazards, it makes sense for cities to have that garbage hauled to a compost plant for *treatment*. This *treatment* reduces the bulk of the original garbage, changes it from a smelly, health-dangerous material to a safe, stable product.

The process of composting has changed the qualities of the original garbage. Now municipal officials have several choices. Here is a list of what those future choices could be:

1. Spread the compost on municipally-owned land.
2. Spread the compost on state-owned land.
3. Spread the compost on federally-owned land.
4. Spread the compost on privately-owned land.
5. Sell whatever amount of compost can be sold.
6. Stockpile as uses 1 through 5 develop.

The application of city-produced compost on state, federal or privately-owned land could be set up as a reimbursable program whereby the city would be allotted so much per ton for having used its compost to improve specified lands.

Recently more and more cooperative-type programs between federal and state, between state and municipality, have worked out quite well. Recreational areas have been developed, libraries have been built, roads have been paved, etc., as a result of such cooperation.

The concentration of people in urban areas has made waste management a serious national problem. I see no reason why a national program of composted garbage distribution could not be legislated and successful. Such a program holds the key to improved waste treatment and utilization in the United States.

* * * * * *

The need for new patterns of cooperation among federal, state, municipal and county agencies to solve the waste crisis is presented clearly by Tom Alexander in an article in the October, 1967 issue of *Fortune* entitled, "Where Will We Put All That Garbage?" Here are some excerpts:

"We need new incentives to change the economics of disposal, new kinds of authorities, new kinds of disposal systems that can be just as flexible as a missile warning network.

". . . Soil scientists agree that heavily farmed land should have some sort of organic humus added periodically. In the land-rich U.S. this is usually supplied by letting land lie fallow for a year and then plowing in a cover crop. But as the world's population grows, fallow land will become more and more of a luxury.

Now it is being suggested that the city should pay the cost of making the compost and perhaps even of plowing it into the farmers' land and be thankful that it has a place to put its refuse. Chicago already helps defray the high cost of disposing of the semi-solid, nutrient-rich 'sludge' residue from its sewage treatment plants by drying and barging it to Florida and selling it for use on citrus groves. Now Chicago is investigating piping the sludge some 90 miles for use on farmed-out lands in Kankakee County, Illinois. If this sludge could be mixed with refuse compost, it would make the compost considerably more valuable as a soil conditioner and nutrient."

Alexander concludes his fine survey report by saying:

"Though much needs to be learned about the costs and practicality of long-distance refuse pipelines, one could envision networks of such pipelines carrying compost and sewage from many sources to marginal agricultural areas. Even though the initial capital costs of such a system might be high, they would probably be offset by low operating costs.

"More important, there would be little noise, odor, unsightliness and inconvenience. Furthermore, the network would double as an irrigation and fertilization system.

"It seems likely that some sort of utility company—wholly private like A.T.&T. or quasi-public like Comsat—would be best suited to plan, build and operate such a regional network. The corporation could be paid a regulated price per ton or per household to get rid of all wastes. Then such a profit-oriented company would try to make what extra money it could through cost cutting, salvage, irrigation charges, composting, or heat recovery.

"But garbage network or no, one thing is clear: Waste disposal will have to be done differently—and soon. The present approaches to waste handling are inadequate, expensive, and wasteful of natural resources. It appears to be only a matter of time before the congested areas of the U.S. will wake up to find garbage on their doorstep unless they reach out to avail themselves of the systems approach to waste disposal."

* * * * * *

What city does not have land to reclaim for an industrial park—or an area that could be developed into a recreational park—or an airport area that needs expansion—or any of a number of such needs?

And the opportunities for sites to benefit from compost application multiply as you dwell on its value to land owned by the state and by the federal government.

In today's world of municipal composting, one gets the definite impression that the easiest thing to do is to put some compost in a bag, give it a catchy name, and whoosh!—instant distribution to farm and garden outlets and instant sales.

Well, today's highly-specialized marketing methods just do not work out that way. Successful sales demand merchandising methods that almost invariably call for large promotional budgets and a large sales staff. To justify that kind of expenditure calls for a complete line of fertilizers and soil conditioners—not a single compost brand.

If compost produced from city garbage is to be sold, it should be sold by existing fertilizer companies, who already have markets set up for their product line. Such firms could buy whatever amounts of compost they want from the city. They could package it and sell it alongside their other products.

I wrote to a few fertilizer companies around the nation, asking if they would be interested in buying and distributing compost produced from city wastes.

Here is one answer from a Missouri company:

"We are formulators and distributors of lawn and garden chemical products and if city wastes in composted form were available on a local and economical basis, we would definitely be interested in distributing such products through retailers to the homeowner."

From the recent surveys and those taken about 6 years ago, as well as from conversations with owners of fertilizer firms, I am convinced that many would include in their line of products a locally-produced compost if it were available.

* * * * * *

Philadelphia and other cities have arranged with railroad companies to set up a "trash train" schedule to transport garbage to strip-mine areas. Pulverizing stations are to be set up to grind garbage for more compact loading of the railroad cars. What these cities and public works officials have not recognized as yet is that for another $2.00 or less per ton, they could be spreading a safe soil-builder on lands instead of dumping untreated garbage.

Many government reports have been issued on the need for improved waste treatment methods and on the need to avoid pollution. If the federal government, perhaps through the Public Health Service's Office of Solid Wastes, would develop a pilot program with several cities, whereby it would guarantee to "take" finished compost—that is, accept it on federal or state lands, I firmly believe that compost plants operating in those cities would become the prototype for successful operation the nation over.

In the long history of municipal waste treatment, it has been accepted practice to spend up to 80 per cent of expenditures on collection (men and trucks) from individual households and transportation to the dumps. To prevent further poisoning of our air, water and land resources, more money must be spent on the other end.

It is time to apply a new philosophy to wastes. Until now, we have stopped short—content to arrange for the development of a process that would convert a waste to an asset. But the asset becomes suspect and remains a burden if there is no place to put it.

The situation would change overnight if—as stated above—we recognize the fact that the city is responsible for garbage—*before and after* composting—and the city can then work with larger government authorities for the benefit of us all.

We must implement a national program for using wastes.

CHAPTER VIII

A Heritage of Waste Leads to the World's Most Disposable Society

"Part of the American heritage is a wastefulness of resources with little concern for any long-range implications of the expedients of the moment. Perhaps the very problems of survival made it inevitable that we should burn the hardwood forests to clear land for farming, throw the topsoil to the winds, and otherwise despoil rather than conserve the resources of North America. The important factor in the context of solid wastes management is that the experience helped to establish a national attitude toward 'unwants' of the moment which is reflected in both the economic and technological aspects of wastes disposal."

The man who spoke those words has spent a good part of his life trying to improve the way our nation handles its wastes. His name is P. H. McGauhey, and he's director of the University of California's Sanitary Engineering Research Laboratory.

Prof. McGauhey stresses that too few levels of our society are concerned with the solid wastes problem. Most of us are willing to let the solution rest with the technicians. But now we must look beyond technical and economic considerations, says Prof. McGauhey, and go right into the areas of public policy, planning and the social goals of our society.

If we really want pollution control, we can no longer tolerate the prevalent attitudes of citizens, engineers, public officials and private entrepreneurs. (There is a swinish side to the most civilized of us, Prof. McGauhey contends.) We must overcome the American tradition of waste. ("Any suggestion that money should be invested in 'waste' for the sake of disposal or any other financially unrewarding goal is likely to be considered absurd, however inclined the citizen may be to object to polluted water, smog, rodents, odors and unsightly debris.")

The result is that we become entrapped in this lowest

common denominator of public spending. We complain —quite eloquently, in fact. (Probably the safest campaign topic for a politician is anti-pollution.) But then we do nothing. We look for the lowest bidder when considering waste management systems, just as if we were buying one more automobile.

But waste management systems are not the same. In the words of McGauhey, we "apologize for the fact that (waste management) systems cost anything whatsoever, often retreating from expedient to expedient on the basis of cost in dollars instead of leading the public to an understanding that solid wastes management is worth whatever it costs within the framework of honest engineering and sound public health practices; and that this cost is part of the price man must pay for the benefits of a modern urban-industrial-agricultural society."

The ideal goal of too many city officials is a policy of no-spending-at-all. And this vote-getting attitude puts waste management into the heart of the political arena.

Few people may say it, but almost all of us act as if conservation were not a practical way of life in the throw-away Twentieth Century. We've unconsciously accepted a golden rule of pollution.

A Mississippi biochemist made headlines recently by suggesting at a federal water pollution hearing that from an economic standpoint, the nation's waterways might be best used as "open sewers and navigable channels." Forget about pollution, he advised, which was an inevitable outgrowth of increasing population and industrialization, and anyway the public is not willing to pay the cost to control water pollution. Why else would so many bond issues for sewage treatment facilities be defeated?

Now, very few of us would say we agree with the biochemist from Mississippi. But too often *our actions come close to his words:*

"Apparently some thoughts are publicly unthinkable —for example that pollution is not necessarily bad, or that the majority of citizens simply do not care whether or not the water is polluted. . . . I live in an air-conditioned house. I drive an air-conditioned car. I work in an air-conditioned office. My environment is awfully artificial and I wouldn't change it for the world. I find it very comfortable."

Do we as citizens of the United States *really* want pollution control? Specifically, have we made it clear we're willing to *pay* for a safer place to live? The answer has got to be NO! The plain fact is that public attitudes and policy present a bigger hurdle to pollution abatement and waste utilization than any so-called lack of technology.

* * * * * *

Royce Hanson, president of the Washington Center for Metropolitan Studies, made these enlightening remarks on "Politics and Trash":

Unfortunately for solid wastes, its management costs more than a street-crossing light or another policeman, but not as much as a nuclear power plant or a major dam. It costs too much to be considered trivial, yet not large enough to be beyond the comprehension of the average householder.

The key to the politics of the system, says Mr. Hanson, is the average household, which we often overlook in our focus on delivery and disposal. It is the household which generates the work and which must be politically satisfied to pay for the technical system.

Continues Mr. Hanson:

"The household is primarily concerned with two politically critical aspects of waste management—getting the stuff off its premises as fast as possible and the neatness of the collection service. From a very practical political as well as sanitary engineering and public health point of view, there may be considerable utility in linking new programs to better household service as well as to grand objectives such as abatement of air pollution and exurban golf courses. Most of us can exist with the city dump's fires, but not with a heap of trash composting on the back step. Aside from the political values, it does seem unfortunate that the *world's most disposable society can't dispose of its throwaways more efficiently.*"

The pressures on city officials to turn to cheap, conformist solutions are almost irresistible. Tradition—even if it's a tradition of waste and pollution—is the safest route to follow.

If we really want pollution control, then you and I must make it clear to the politicians of our nation—especially the city officials—that we are willing to pay for treating wastes because we know we *benefit* by paying those costs.

In a campaign speech about 135 years ago, Abraham Lincoln said:

"There cannot justly be any objection to . . . good things, providing they cost nothing. The only objection is to pay for them."

With the ever-increasing problem of pollution, the cost must be considered within our ability to pay.

* * * * * *

So much pollution has fallen on the world we live in that Julian Huxley, the noted British biologist, suggests we call our environment the "effluent" society instead of the affluent society of yesteryear.

Dr. Luther L. Terry, former United States Surgeon General, went back 2,000 years to cite a truth uttered by the philosopher Lucretius:

"Nature resolves everything into its component elements, but annihilates nothing." Continued Dr. Terry:

"This is the scientific truth which the community of men must accept today in approaching the problems of air and water pollution and in devising protection against radiation. It manifests itself, too, in the mysterious biochemical reactions which take place within the human body as we handle and use some of the thousands of new products and processes which are the mark of modern industrial life.

". . . The urgency of our task is heightened by the fact that the problems are multiplying so rapidly.

". . . There is now nearly six times as much pollution in our rivers, streams, and lakes as 60 years ago, and the amount is still increasing. An expanding population increases the demand for a fresh water supply and, at the same time, increases the volume of waste. The crowding of people into urban centers intensifies the problems of waste disposal. Application of commercial fertilizers, and use of a vast array of new herbicides and insecticides, contribute to pollution.

". . . We are in danger, on the one hand, of creating an incredible disharmony in nature which will ultimately degrade and enslave us. Or we can create an environment which can enrich our lives, our society, and our individual well-being. It is for our generation to decide."

There's no doubt. We've got to overcome our heritage of waste. If we can begin to use our garbage, we'll have made a tremendous contribution to our nation's future.

CHAPTER IX

Building A Bridge between Today's Wastes and Tomorrow's Needs

I hope that as you read these pages you will understand that this plea and plan for waste utilization does not originate with me or any single group of individuals. During the years I have edited Compost Science, I have met and corresponded with innumerable persons who have shared these views. These people have ranged from experienced scientists in all fields to inexperienced laymen in all fields.

One group—and I'm really not certain if it's fair to refer to the individuals as a group—has been particularly significant in the field of waste utilization. They are all in California—some with the California

State Department of Health, some with the University of California's Sanitary Engineering Research Laboratory.

Much of the early reliable data on municipal composting in this country appeared in the University of California's technical bulletin, *Reclamation of Municipal Refuse by Composting*. Dr. McGauhey and Dr. Clarence Golueke, who has served on the editorial board of Compost Science since its inception, have been constant and consistent in their goals for recycling wastes. The same qualities have marked the writings of Richard Peters and Frank M. Stead of the California State Department of Public Health.

* * * * * *

Back in Autumn, 1965, Frank Stead had an article in Compost Science on "Social and Legal Implications of Organic Waste Management." This paper was originally presented by Mr. Stead at a poultry waste management symposium in Lincoln, Nebraska, when he was Chief of California's Division of Environmental Sanitation.

I've excerpted much of this paper, as Mr. Stead has so clearly and eloquently presented the case for our adopting a *positive* program to manage our environmental and health problems instead of practicing a hit-or-miss plan of containment. He says "Our past negative role of merely stopping or even preventing something has furnished an unsatisfactory background for public understanding of today's problems." We must prepare ourselves individually and as a society to cope with new elements in a changing world.

The new frontier is not represented by the clearing of the land and the exploitation of resources. "Our client is man," but we must allow him to "inherit the earth" by developing a model that will help achieve that objective.

The transition we must make, according to Mr. Stead, is a shift from protection of people to protection and wise management of natural resources.

> "The three basic natural resources of man are air, land, and water. Man lives at the interface of land and air, and, of necessity, in the presence of water. But man needs more than air to breathe, water to drink, and barren platform to stand on. The ultimate and fundamental resource is living matter—plant and animal.
>
> *"The task of managing man's environment then really boils down to a management of organic material, the substance of all living things, and the surplus of which we are wont to call organic wastes."*

Mr. Stead now develops the concept of organic material as a resource, the potential of organic wastes, and its inherent characteristics which enable it to be converted and transformed for uses not apparent on casual observation.

* * * * * *

> The first characteristic of organic matter that strikes one is its ability to store energy. Our whole economy in the United States is based upon the

use of fossil fuels as the principal energy source. Coal and petroleum may be thought of as stored-up solar energy which may be released thermally by burning in a flame. Similarly, plant and animal protein, fat and carbohydrate represent stored up solar energy which as food furnish the energy to supply the human body through the process of chemical metabolism.

Secondly, save only for minerals, organic matter with its infinite variety in nature makes up all of the material in solid form used by man for his shelter, his clothing, his tools, and his structural materials, as well as his paintings and other cherished possessions.

The miracle of organic matter, however, is that it is the essence of all living things, both plant and animal, ranging from algae to oak trees and from mosquitoes to elephants, even including man himself with his unbelievably complex physical and mental organization. It is indeed the very stuff of life.

This phenomenon would be awe-inspiring enough if plants and animals were created out of specialized new materials which lasted the life of the organism and disappeared, but the real wonder is tied up in the phenomenon of conservation of material: That is, our earth system has a fixed supply of organic material and this supply is cycled over and over again through soil, plants, animals, and waste products. In other words, atoms of carbon, hydrogen, oxygen, nitrogen, sulfur, and a few metals and halogens may combine once as seed of a plant and later be incorporated in an animal organism. This unlimited power of organic material to synthesize and decompose only to resynthesize is the very secret of life.

* * * * * *

With these marvelous occurrences going on around him in nature, it is only natural that man should study them. Early observation taught him that such organic wastes as animal manures would spontaneously decompose to produce a humus which, when returned to soil, greatly aided the subsequent growth of plants in the soil. Study also indicated that the energy for this transformation was in the animal manure, but that the conversion took place only if bacteria were present. Later, he noted that the benefits of the humus were both as a physical soil conditioner and as a source of specific needed chemical plant nutrients, principally nitrogen, phosphorus, and potash. With a little experimentation he learned that other organic wastes could be converted in the same manner but that, sometimes, very foul odors were produced. Finally, he discovered that if carried on in the presence of an excess of oxygen, no nuisance was produced and the plant nutrients suffered a minimum loss as volatile gases. Thus, a century ago, man learned how to deliberately transform organic matter from a waste product to a useful substance.

With these exciting possibilities just around the corner, where do we stand today in our management of solid organic wastes as a resource? Household garbage goes to the sewers through garbage grinders, is spread on open dumps along with rubbish to constitute an eyesore, an odor nuisance, and a potential hazard to health, or at best, is placed in landfills without preliminary processing so that great shrinkage occurs and the resulting land has only very limited usefulness for ten or twenty years. Animal manures from such commercial animal operations as dairies, hog farms, cattle feed yards, and poultry ranches are still handled in accordance with practices developed when these activities were isolated in rural territory, and as a result, almost without exception, now present a serious nuisance problem to adjacent residential areas as well as a serious loss of resource. Wastes from the fruit and vegetable processing industry, totaling many hundreds of tons per day, are sent to dumps or to already overtaxed sewage treatment plants with no effort at conservation. Finally, sewage solids, separated from sewage and partially stabilized by anaerobic digestion at great cost, are, in the major metropolitan areas, discharged to the ocean not only at a total loss of this resource, but at the expense of a potential threat to the biological balance on the continental shelf.

* * * * * *

What is the explanation for this seeming blindness on our part in management of a resource? I believe it stems from our unquestioning acceptance of a philosophy developed fifty years ago under a far different set of conditions. This philosophy may be expressed in terms of the following deeply cherished basic assumptions:

1. Conservation is unnatural; natural resources are inexhaustible and should be used once and thrown away.
2. Rivers, bays, and oceans are the natural sewers and dumping grounds for wastes.
3. Waste organic matter can be converted into nothing except a low grade soil amendment with minimal plant food value.
4. With respect to nuisances to surrounding inhabitants "first in time, means first in right."

None of these assumptions are axioms and I believe none of them are sound.

The first and most obvious step to be taken is to apply the aerobic stabilization processes, referred to earlier, to convert putrescible and bulky organic wastes into a useable agricultural commodity. I refer, of course, to the aerobic composting process which is capable of reducing virtually all types of solid organic refuse into a valuable humus if a few clearly defined variables are controlled.

One can easily visualize the transformation of present sanitary landfill operations into sites where all types of waste organic matter—garbage rubbish, food plant wastes, animal and poultry manures, and raw or digested sewage sludge—would be reduced to optimum size, blended in optimum proportions of nitrogen and carbon and quickly converted in an aerobic environment to a nuisance-free compost. Such a plan would virtually solve the land requirement problem which now plagues our cities since, in the early years the compost could be placed in a compact fill occupying only a fraction of the space required for raw refuse. Later, when an agricultural demand for this material has been established, this material could be mined and transported away and the site could become a fixed processing operation, namely, a new industry

Finally, here are some comments by Mr. Stead regarding organic wastes as a savings account:

We can determine now that we must adopt a new concept—one with a prospect of permanent success. The key to the prospect is the concept of environmental health (as distinguished from environmental sanitation), which consists of the planned management of natural resources—air, water and land—as a total system. In some areas this can be accomplished on a state-wide basis, and in other areas it will be on a multi-state basis.

The geographical common denominator is set by the topography, not by the political boundary. In the case of solid wastes, we must give some thought to converting useless organic materials to a resource of value. We must think of organic wastes as a savings account—a working supply of basic materials, not only for the production of fabrics, structural materials and chemicals, but also as a source of carbon, hydrogen, oxygen, sulphur and a few other materials that have the ability to combine, to form plants and animal materials, and which can be broken down and reprocessed.

Let us now return to the present and consider the changes in direction that can be brought about under the existing organization. These are the things I foresee: At the beginning of the transition, let us say for the first five years, I see composting and the placing of the compost in a compact fill.

Secondly, I see the combining of all of the organic wastes that I have mentioned—agricultural, industrial and household—to produce a consolidated, stabilized end product that can in turn be used as a raw material for other useful products.

CHAPTER X

What to Do When the Junk Dealers Don't Want It

Salvage in one form or another is the heart of the solution to our waste problem. If we can't destroy our wastes without pollution—and we certainly have not been able to—then we must salvage them . . . use them . . . recycle them.

Throughout the pages you have read so far, the problem under discussion has been *organic* solid wastes.

I have concentrated on that part of the waste problem, because composting offers the best solution for organic wastes.

But mineral wastes also cause a great deal of havoc. Auto graveyards, beer cans and bottles constantly plague us. Each year, we produce more than 48 billion metal cans, 26 billion bottles and jars, 65 billion metal and plastic caps; we junk millions of cars. While composting is not applicable to the auto graveyard, the economic principles of salvage remain the same for both organic and mineral wastes.

Unfortunately our policy for salvaging wastes has been characterized by a "laissez faire" attitude reminiscent of the extreme in 19th century philosophy. We must realize that the salvage market (that is, how we manage to recycle all our wastes) is as much a part of successful refuse disposal as collection and treatment.

As it now stands, junk dealers may be able to recycle (that is, sell) the chrome strips of cars, but they can't find buyers for the other parts. Does that mean we should rest happy in the knowledge that de-chromed automobiles will pile up on our hillsides? Of course not! We can and must develop ways to use the junk that the junk dealers don't want. This chapter tries to analyze how it could be done.

* * * * * *

"There are two distinct kinds of economic issues involved in the solid waste problem," writes William A. Vogely, assistant director—Mineral Resource Development of the U.S. Bureau of Mines. "The first has to do with the conservation of the valuable materials obtained in the waste and the second has to do with the handling of the external cost of waste generation."

When it comes to mineral wastes, such as that produced in mining operations, the choice most often is to have the pile used as aggregate (as fill under roadbeds, etc.) rather than try to extract its minerals. In other words the company just wants to get rid of the stuff as quickly and cheaply as it can. The directors of the company do not consider it their responsibility that whatever minerals are in the wastepile are destroyed forever by putting them under a new highway. The company and its stockholders are not in business to sell "left-overs" or commit themselves to utilizing minerals in wastes if it's not readily profitable. A company looks at it this way: it has tons of material to be removed; it has an offer from a road-building company who will cart it away each week; it may be years before someone comes along who can make use of the mineral residue; so he makes a simple decision to have the highwayman haul it away.

Conservation runs a poor second when traditional economics is in the race.

During the last few years, a new force has entered the economics scene with the advent of stricter pollution laws. Dr. Vogely calls this the "Social Cost Issue."

In the past, waste products were piled immediately adjacent to the mine, mill or plant—wherever disposal was cheapest and didn't interfere with company operations. Liquid wastes, of course, were pumped directly into waterways. Any company official who tried to reduce the "social cost" of his waste would suffer a competitive disadvantage since his costs would be higher than others.

The new anti-pollution laws, however, tend to equalize the costs to all firms and thus penalize none. (It should be pointed out that these increased costs are passed along to the consumer, thereby—in Dr. Vogely's words—"falling upon society as a whole.") The new laws often make it cheaper for a company to treat its waste in a way that it can be used, if that turns out to be cheaper in the long run than other alternatives.

What happens to Chicago sludge illustrates this point. When the city dewaters the sludge and incinerates it, the cost per ton is $57. When it dries it and sells it to a fertilizer distributor, its gross cost per ton is $60 less $15 income from the sale—or $45. So the city actually saves $12 per ton by treating it instead of burning it.

Whether the waste comes from mines or humans, treatment costs money. But utilization can mean a lower cost compared to alternative methods.

* * * * * *

In an analysis of the recovery of mineral scrap, Dr. Vogely traces the life cycle of an automobile as it passes from the new car dealer through its vicissitudes as a transportation vehicle and onto the auto junkyard.

The last owner of a retired car has 3 alternatives: abandon it on public or private property; take it to an auto wrecker, or take it directly to the scrap processors (which is just about never done). The total number of cars abandoned depends on the value of the cars to wreckers or processors compared with the ease of abandonment.

As shown in the Bureau of Mines Survey for 1965, 20 per cent of the junked autos were abandoned in graveyards, or on public or private property. Eighty per cent of the total junked autos were in the "industrial stream," with 73 per cent being in auto wrecker yards. The analyses indicates that moving of the hulks from the auto wrecker yards is *primarily an economic problem.*

The hulks will move as their value is raised and, in fact, one can conceive of a value high enough to cause the auto wreckers to change their methods of doing business so as to store parts in inventory accumulations rather than in used hulks.

* * * * * *

Concerning the economic factors of mineral waste utilization, Dr. Vogely draws these conclusions:

> The process of producing minerals from the earth, processing them, manufacturing products and disposing or reusing after manufacturing is one which generates waste and spillover damages at many points. The economic system does not provide automatic incentives to minimize the waste or spillover damages.
>
> There seems to be much to be gained from looking at the entire life cycle of a mineral resource

as a system. Design mining so as to prevent social cost and to provide for the fullest recovery over time of the mineral constituents. Provide for the manufacture of articles in ways which will facilitate the recovery and reuse of the mineral constituents. Provide in junk collection facilities ways and means for economically segregating materials. In this way, reuse and prevention of waste can materially increase mineral supplies and postpone the time of absolute scarcity.

Those principles of utilization must be met—if resources are going to be created from wastes wherever they occur.

* * * * * *

Currently about one-fourth of the 165 million tons of municipal solid wastes collected each year are disposed of by incineration. But even incineration does not destroy everything, and scientists are discovering that the residue may be valuable.

Carl Rampacek, Research Director at the College Park, Md., Metallurgy Research Center of the Dept. of the Interior, says that "if all refuse were properly incinerated, the residues would contain more than 11 million tons of iron and nearly a million tons of nonferrous metals including aluminum, copper, lead, zinc and tin. Based on current scrap prices, the combined monetary value of these metals would amount to over $1 billion."

As Mr. Rampacek describes them, incinerator residues are a complex, soaking-wet mixture of metals, glass, dirt and stones, charred and unburned paper and ash containing mineral oxides. But little was known about their exact composition, so the researchers had to establish an "accurate definition of the waste from the standpoint of it being a *raw material for the metallurgist*."

The greatest variations were caused by the degree of burnout which is largely influenced by moisture content of incoming refuse, capacity at which furnaces must operate, and the state of repair of the furnaces. But composition data remained surprisingly consistent, considering those factors, and the mixed nature of the original refuse. For example, iron materials consisting of tin cans, mill scale, wire and massive iron which can be separated easily, accounted for about 30 per cent of the total residues and nearly one-half of the residues in glass.

Here's how Mr. Rampacek views the future in metal recovery from incinerator residues:

> "While progress thus far in this work has been encouraging, a great deal remains to be done to reach the ultimate objective of converting these materials into products suitable for recycling. Copper and tin must be recovered from the iron fractions, preferably by leaching without loss of iron. If this can be accomplished, maximum copper and tin recovery will be realized and the iron remaining will be the highest quality steel scrap.
>
> ". . . Metallurgists rarely encounter such complex mixtures, but it is believed that most of the necessary technology is available and the problem

resolves itself to wise application of the information on hand.

". . . As progress advances on these most urgent problems, attention will be turned to the minor problems. Research will be conducted on recovery of all of the mineral values contained in the fine ash fractions, including the oxides of many metals. There are definite possibilities that good use can be found for the pulverized glass fraction too. Consideration is even being given to developing methods for effective utilization of the carbonaceous slimes and table tailings. *Conceivably, uses can be found for all materials contained in incinerator residues and this so-called solid waste will become a desirable raw material that will no longer be wasted as landfill.*"

* * * * * *

In early 1968, scientists at the Bureau of Mines revealed that they had found gold and silver in the powdered remains of Washington's dumps. They estimated the value at $14 per ton, and projected the value on a national basis to $7 million annually. Most of the silver and gold comes from photographic chemicals and films, old coins, costume jewelry and silverware, and solder in electronic equipment.

According to Bureau of Mines director, Walter R. Hibbard, the 500,000 tons of fly ash which now cost about $1.50 per ton to get rid of from incinerator plants each year indicate "the potential value of what we might call the nation's above-ground bonanza—its

refuse piles." A colleague of his notes that two or three centuries from now, our "dumps will be mines."

But actually, if we set our minds to it, we won't have to live with bigger and bigger dumps for centuries to come. We can begin to use those valuable materials before they lie around that long. We can develop sorting techniques that make extraction economical.

* * * * * *

> "I firmly believe that practically all waste is salvageable or potentially useful, and, if we forget our 'traditional economy,' there is no problem in recovering most of the waste in this country."

Those words were spoken by a man in the salvage business—Lyle Randles, Jr., of the Los Angeles By-Products Company. His firm pays the City of Los Angeles $11.26 per net ton for noncombustible rubbish (cans and other metals). Based on a 1963 contract, the City would get back about $600,000 annually.

Houston, Texas, is another city that sells salvage to

local companies. Tin cans salvaged from the city's incinerator plants bring $11 per ton. That city, like Chicago, also sells its treated sludge to a fertilizer broker.

The point is that salvage is too basic to the solution of our waste treatment problems to rely on the forces of our "traditional economy." At the present time, the scrap and salvage industry is large and complex, and there are associations of scrap (secondary material) dealers in such places as Washington and New York that coordinate the activities of salvage brokers. The items they've been dealing in include newsprint, corrugated cardboard, rags, ferrous metal, cans, nonferrous metal and glass.

The "normal" market for any and all of these materials fluctuates widely depending upon time and location. Baled newsprint may sell for $12 to $15 per ton, and baled corrugated boxes from $7 to $12 per ton. Mixed rags vary from $2 to $30 per ton. Tin cans bring good prices if they're close enough to the copper smelting industry in the western states. Tramp metal, properly baled, can bring $25 per ton. But these prices hold up for relatively small amounts of "junk."

When the salvage market operates successfully now, the effectiveness of recycling wastes is clearly demonstrated. It's quite evident that the traditional salvage market is based on the law of supply and demand. The supply of material for salvage would be increased tremendously if cities were to embark on a full-scale program of reclaiming useful materials from wastes. The traditional economy simply can't stand the success of a waste recycling program, since supply would quickly overwhelm the current demand. *We need to improve the economics for salvage!*

We cannot as a responsible nation afford to allow junk dealers to determine if an auto should wind up in a graveyard, or if its component parts should be salvaged. We should not allow junk dealers to decide if all tin cans should be re-used . . . or all bottles . . . or all newsprint . . . or all corrugated boxes.

Right now, if junk dealers say "We don't want it," that "it" becomes a blight—a hazard that is ugly, unhealthy and degrading. All because a junk dealer says "No!"

Junk dealers are businessmen, out to make a profit. Like all businessmen, they deserve the profits they earn. They do not create or destroy a demand for the materials they serve as middleman for. They do reflect normal demands of industry.

But today's waste problems are *abnormal,* and need *abnormal* handling. That is why we cannot leave it up to the junk dealer to determine what junk we should use and what junk we should dump. We need a new way to "regulate" the junk so that it eventually returns to the mainstream of our economy.

If 1 million cars are ready for the graveyard, we must be ready to provide incentives so that the potential mineral values in all 1 million cars are used. It is not acceptable to have 100,000 rear fenders restored and the rest strewn about; or 200,000 tires and the rest smoldering; or chassis after chassis feebly camouflaged by hedges because Detroit finds it cheaper to make new ones.

There's no satisfactory place to store unused "junk"; the only thing to do is use it.

* * * * * *

One final comment on junk dealers. The Salvage Industry is represented by the National Association of Secondary Material Industries. This consists of some 600 of the leading firms in the nation "concerned with the collection, preparation and utilization of the many types of secondary materials that have a vital place in our economy."

Back in 1965, prior to passage of the Solid Waste Act, a representative of the Salvage Industry appeared before the Subcommittee on Public Health and Welfare which was holding hearings on the proposed legislation. Just about all the testimony given stressed the need for research into the solid waste field and the imminent dangers we faced if new solutions were not achieved. But the spokesman for the Salvage Industry had a far different message.

He expressed serious concern because the proposed Act did not draw a clear line between the materials of economic substance handled by his industry and that which is truly "waste." He stressed that: "Governmental assistance certainly should be directed clearly toward the explicit 'solid waste' areas where the absence of economic value results in the lack of normal disposal channels." His industry wanted to make certain that the solid waste referred to in the new law would exempt any wastes that had "direct commercial scrap value in their present form."

"Can you imagine what our cities and countryside would be like had not this industry pioneered and developed means of collecting, processing, and utilizing secondary materials? . . . These growth factors and new techniques truly make our industry more important than ever as an industrial reclaimer, a direct adversary of 'waste.'

"It is, therefore, our contention that those who are concerned with minimizing 'solid waste' should also be interested in protecting the interests of the secondary materials industry in view of the valuable functions it performs.

". . . The economic value of the overwhelming proportion of the secondary materials handled by our industry, in our opinion require no Federal program to cope with this. The material has economic value. It flows in normal ways back into industry. We do not feel that the funds of the Federal Government should be placed in the hands of those who would be in competition with our industry.

". . . At one time, to cite an example, the Department of Defense was operating an aluminum sweating operation in direct competition with our industry and later legislation was passed to limit this to situations which were only in the national interest. The problem was resolved. There are State and city governments trying to solve the solid-waste problems who are also expanding beyond what we consider the disposal of solid wastes."

It must be remembered that the Solid Waste Act of 1965, while a great step forward in encouraging new solutions, primarily provided for research. It was no specific attempt to get the federal government involved in the utilization and sale—or even necessarily consumption—of any product generated by wastes.

It does not seem farfetched to anticipate that much more opposition can be expected from industries with products similar to those which might be generated from wastes.

Future legislation will have to indicate how wastes can be recycled back into the economy without creating unfair competition with private enterprise. However, if we are agreed that the ultimate disposal method must be utilization, then fear of competition with private industry cannot stop this trend.

CHAPTER XI

Put Government into Wastes --Not Vice Versa

For years, reformers have been shouting about the need to get the waste out of state and federal governments. For my part, not enough attention has been given to stressing the opposite route—getting government into the waste picture.

The persons you vote into office in local elections—the mayor, councilmen et al—are the ones who have the primary responsibility for deciding whether to use public trucks to pick up the garbage or hire some outfit to do the job. Local officials will decide how to treat the garbage and where to locate the plant. But I believe your local men need help when it comes to figuring out how to get rid of the stuff. State and federal agencies are the most logical sources for aid.

Larger governmental authorities can finance large-scale experiments that will lead to dramatic solutions for many communities facing a common problem. The federal government in cooperation with state agencies can guarantee that local decisions affecting pollution are made with the fullest possible knowledge of their pollution consequences. They can encourage the setting up of super-authorities to deal with super-pollution "problem sheds." Such problem sheds are not necessarily bounded by county or state lines, since man-made pollution spreads across any man-made boundaries. Only by federal and state government initiative will a national system of waste management come about.

* * * * * *

Although pollution like education has traditionally been considered a state and local matter, the federal government's role in setting standards in matters which directly affect the strength of our nation has been clearly established.

For example, most U.S. citizens maintain that educational standards for their state rest with their state officials. Yet we have come to recognize, and take for granted, the vital need for a federal subsidy for education. All the problems of the teacher exodus, inadequate classroom facilities, and teaching aids could not be met through local tax resources.

In 1965, President Johnson based his education programs around the needs of poor pupils and schools which served them, building support from the public concern over the problem of poverty in an affluent society.

"Poverty," said the President, "has many roots, but the tap root is ignorance." Just like poverty, pollution also has many roots, but the tap root is hopelessness. Only the federal government has sufficient resources not to be overwhelmed, not always to be on the defensive but to launch a frontal attack on pollution.

The Government has a role in setting standards in this country on matters which directly affect the strength of our nation. Certainly the destruction, or hopefully the potential *construction,* of our natural resources fits this category.

Today, our municipal dumps joined by factories, cars and furnaces throw off more than 150 million tons of pollutants annually into the air we breathe—almost one-half million tons every day. Each day the nation produces enough trash to fill a freight train more than 150 miles long. New York City alone collects enough refuse to fill a 7-mile-long train daily. Citizens of the San Francisco Bay area will throw away enough refuse between now and the year 2,000 to cover a 50-square-mile area to 20 feet deep. To paraphrase one Public Health Service official, when we're not drowning in our

own sewage, we're smothering ourselves under a blanket of garbage.

For years, this deliberate pollution has worried researchers and progressive legislators. New Orleans scientists found that tiny particles wafted from smoldering city garbage dumps caused asthma outbreaks that sent thousands of wheezing victims to hospitals. A New York research team revealed that intense pollution from refuse burning and coal smoke filled hospitals with cases of heart illness and upper respiratory infection. Other researchers have proven the relationship between urban air pollutants and cancer.

You just can't get away from the fact that garbage, as it is now being mishandled, takes away from the strength of America. This being the case, the federal government has a constitutional right to help clean up the mess.

In a Senate debate on school aid some years ago, Claude Pepper of Florida made this observation on the constitutional authority for such action:

"It seems to me that there is ample authority in more than one section of the Constitution for enactment of this (general aid) legislation. . . . I refer to the provision which gives Congress the power to do whatever is necessary to carry out the explicit powers conferred upon Congress by the Constitution. . . . I did not see anything in the Constitution which specifically gives Congress the power to provide for the public health, but if we had a nation of disabled people, if we had a nation of citizens smitten by disease, we certainly would not have a strong republic. If we had to base the (general aid) legislation upon the power of the Congress to provide for the common defense as well as for the general welfare, I think we would have the authority under the Constitution to aid the states in providing for healthy, strong, skilled and intelligent citizenry."

Certainly this same reasoning leaves no doubt about the federal government's authority under the Constitution to set standards for waste treatment and utilization.

* * * * * *

Hundreds and hundreds of cities in the U.S. use their water supply, and then pass their sewage into the nearest waterway. Everything goes untreated into the river which eventually becomes the drinking water source for the city downstream.

Raw sewage sludge continues to be a major cause of

water pollution. In reviewing the enormous problem, Senator Gaylord Nelson revealed that the federal government pays 30 per cent of the cost of sewage treatment plants. He thinks this figure should be much higher. The federal government is already paying up to 90 per cent of the cost of federal highways. Under what conceivable perversion of priorities could highways be thought of as more important than clean drinking water? Says Senator Nelson:

> "If I were to list the 10 most important problems in the country, I would put on the top as the most important over the long pull the issue of the quality of the environment and the status of the resources in which we live. . . . If we continue at the present pace of pollution, in 25 to 30 years for all practical purposes, there will be no nonpolluted river or lake in America. . . . In my judgment, the answer is for Congress to raise matching funds to 90 per cent of the cost of sewage treatment plants and have every single municipality in America be required to install treatment plants."

Sen. Nelson's mention of federal support for highways is an excellent analogy, for if federal *waste utilization* laws could ever approach the highway program, the silver linings would be aglow.

* * * * * *

How did all this highway-building subsidy get its start? How does it happen that nobody complains of the federal government's role in highway building these days? Did Americans always take it for granted that the federal bureaucrats should direct their highway construction programs by controlling the purse strings?

Not on your life! Federal aid to road construction failed consistently in earlier days, because there was a fairly general opinion that this would be unconstitutional. The states, cities and towns, it was felt, had to provide their own highways.

But by 1802, when Ohio was admitted to the union, a provision was made resulting in the appropriation of five per cent of the new proceeds from the sale of public lands in Ohio for building roads in and to that state.

The act provided that roads would be laid out under authority of Congress. By 1811, construction began on the first 10 miles of the Cumberland Road eventually going from Cumberland, Maryland, to St. Louis, Missouri. It was paid for mainly by appropriations from sale of public lands which amounted to around 7 million dollars. By 1853, the entire road had been given to states in which it was located.

Organized bicycle riders provided the first popular demand for good roads in the United States. The National Good Roads Association was founded in Chicago in 1890. New Jersey passed its state-aid law in 1891 and the other states quickly followed.

Soon the clamor from owners of automobiles led to the first Federal Aid Act in 1916, which was funded by poll taxes. In 1921, Congress set up the U.S. Bureau of Public Roads giving financial help to the states, just as the states gave aid to the counties. The funds increased from less than $100 million annually until they reached $330 million in 1937.

Today the federal subsidy of roads is taken for granted. Before the money is allocated, however, state highway departments must satisfy certain standards for highway construction determined by the federal government. This seems most reasonable. If the federal government gives the money, it requires that the money not be wasted on inferior highways. Standards range from construction materials to safety measures (shoulder width, guard rails, etc.) as well as the "Beautify America" campaign now underway.

How good would our roads be today if the federal government were not directly involved in financing them? Rich states might be able to provide fairly good roads. Poorer states would have to be content with fewer and less satisfactory roads. There would be no possibility of crossing the country in a few days on superlatively fine roads as any American can do today with no trouble at all.

Today anyone who has studied the extent and contradictions of the solid wastes disposal problem would surely qualify it as at least as big a problem as the transportation problem.

Yet we have little help from the federal government and we have a fairly general attitude, among officials, that the federal government should keep its hands off this problem. It's a local problem, these people maintain, and therefore should be solved at the local level.

But solid waste management cannot be solved at the local level without help from the federal government. No matter how effective the waste treatment, there is always a residue. To a city, that residue is a problem to get rid of off city limits. To a nation, that residue is a resource to be used within the continental limits. As population grows and cities expand, so does this enormous problem of solid wastes. Once the stuff is in your eyes, not even closing them brings relief.

* * * * * *

Carriages, buggies and early automobiles were able to make out somehow on roads that were little more than cowpaths, thrown together of mud, clay, logs, bricks, stones—anything that could be used. But such roads are obviously a complete impossibility in a society that turns out millions of cars each year. The problem of providing roads to handle automotive developments was simply too big for local governments.

In just the same way, the problems of waste disposal are already too large for local governments to handle without help from the federal government.

Our population pressures are so great that there is almost literally no place available in the neighborhood of most towns and cities where wastes can be disposed of without despoiling land that might be valuable some day, without creating health hazards to local citizens or to the citizens in some neighboring community, without befouling natural resources that really belong to all the people of this country. Even in the few cities where there is still some room, time is short as population growth will soon force a look for new solutions. Then, too, no matter how effective incineration or composting may be, there is bound to be some residue. In other words, a city must be able not only to treat its wastes.

It must also have the capability to *use* the treated wastes—or they too become a liability which has to be stored somewhere or transported somewhere. Within most city governments, as they exist today, such use cannot be arranged for.

The city neither has the money to finance a "closed circuit" of recycling wastes, nor the physical land areas to arrange for this degree of waste utilization.

In the debate over whether cities must relinquish authority over waste treatment as the federal government increases its responsibility, I think that only a straw-man, not a valued American heritage, is being destroyed. In other words, we can agree on the need for independent, self-reliant viable cities, but cities are not *giving anything up* in this area of waste-treatment-utilization that is so vital for our collective welfare. How is it possible to give something up if after all these years most U.S. cities haven't touched this area of public responsibility with a 10-foot pole!

* * * * * *

One market analyst, William H. Ducker, observes that more and more large manufacturers now recognize that the field of environmental pollution of necessity must fall within the jurisdiction of the Federal Government, even though most of these men are philosophically opposed to any increase in Federal regulation.

"They are honest enough to admit, however, that it will take Federal Government action to remove antipollution equipment from the competitive environment. For example, no one steel company is likely to spend several millions of dollars on pollution control when its competition does not have to do so. Along the same line, if a community or group of communities along, say, the Detroit River/Lake Erie waterway, forces manufacturers to comply with rigid pollution specifications, these plants would be at a competitive disadvantage unless other communities, such as Gary, Indiana, also forced compliance.

"If, however, the Federal Government set the same specifications for an entire industry regardless of its geographic location, each firm within that industry would theoretically have to spend about the same relative amount of money. While industry prices would be forced up somewhat to cover the added cost, no one group of plants would have a cost advantage over another group."

CHAPTER XII

Making A Buck Wherever You Can

I used to live in northern New Jersey, and for years when anyone mentioned pollution, I thought of the smells and smoke that spewed out of the companies along U.S. Route 1. For as long as possible, industry ignored its wastes.

If you drive by that same area today on the New Jersey Turnpike—somewhere between the Newark Airport and Lincoln Tunnel, you'll get the feeling that those industries still continue to ignore their wastes. But, little by little, progress is being made to force companies to abide by pollution laws and install control equipment.

It's not that company executives like to make a mess out of air and water; it's simply that there's no obvious profit in controlling pollution. If a company buys some expensive equipment to filter out the waste particles before it goes into streams or atmosphere, it winds up being stuck with a residue it never had to worry about before.

The more imaginative companies have finally quit complaining about the pollution laws that forced them to stop spewing their wastes out. They have actually applied their aggressive marketing know-how to figure out how to profit from what had been an unwanted by-product.

> "When a company looks at pollution control as a thorn in its side, it may be missing the boat," declares Robert J. Anderson, a partner in a Maumee, Ohio, grain, fertilizer, feed and equipment concern. "When you get down and work at it, you can often turn a control problem into a profit." Filtering devices in the grain elevators of Anderson's company collect grain dust and compress it into pellets. By selling the pellets as cattle feed, the concern expects to earn back the $750,000 cost of the dust-collecting equipment within 5 years.

I don't believe that even the imaginative companies

would have made the effort needed to find a market if a government pollution agency had not gotten on their backs. First came the public agency setting a standard that meant the company could no longer continue a practice that resulted in pollution. Without the government standards, the company's only justification for expenditures for pollution control equipment to their stockholders would have been that it wanted to be nice to fishermen and bird-watchers and nature-lovers . . . or "just for nice," as the Pennsylvania Dutch say.

But "just for nice" is a weak explanation in business. Complying with federal regulations, however, is a perfectly-understood explanation. Stockholders may complain about the bureaucracy in Washington that forces production costs to go up, but that kind of complaining doesn't get division managers fired. Besides, as long as the entire industry must increase its costs by having to install anti-pollution devices, then the added expense will most likely be passed on to the consumer.

Private firms have a big advantage over cities when it comes to offsetting the costs of pollution control equipment. Here's what I mean: Most control equipment by its very nature retains the material that formerly caused the pollution. Some kind of filtering device, whether in a smoke stack or a drain, serves as a collector. A residue is obtained, a by-product you might say. Unlike municipalities, companies are accustomed to analyzing the profit potential in all by-products. The research staff sets to work to develop that potential. Eureka! Another success!

Public wastes and private wastes are making a mess out of our country. But private companies at least have the incentive and capability to put a salvaged waste into production economically. Take the case of the Great Northern Paper Company at its plant in Millinocket, Maine. Required to clean up the Penobscot River, Great Northern installed a new recovery boiler for the chemicals used at its sulphite mill. The chairman of the company reported to stockholders that in addition to fulfilling the need for water improvement, the Company expects to obtain a significant monetary return on its investment, since the recovery process will enable Great Northern to reclaim between 70 and 80 per cent of its pulping chemicals and to generate increased power from the burning of waste liquor as fuel.

New Scientist, for November 2, 1967, reports on methods of recovering sulfur from chimney and flue gases. Supplies of sulfur are dwindling fast and it is an important ingredient in commercial fertilizer. On the other hand, it is one of the most objectionable ingredients in air pollution.

"Remove the sulfur either from the fuel or from the flue-gases after it is burned, and not only have you at once made the world a healthier place to live, but sulfur will never again be in short supply and fertilizer will be cheap and plentiful," says the *New Scientist* article.

Two new methods of obtaining valuable food protein from what was formerly difficult-to-dispose-of waste are described in *Municipal Engineering* for February 2, 1968. A Canadian firm is developing a method of using poultry-packing waste to make high-protein animal feed. The new idea involves macerating, drying and digesting the waste in heated tanks to produce a crystallized high-protein powder which can then be fed to other poultry, either alone or mixed with other feed.

Manufacturers of cod liver oil have produced troubles for sewage disposal plants, since the extraction process they use results in a run-off of a protein-rich fluid which clogs sewage works. It is thus both a disposal problem and a waste of protein, which is the most expensive of all food elements and the most essential. A process has now been developed to produce from the run-off a precipitate that can then be spray-dried, to yield a protein powder that can become the basis of animal foods.

One pollution problem which seemed almost insurmountable later proved to be a boon to a new industry. A Virginia plant manufactured smokeless powder during World War I and found itself with enormous quantities of something called nitre cake for which it could find no disposal method. It was decided to bury it, rather than cause an air pollution or water pollution problem with its disposal. After the end of the war a paper plant was established in the same neighborhood and the nitre cake proved to be useful in the paper-making process. It was dug up and used.

The story of cotton by-products shows perhaps better than any other what can be done with substances formerly thought of as useless. When the cotton gin was invented, what had been no problem at all to individual farmers—disposal of the cotton seed—became a gigantic headache to the owners of cotton gins, as the mountains of seed piled higher and higher.

Over the years, uses were found for every part of the seed and the lint that clung to it, so that eventually farmers were raising cotton because of the profits from the by-products, long after the synthetic fiber industry had made cotton-growing less profitable. The lint is now processed into many different products; the seed is separated into hull and meat. The hull has many different commercial uses. The seed is processed into salad oil and soap. The cake of cottonseed that remains after everything else of value has been removed is used as fertilizer or fed to cattle. This triumph of turning to profitable use by-products which formerly presented immense disposal problems is perhaps the best example of the kind of solutions that can be worked out to replace the scandalously wasteful and careless ways we are presently using in many industries.

A Japanese inventor has developed a huge machine which he claims will, under enormous pressure, make building blocks of any and all garbage and refuse. You can feed into this enormous hydraulic press anything from refrigerators or sofas down to all the contents of garbage cans. Everything is shredded, pulverized, sterilized by heat and formed into cubes weighing about 5-7 tons, which can then be used for building, landfill or almost anything else. They can be encased in cement, asphalt or plastic. He calls it a "Garbecue."

Almost everybody who writes on the subject, however, ends with words that sound prophetic and pleasant. The *Readers Digest* puts it this way: "Many scientists believe that eventually we must return to the soil much of the organic matter that we harvest from fields and forests." Harold Gilliam in the *Audubon* study, said: "There is urgent need for research on the economic feasibility of such rational methods as composting—reusing the refuse for soil enrichment and other constructive purposes."

A Committee on Environmental Alteration established by the prestigious American Association for the Advancement of Science puts it this way:

"It seems clear that the collection and containment not only of chemical but also of other waste materials at or near their source in industrial, municipal, agricultural, and other systems are generally preferable to discharging such materials into streams, lakes, oceans, and atmosphere, or spreading them on the landscape.

Furthermore, in many instances the recycling of these materials proves to be desirable, not only in the social sense of reducing hazards to the general public and the surrounding region, but also in the economic sense of reducing the required input of a diluting medium for industrial plants or other activities. Frequently, valuable by-products can be extracted in the course of such containment and recycling operations. Even in cases where it is not known how to use waste material or whether it is harmful, prudence and good sense indicate that, when feasible, the material should be stored, converted or otherwise kept from being discharged into the environment.

"We recommend that containment and recycling possibilities be kept under continuous review by appropriate government, industrial, agricultural, and other agencies to encourage the substitution of biodegradable chemicals and containers, to sponsor the redesign of industrial processes to embrace the recycling and by-product use of waste materials, and otherwise to exploit the use of technology for dealing with these problems."

* * * * * *

You can readily see that a city or any governmental agency faced with a steady supply of salvaged wastes would have more trouble than private industry.

Since a city is unable to mount an intensive and imaginative campaign for marketing its salvaged wastes, it is forced to rely on local "traditional" market demand. In some cases, a market does exist since the salvaged waste can be readily used as raw material for industrial processing. Houston, Texas, and some western cities have entered into contracts which bring those municipalities substantial revenue from the sale of metals salvaged at incinerator plants. The Proler Steel Corporation pays Houston $13.00 per ton for tin cans, magnetically separated and flattened at the Houston incinerator.

But other cities with similar materials are not so fortunate. The tin cans—possessing the same characteristics as that bringing $13.00 per ton in Houston—must be dumped. Their mineral values are mostly lost to society. *With most wastes, pollution begins where resource value ends.*

Again and again, we see the critical decision of salvage and recycling made on the basis of this short-term thinking. A new force representing the economics of resource conservation and environmental health must be brought to bear on traditional economics.

CHAPTER XIII

Selling Composted Garbage Like Toothpaste

Whenever possible, cities should try to get whatever income they can from their treated wastes. As I have tried to make clear, a city should not base its treatment method upon a commercial salvage operation, but that does not mean it should be meek about exploring markets.

Salvage income can be available for cans, tires, etc., and certainly for composted garbage. Not all the compost produced by municipal treatment plants need be sold. Indeed most of it can be stockpiled safely in a minimum area. However, a substantial percentage, representing thousands of tons annually, could be sold to large fertilizer firms.

Just imagine what would happen if the largest fertilizer company in the U.S. decided to advertise a compost brand name the way Colgate pushes its toothpaste, the way Ballantine raves about premium beer, or General Motors sells sports cars? Suppose this leading fertilizer company throws caution to the wind and decides to use its full nation-wide distribution power to get the compost story across to gardeners and farmers.

The sale of organic soil conditioners is a big business even before the entry of our mythical Number One. According to the United States Department of Agriculture, more than 25,000 tons of compost are already being sold, along with some 340,000 tons of dried manures and 120,000 tons of sewage sludge.

Production of *Milorganite,* Milwaukee's granular activated sewage sludge, has increased to about 80,000 tons yearly. Annual sales volume is $3,500,000. "We still sell everything we produce," reports C. G. Wilson, Sales Manager and Head Agronomist of the Milwaukee Sewerage Commission.

Available in every state, *Milorganite* is used extensively in golf courses, lawns, city parks, as well as gardens and orchards. It is packaged only in 50-pound bags.

From Mobile, Alabama, Public Works Commissioner Lambert Mims writes that the city compost plant is operating and "we are manufacturing a very fine product marketed in the Southeast under the label of *Mobile-Aid Organic Soil Conditioner.*" The compost is packaged in 25- and 50-pound bags or sold in bulk to commercial users.

The Fairfield Digester System is used in Altoona, Pennsylvania, to turn out the *Fairfield Organic Humus Builder*—"a sanitary soil builder converted from fresh garbage . . . A mechanically-controlled environment contributes to the uniform effectiveness of the process."

Alive is the brand-name given to compost produced at the Metroganic plants in Houston, Texas, and Gainesville, Florida. This company hopes to market it nationally, and has had tests of the compost conducted at Texas A & M University. In germination and early growth of tomato and pepper seedlings, the compost has been superior to sphagnum peat.

Finished compost from the Johnson City, Tennessee, plant will be used and marketed in a variety of ways—in greenhouse experiments; on bare areas such as highway cuts and stripmine spoil banks to control erosion; as well as on gardens, parks, lawns, golf courses and truck farms.

In Los Angeles County, there are some 115,000 beef and dairy cattle and 3 million pen-caged, egg-laying hens. The Public Health Service has announced that a $90,000 grant to study improved sanitary methods of composting manure has been awarded the city of Cerritos, with Charles Senn of the University of California at Los Angeles directing the one-year investigation.

"The study," says Richard Vaughan, Chief of the PHS Solid Wastes Program, "will aim at developing new techniques in converting animal wastes into soil conditioners or fertilizers. Currently, untreated manure is hauled from the corral sites and stockpiled near urban areas in the two California counties for as long as a year before it is composted."

One stockpile in Cerritos, near Los Angeles, is 30 to 40 feet high and contains over 10 million cubic feet of manure. The result is often a fly problem, seepage to pollute ground waters, and dust and odors.

The study proposes to develop a method to permit prompt composting on the feedlot sites. Concludes Mr. Vaughan: "Improved composting of animal waste can assist such large dairy states as Wisconsin, Minnesota, New York, Pennsylvania, Ohio and Michigan. Other states—Iowa, Nebraska, Illinois, Texas, Colorado and Kansas—that raise beef cattle in concentrated feedlots will also have an interest in the compost research."

An article on "Manure Management, Costs and

Product Forms" by J. Van Dam and C. A. Perry, Los Angeles County farm advisers, shows that a good market exists for composted manure, both packaged and in bulk, all within a competitive price range. They observe: "By approaching the manure-removal and disposal problem with positive thinking, a livestock feedlot operator could realize a profit from the sale of manure."

As you can see, more organized research is being done on composting, and more organic soil conditioners are reaching the market.

* * * * * *

Just think what would happen as a result of a massive, coast-to-coast advertising campaign!

Years ago, fertilizer and soil conditioners were almost exclusively products for farmers. Not so anymore. Today, the U.S. non-farm fertilizer market exceeds $300 million—more than 50 per cent going to home gardeners and the balance to professional landscapers, golf courses, parks, etc.

Non-farm fertilizer use has grown at the rate of about 10 per cent per year since 1960, with major growth occurring in California, Texas, Florida, Northeast and the East North Central areas. An industry representative estimates that about 25 per cent is sold in garden centers, 20 per cent in hardware stores, 10 per cent in discount and department stores.

On a tonnage basis, organic fertilizers (including manure, etc.) comprise one-third of the fertilizers sold.

U.S. production of peat has been climbing in recent

years to meet demand from home gardeners who use it as a soil conditioner. The U.S. Bureau of Mines says about 600,000 tons of peat, worth more than $5 million at wholesale, are harvested annually, triple the 210,000 harvested in 1952.

Compost has been shown time and again in agricultural experiment station tests to out-perform peat in beneficial effects on soil and plant growth.

* * * * * *

Again, let's contemplate what would happen if the General Motors of the fertilizer industry is bagging compost at its plants throughout the U.S.

1. *Distribution* — Many processing and packaging plants located in different areas of the country could be shipped compost from nearby composting installations (city treatment plants). Transportation costs for bulk shipment (from refuse compost to fertilizer plant) would thus be minimal. Sales of bagged compost would be similar to distribution of the firm's other fertilizers and soil conditioners.

2. *Marketing* — As far as the municipality in charge of the refuse compost plant is concerned, its worries over disposing of much of the end product no longer exist. The fertilizer company has a contract to take all the compost produced at a fixed price. A single trade name will permit effective national advertising.

The fertilizer company will use a large percentage of the compost as filler for its regular fertilizer line, in addition to marketing the compost as a completely organic soil conditioner. Its distribution channels to garden centers, farm co-ops, supermarkets, discount stores, etc., can easily absorb the new line.

The firm's agronomists can arrange large-scale tests; its budget can afford to sponsor research grants to well-known agricultural experiment stations; industry trade journals will feature articles "announcing" a great new product. (The comment has been made, I believe accurately, that agricultural chemical trade journals are glad to write about almost any fertilizer or conditioner that can be put in a bag and sold—*after* it's been put in a bag.)

3. *Spreading Equipment* — Because compost will be available at economical prices, farm and garden equipment manufacturers will improve the type of spreaders now available for applying humus. This new equipment will make compost spreading a simple and economical task for large-scale farmers, clearly worthwhile in terms of soil fertility and increased yields.

CHAPTER XIV

Improving Soils to Feed People

You don't have to be an organic gardener to know that humus helps to make soils productive. One soils science textbook after another spells out the need to maintain a high organic nitrogen content in the soil.

The rich, black texture of a fertile soil comes from its organic matter content. Rain can penetrate a humus-rich soil, providing vital water to plant roots, rather than flowing in flood-like fashion away. Tilling can be done more easily when a heavy clay soil is loosened up by organic material. Soil nitrogen is released from organic residues, and minerals like phosphorus and potassium are helped to become more available to plants through the action of organic acids from decomposing humus.

Years ago, when farm animals were plentiful and

tractors scarce, much organic matter was returned to soils via animal manures. But today's machines leave fumes, not manure, in their tracks; fertilizers build up the harvests—taking more humus out of the soil. If unchecked, the rapid depletion of humus means that topsoil becomes less and less fertile.

Agronomists have warned farmers about a short-sighted, abusive land policy. But farmers have enough troubles without worrying about fertility in years to come. Their main concern is how to make a living now.

Our nation has been blessed with an abundance of land resources. It's estimated that the U.S. has about 638 million acres of non-federal land suitable for regular cultivation and another 169 million acres acceptable for limited cultivation. In addition, there are 644 million acres not considered suitable, plus another 761 million acres of federally-owned land mostly in forests and range.

What would happen if our legislators, convinced that composting provided a sensible and economical solution to refuse disposal, set up an agricultural program to encourage application of compost on these lands?

Every year, one million acres of land in this country are converted from farms to hot dog stands, gas stations, suburban houses, shopping centers and everything else that goes with urban sprawl. This overwhelming trend toward urbanization brings all us garbage-makers into a tight little web. The resulting problems to humans are by now well-known. The problem to the land is finally getting some attention.

Prof. Ian McHarg, chairman of the University of

Pennsylvania's Department of Landscape Architecture, describes the modern metropolis "as sterilizing and waterproofing thousands of square miles of land." As examples of what to do to stop fouling our rivers and polluting our atmosphere, Prof. McHarg cites the way the Japanese in the past thousand years have enriched their land while farming it intensely, and the way the Dutch have "created" Holland.

> "If you want a closer example, take the Amish, Mennonite and Pennsylvania Dutch farmers . . . they know where their food comes from. They know a crop is directly related to the organic materials, water, and sunlight that go into the soil."

Dr. S. J. Toth of Rutgers University's Department of Soils wrote in Compost Science that soils in the Northeast particularly need dressings of organic matter to maintain their high productivity. Here's what he had to say about conditions in New Jersey:

> "It is surprising to find in a state as highly industrialized as in New Jersey, for example, that stores of organic residues are available in amounts sufficient to add one ton of organic matter to all the cultivated land in the state. The supplies are roughly between 1½ and 2 million tons.
>
> "The sources of this supply are garbage (1,000,000 tons), sewage sludge (200,000 tons) and industry (500,000 tons). Approximately half of the industrial wastes represent manure and bedding from the poultry industry and the rest is organic industrial wastes of various types, mainly from the instant coffee and tea, antibiotic, botanical drug

and spice industries. The supply of these organic industrial wastes is sufficient to dress one-quarter million acres with one ton of organic matter per acre."

Robert Rodale, editor of *Organic Gardening and Farming* magazine, offers these observations:

Compost has traditionally been a home-made product. The majority of gardeners have at some time made compost, and a large number make and use it regularly. But there is a need for compost far in excess of the amount now being made.

The chief value of compost is its ability to improve the physical and biological structure of the soil. Humus in the soil holds moisture, therefore enabling plants to feed and grow during periods of drought. It provides the proper environment and nutrients for the many types of microorganisms that improve the soil through the increase of their populations.

It is a general rule that compost (and all forms of humus soil conditioners) becomes more important as soil cultivation becomes more intense. The average home vegetable garden is an example of just about the most intensive type of agriculture practiced in the U.S. More food is produced per square foot in a well-cared-for garden than in any form of modern agriculture, chiefly because rows and plants are spaced so closely together. A lush lawn is also an example of a very intensive type of culture, as are flower beds, shrub borders and foundation plantings. In such situations, large amounts of humus are required to insure not only that there will be proper drainage through wet seasons, but that the soil will remain productive through drought. In intensive culture, humus also provides a reservoir for nutrients. Nitrogen, for example, has an affinity for humus, and nitrogen management is extremely difficult in gardens with soil of low humus content.

The humus needs of gardeners and landscapers are largely being filled (though usually inadequately) by home-made compost, peat and local products such as cocoa bean hulls, gin trash, sawdust and pine needles. Vast quantities of those organic products are being used as soil conditioners. Peat is the currently used organic product most similar to commercial compost. It is estimated that 700,000 tons of imported and domestic peat are used each year.

Compost is usually sold in competition with peat. However, compost has an additional distinction which makes it of particular value to gardeners. It is one of the few fertilizer materials which can be placed in direct contact with plant roots without causing harm to the plant. Compost is therefore a foolproof product, a feature much appreciated by gardeners.

* * * * * *

Reclaiming areas destroyed by surface mining for agricultural and forestry purposes is of vital importance in the brown coal surface mining areas of the German Federal Republic, writes Prof. H. Kick of the Agricultural Chemistry Institute, Bonn University.

> "The results obtained so far in our experiments and those obtained elsewhere show that garbage composts, garbage sludge composts and also sewage

sludge composts can be utilized for soil amelioration purposes advantageously from the economic point of view, provided the cost of transportation and the cost for the distribution of the material can be kept within acceptable limits."

Compost made from our city wastes could be used as effectively to reclaim mining lands in the United States.

Yes, compost could be a tremendous force in building soil fertility throughout the world and thereby help to feed the world's population. Interestingly enough, a British anthropologist theorizes that the value of compost was evident before it ever reached the soil. According to D. C. Arnold, compost was the world's first source of beneficial fire for mankind. He theorizes that the idea of using fire as a tool probably came to early man while he was watching smoke and flames rise from a pile of matted vegetation he had used for a bed.

Backyard gardeners and today's researchers are well aware of the heat generated by decomposing material in a compost heap. The intense digestion of the organic materials causes temperatures to reach 180° F. and higher. So the reasoning behind Dr. Arnold's theory is easy to understand.

"Accumulation of sun-dried vegetation in a bed, perhaps a communal one, could lead to the 'compost heap' situation, presenting man with a small-scale source of heat. Occasionally," writes Dr. Arnold, "no doubt such beds would burst into flame, providing at once the source and the idea of a blazing fire."

In support of his theory, Dr. Arnold cites the regard with which the active compost heap was held—"a most precious family possession." He believes that the care some primitive races give their existing fires today could have derived from that compost heap tradition of their ancestors.

From that early lesson in fire-building, mankind learned other uses for the decaying organic material we have come to call *humus*. We have learned that the "fire" it generates in the soil is largely responsible for the fertility of our land and the successful development of plants.

Over 150 years ago, the founder of the "humus theory," Albert Thaer, became convinced that humus was the only direct nutrient source for plants. His idea was generally accepted until Baron Von Leibig came along with his view that minerals were the only important food. Leibig's experiments had a tremendous effect on soil science and our artificial fertilizer policy of today is directly descended from his work. After Leibig, no one gave much thought to the possibility that humus was taken up directly by the plant as food. People did recognize, though, the value of humus in holding moisture, feeding microorganisms and in supplying the agents that make locked-up soil minerals available to plants.

The relationship between plant growth and compost is so interwoven and complex that it will be many more years (if ever) until scientific research comes up with the full explanation. Recently, for example, a South African agronomist has shown that there is some truth in Thaer's old humus theory—that plants do take up certain portions of humus directly as food and that even in minute quantities, these humus plant foods have considerable effect.

Here briefly is a review of how soil fertility and plant development depend on compost and other organic materials:

Improving soil physical condition. Soil aggregation, moisture-holding capacity, stopping runoff, and optimum soil temperature—all are greatly helped by organic matter directly or by by-products of organic matter decomposition.

Increasing microbial population. While partially decomposed organic material has more effect on the activities of microorganisms than well-decomposed material, both encourage microbial activity.

Exchange capacity increases and leaching of plant nutrients decreases. Availability of phosphorus is greatly increased by organic matter in soil, while nearly all of the nitrogen in the soil becomes available for plant use as a result of decomposing organic matter.

Increased crop yields. Greater harvests are obtained because of improved physical and biological action of soil.

* * * * * *

More scientists are becoming increasingly skeptical about the over-reliance upon artificial fertilizers. The runoff of artificial fertilizers from farm fields into streams, lakes and rivers pollutes our nation's waterways as much as city wastes dumped directly into them.

Dr. Barry Commoner, Chairman of the Botany Department and Director of the Center for the Biology of Natural Systems at Washington University in St. Louis, warned that extensive use of inorganic fertilizers containing nitrates could accumulate to dangerous levels in foods, particularly certain baby foods.

In Lake Erie, says Dr. Commoner, pollution by nitrates from inorganic fertilizers probably equals pollution by sewage. The leaching of inorganic fertilizers into waterways encourages growth of "algae blooms," which make huge "cesspools of the nation's lakes." Algae growth removes oxygen from water, making it less able to support animal life.

Dr. Commoner suggested that "in the long run, the problem of nitrate pollution will require a *fundamental revision* of the entire economy of agricultural production in this country." He said that the only way he saw out of the problem was to return to the use of the more costly organic fertilizers such as manure and compost, which are very low in nitrate content.

"Right now, this is not considered practical. It is far too costly, I realize. But if it is the only solution—and I think it is—then we are going to have to find a way to *make* it practical!"

Now we have still another reason to turn garbage and other wastes into a never-ending source of soil-building material. Reducing the volume of artificial fertilizer runoff into our waterways is yet another benefit.

CHAPTER XV

Don't Forget The Garbage, Dear

I have a friend who spends much of his professional and spare time as a researcher checking how garbage will some day travel through pipes. But he still gripes about his wife's parting message to put the garbage out before he leaves for the laboratory.

For most of us our concern with garbage ends when

the bag goes down the apartment house chute or the can is picked up by the collectors. But actually that just ends your responsibility for your *personal* garbage—*the best is yet to come!* Right now, you're "hearing" from garbage you've never met personally—don't know where it came from, let alone what or whose it was. You not only hear about it—you smell it, breathe it, drink it!

How many more years must the cumbersome and chaotic collection and disposal of garbage continue? How much longer will the Space Age put up with diesel-powered garbage trucks—whether they be red, yellow or green—as significant progress over the horse-drawn garbage wagons of yesteryear?

In February, 1968, New York City residents saw some 100,000 tons of garbage pile up in the streets during a nine-day sanitation strike. The way we're going now, a similar strike in the year 2,000 may result in 500,000 tons of refuse.

Is this the scene we must look forward to?

* * * * * *

The literature of garbage treatment in the U.S. is mighty grim stuff to plow through. Many officials in the Public Health Service as well as public works officials all over the country will reel off one insurmountable problem after another. The only changes that might take place are for the worse. More people, more garbage, more plastic, more pollution.

Is there a Utopian method for garbage? What would it be like? I don't care if it's expensive, impractical, untried, unsound—even "impossible!" The point is if you don't have an ideal plan of what you'd like to do with garbage, then the city streets—with their rows of trash cans . . . the city dumps—with crud all over the place . . . the nation's waterways—with death built into them . . . become our norm.

* * * * * *

Here we are in Utopia. And where did all the garbage go?

The Utopia we see has an urban, not rural, setting. "Indeed the first utopia was the city itself," writes Lewis Mumford, who also points out that utopias from Plato to Bellamy have been visualized largely in terms of the city. The Greeks never could conceive of a human commonwealth except in the concrete form of a city. Alexander spent much of his energy building cities. Through the ages, writers and philosophers recognized the city as a reflector of the complexities of society within a frame that respected the human scale. "To an extraordinary extent," says Mumford, "the archetypal city placed the stamp of divine order and human purpose on all its institutions, transforming ritual into drama, custom and caprice into formal law, and empirical knowledge spotted with superstition into exact astronomical observation and fine mathematical calculation."

Since our utopia will be found in a city, the citizens cannot bury their garbage in Thoreau-like style under the forest floor. Instead we must envision how city officials will bring "divine order and human purpose" to the earthly task of garbage disposal.

Interestingly enough, conservationist Paul B. Sears cites the parent-child relationship of the city to the farm. "Save for the small voice of the rural romantics, utopia is the apotheosis of the city. This was true even in the days when the Hebrew prophets were thundering against the vices of urban life. . . . But it is well to recall, too, that the city is the child of agriculture. Cities were impossible until men and women were released from the constant quest for food."

It's dangerous, fallacious, and plain stupid to think of the city as being isolated from its surrounding country. Sears describes it this way:

"The country is still its (the city's) nutrient base, strikingly so in regard to its increasing thirst for water, for which the city's tubular tentacles reach out further and further. Its economic health is tied more closely to that of the rural region than is commonly realized. . . . *There can be no effective movement toward utopia without including the entire ecosystem of community and environment."*

* * * * * *

The shapers of tomorrow—some leading members of the "think tank," in fact—have been promoting a bubble-encased city as the dream of the future. What better way to manipulate our environment, they theorize, than by using technology to keep the outside outside, and the inside inside.

Plans which have been developed for the Experimental City call for its being enclosed in a series of domes two miles in diameter. Each dome would be a mile high at its apex and six of these domes will accommodate about a quarter of a million people—approximately the ideal size for such a city according to the thinkers.

Athelstan Spilhaus, president of the Franklin Institute in Philadelphia and chairman of a Committee on Pollution appointed by the National Academy of Sciences, has described the "metaphysical" city. It would permit no discharge of smoke or fumes into the atmosphere. Vehicles entering the city would travel by tunnel whose air would be drawn out through "fume sewers" to air-cleaning plants. Automated rail systems would take residents from their homes to the desired destination.

All wires, pipes and other utilities would be routed through underground tunnels and ducts. *All garbage and trash would be disposed of by grinding and piping to a place where it can be treated.*

Such futuristic designers as Buckminster Fuller and Air Force Systems head General Bernard Schriever have worked with Dr. Spilhaus on the domed city—the first of which is planned for northern Minnesota. The scientists envision a first generation of 50 Experimental Cities, one per state, to test new ideas in adaptation to local conditions and needs. Each city would be surrounded by about 40,000 acres of open country—presumably so the quarter of a million residents inside the dome could hear crickets chirping or birds singing through a broken pane.

The Ministry of Building and Power in Russia also has plans for an enclosed community. The town, Snezhnogorsk, is planned to house 1,500 power station workers and will be 600 meters wide and 20 meters high. The living quarters will form the outside wall of the town and will face inside to a domed garden with an artificial climate heated by means of infra-red rays.

When I first heard about plans for domed cities, I couldn't help but think of Rachel Carson's introduction to *Silent Spring.*

> "On the mornings that had once throbbed with the dawn chorus of robins, catbirds, doves, jays, wrens and scores of other bird voices there was now no sound; only silence lay over the fields and woods and marsh."

In a domed city, man would create his own silent spring—not by misuse of pesticides but by setting up a micro-climate and an environmental isolation so absolute that he had created a "Silent Year."

This is not to make light of the goals of the domed city. The dream of a smokeless, noiseless, trafficless city is one we should strive for. But I doubt if it's worth achieving at the price of stifled, uniform, physically-penned-up individuals living in an encased "island of purity."

* * * * * *

Paul Sears has speculated on the possibility of human beings becoming fossilized before our time. "There is good reason to believe that people could be conditioned to adjust themselves to a highly artificial, technologically controlled environment such as the city under a plastic bubble which some envision. Such an arrangement, however, would not only congeal the pattern of living, but be vulnerable on two counts. The greater our dependence upon an elaborate chain of technology, the more liable we become to disaster through failure of any link. And the more restricted our range of experience, even though physical needs are met, the greater our loss of flexibility to meet emergency. The too-sheltered child is an example. So is battery-grown poultry. These birds, raised under completely controlled conditions, must be protected against sudden noises or even the presence of a stranger. Otherwise they pile up in a corner and smother each other."

True, the city has become a sprawling monster. In our attempts to control it, we're tempted to try anything to stop the crime in the streets . . . the contamination of our water and air . . . loss of open spaces.

The temptation to "think big" is the best thing that happened to the pollution problem in the United States. If we didn't have men thinking about domed cities, we would never get anyone to consider getting rid of garbage cans, and open burning dumps, and automobile junkyards, and recycling wastes. And that's what composting is all about. And that's also why a combination of pipelines to carry garbage out of cities and composting plants in rural areas to treat the waste has a Utopian quality.

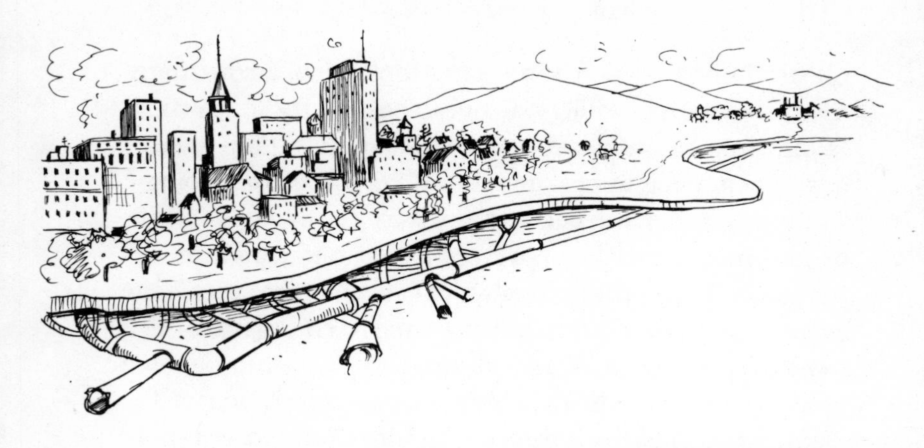

CHAPTER XVI

Piping City Garbage into Rural Compost Plants

In most cities, you can tell your slum areas by the overturned garbage cans in the streets. And any street that's good enough for garbage is good enough for poverty, sickness and mugging. Someday in the future, we shall have the means to get trash out of the city limits without resorting to garbage cans or garbage trucks.

Pipelines will be the answer—underground pipes. Each apartment and house will connect with a network of pipes to convey refuse quickly and economically away from crowded urban centers to treatment plants in the country.

Surely, it's ridiculous for modern-day apartment dwellers like those living in Habitat to be forced to keep a garbage bag under the kitchen sink or to tolerate a basement incinerator sending forth wispy orange peel chars that blanket the neighborhood. And it is just as ludicrous picturing the flight of the garbage-carrier in the year 2,000, dashing to the trash barrel out back, hoping the bag won't bust before he reaches it.

Furthermore, if city traffic is ever to be eased and if street noise is ever to subside, can it be done with the continuation of present methods for garbage pick-up?

In an age when we're willing to invest many millions to get a few hundred people flashing their way cross country in a supersonic transport, there's no need to continue to put up with draggy transport methods for wastes. Garbage trucks, horse-drawn or diesel-powered, have been an accepted phenomenon of urban living for many, many years. They clearly did a commendable job in the past, but operating on today's city streets, they symbolize the archaic methods used in the handling of organic wastes in this country.

As long as men have to race along the sidewalks emptying cans into a slowly-moving cavity, we're heading in the wrong direction right at the start. We'll never catch up in the utopia-chaos race. At the current rate at which wastes are produced, it's almost impossible to develop sensible, sanitary and economical treatment methods using "ancient" transportation media like garbage trucks to haul wastes to the treatment area.

A system of pipeline transportation makes a great deal of sense for moving wastes from where they are produced to where they will be treated.

By the time pipes replace trucks, we'll undoubtedly have changed our garbage attitudes as well as our garbage transportation methods. We will fully realize that the waste treatment process must be constructive, not destructive.

How much longer can we continue to be wasteful with wastes is the question. How much longer can we pollute without, in the words of one scientist, "setting a time bomb that will explode in the face of society anywhere from a month to a generation in the future?"

I think semantics have played a major role in blocking the development of new uses for organic wastes. In the future, we no longer will have "garbage or sewage sludge"; we'll have a more acceptable term, *"Organic Ore."* After this material is "mined"—that is, ground and collected via pipeline—it's ready for the compost plant.

The compost plant will be located in the heart of the farm belt, and the pipelines will take the organic ore right there. Using technology already developed at plants throughout the world, the wastes are stabilized by controlled treatment methods into a humus-rich, hygienic material.

Whatever use is chosen—be it soil conditioning, land reclaiming, etc., the stockpile is there to be recycled like any other natural resource.

(*The following is based on a report by Judith Fowler which originally appeared in the Pennsylvania Gazette and subsequently appeared in the Summer,* 1968 *issue of Compost Science.*)

Dr. Iraj Zandi, associate professor in the University of Pennsylvania's Towne School of Engineering, is a man who is trying to develop a better way for urban wastes to travel from source to utilization site. He takes a dim view of truck collection and believes "no management miracles" or refinement of the present solid waste handling techniques will eliminate all the objectional aspects.

"You can put the men in nice, clean uniforms, you

can get the latest trucks and equipment, but the fact remains they are doing a job that shouldn't be done by hand. The system of collection remains the same and there needs to be a basic change in the system," he believes.

Some of the arguments Dr. Zandi uses against truck collections are the obvious ones of its being inefficient, unsanitary, unsightly, and inconvenient. But truck collection also ties up already overtaxed streets, often during prime afternoon hours. And collection stations, incinerator facilities, and dumping grounds represent mismangement of valuable lands in cities where land comes at a premium, not to mention the ability of these eyesores to extend their worthlessness to surrounding properties.

Collection and removal also demand vast resources of unskilled labor that could be redirected into other, more healthy employment. Also of concern to Zandi is that the method "creates a potential health hazard, contributes to air pollution when incinerated, and adds to water and land pollution when buried."

He believes the truck collection method has to be changed and will be because "obviously, the steady growth of cities combined with a more waste producing, affluent society, will only intensify all these unfavorable characteristics." He also believes that he and his crew of engineers have documented a workable solution that in the long run will prove more economical than truck collection.

The pipeline method, which Dr. Zandi adapted from his knowledge of transport of other material via pipeline and combined with the concept of a slurry, would work something like this:

Homes and businesses would be equipped with some type of solid waste crusher machines, similar to sink disposal units but larger in size, which would be able to pulverize garbage into small pieces. Then small, high-pressure pipelines, designed to transport the ground-up garbage, would be snaked through existing sanitary sewer lines.

Home owners and small businesses would empty garbage into chutes leading down to the pulverizers. After the garbage was reduced, it would empty into pipelines and be carried away. Apartments and large businesses would have a more complicated chute system and need larger pulverizing equipment, but would work in basically the same way. (Businesses with large bulk items would continue to dispose of them through junk dealers.) The pipes would lead to central collection points for treatment or could be carried miles away from the city to be reclaimed, diverted for compost.

Because the system would be on-site and could be used as many hours a day as it took to rid the site of its garbage, there would be no more problems of delayed pickup, odor, flies, or rats and less chance of fire breaking out in piled-up garbage and refuse.

The idea behind the plan is an application of a law of fluid mechanics: If a liquid (slurry) or gas (pressurized air) traveling at a high rate of speed is able to entrap solid material and carry it along, why can't waste water mixed with crushed solid waste traveling at a high rate of speed entrap the solid waste particles and carry them along?

Dr. Zandi notes that industry has used the slurry method for transporting coal, iron ore tailing, gold slim, and fly ash. In Germany, three million tons of coal a year is moved via pipeline. In the U.S., gilsonite, a tradename for a black, lustrous asphalt, is carried the 72 miles from Bonanza, Utah, to Grand Junction, Colo., at the rate of 850 tons a day.

The pneumatic system would use suction to collect all waste in the study area at four main points, where waste water from the sewage system would be mixed in, the waste fed through the crusher and then pumped away.

Based on all the data, Dr. Zandi estimates the people of Philadelphia will spend some $90.6 million over the next 50 years to collect and remove solid waste in the study area to a distance of 50 miles. The distance is important in the light of moves made by the city to investigate and utilize opportunities to dump trash into abandoned strip mines in the northeastern portion of the state via Reading Railroad transport.

Dr. Zandi's statistics show that it would take $88.3 million to put a completely closed, automatic piping system into operation in the study area. This is $1.46 million less than what was estimated for truck collection 50 years from now. Initial installation cost was estimated at $67.34 million.

* * * * * *

(*Dr. Zandi prepared a report of his pipeline*

research for Compost Science. Following are excerpts:)

"In the walls of the cubicle were three orifices. To the right for written messages; to the left, a larger one for newspapers; and in the side wall, within easy reach of Winston's arm, a large oblong slit protected by a wire grating. This last was for the disposal of waste paper. Similar slits existed in thousands or tens of thousands throughout the building, not only in every room but at short intervals in every corridor. For some reason they were nicknamed memory holes. When one knew that any document was due for destruction, or even when one saw a scrap of waste paper lying about, it was an automatic action to lift the flap of the nearest memory hole and drop it in, where upon it would be whirled away on a current of warm air to the enormous furnaces which were hidden somewhere in the recesses of the building."

And so wrote George Orwell in his political novel entitled "1984." Should Mr. Orwell be alive today and could visit Sundeberg, a suburb of Stockholm, Sweden, indeed, he would be pleasantly surprised to find that his prophetic concept of MEMORY HOLES is not any longer a part of a fiction story—but is a reality.

In Sundeberg, the solid waste (all kinds) of 250 apartments is swept along in an air stream of about 90 feet per second mean velocity in underground pipes ranging in diameter from 20 to 24 inches.

(Dr. Zandi concluded his paper by speculating on the possibility of composting wastes

while they are in transit in pipes and also the social consequences of pipeline transport.)

The most challenging of possibilities is the utilization of pipeline itself as a biological reactor for the stabilization of organic components of solid waste slurry to a sufficiently stable state for direct application on land for the purpose of soil buildup and irrigation. Assuming a mean velocity of 3 feet per second in pipeline and a hundred miles transportation distance for detention time in pipeline is about 50 hours—plenty of time for decomposition of organic matter if suitable environmental conditions can be obtained throughout the line. Maybe seedings with microorganisms, addition of some nutrient material and sewage sludges, venting of gaseous products, or oxygenation is required, but the potentialities are great.

It is very difficult, if not impossible to conceive all social implications of a pneumo-slurry system of solid waste collection and removal. However, it is not hard to sense the pleasure of living in a community where one does not have to worry about garbage cans being spilled over, the collection day forgotten, and garbage collectors striking.

How would city slums look with no trash around, flies missing and rats scarce? What kind of air would we breath when thousands of home incinerators in center cities are eliminated? What kind of real productive work would be accomplished by hundreds of thousands of men whose job is now to collect the waste? What will be the effect of new job openings for manufacturing, installing and operating the pneumo-slurry system on the society? These and many other questions, some quite disturbing, should be answered as best as is possible—before collection and removal of solid waste can become accepted.

The level of investment required for pneumo-slurry system is quite high. When investment in technological development is very high a wrong judgement can be extremely expensive. This is why it is so important that a thorough study of all socio-economical and technical aspects of the problem be undertaken prior to any commitment of time and money. Pipeline transport of solid waste is not meant to be a "cure-all" solution. Of course, there will be conditions when pipeline is not at all possible. It should be noted, however, than an inherent advantage of a pipeline system is the scale economy which may be achieved. The regional problems may be handled by one authority—and this very well may be a profit making organization. Many smaller communities which are presently unable to effectively dispose of their solid waste easily can join the larger cities. From the change to new technology comes the need and the opportunity for large business or Governmental organizations. It alone can mobilize the requisite skills, and commit the required magnitude of capital.

When one looks to the crystal ball for an insight into the future and ponders whether or not pipeline transport of solid waste would ever come to pass as a reality, it will be helpful if he remembers the histories of water distribution and waste water collection systems. It is a consequence or affluency and technology that better, more efficient, and reposeful systems will eventually replace the older ones.

In the words of John Galbraith in *The New Industrial State,* "The high production and income which are the fruits of advanced technology and expansive organization remove a very large part of the population from the compulsions and pressures of physical want. In consequence their economic behavior becomes in some measure malleable. No hungry man who is also sober can be persuaded to use his last dollar for anything but food. But a well-fed, well-clad, well-sheltered and otherwise well-tended person can be persuaded as between an electric razor and an electric toothbrush" —or a pneumo-slurry system.

CHAPTER XVII

New Laws Could Be A Big Help

Throughout our nation's history, very little pollution has been abated without legislation or litigation or the threat of such action. Prime areas of concern to public health officials have been a consistent (hopefully pure) water supply, sewage disposal and refuse collection.

Some early attempts at anti-pollution legislation sound rather funny by contemporary standards. In 1797, for example, Pennsylvania still had a law regulating the "importation of Germans" which provided that the sailors at sea weekly burned charcoal or tobacco between decks and washed down the ship with vinegar.

But there was nothing humorous about the epidemics which hit American cities. When more than 5,000 people died in Memphis during an 1879 epidemic, national attention was focused on sewage disposal—or lack of it. The National Board of Health created by Congress exposed shocking conditions.

National boards or panels or similar groups, appointed during the last 90 years, have continued to expose shocking conditions.

* * * * * *

Studies of water pollution problems as they affected public health were authorized by the Public Health Service Act of 1912. But specific legislation bringing the federal government into the pollution arena was not passed until after World War II. The Federal Water Pollution Control Act has become the basic statutory authority for a federal program for the prevention, control and abatement of water pollution.

The Act declared that states and local governments have the chief responsibility for preventing and controlling water pollution. The federal role is to assist the states in fulfilling this responsibility and to work with the states in solving interstate pollution problems.

Many people in and out of government continue to debate the federal role in pollution control. In 1965 hearings on pollution control legislation, Congressman

William Springer of Illinois continually interrupted witnesses who stressed pollution problems to make the point that: "If local waste isn't a local problem, then I don't know of any local problem that you have."

One of the key provisions of the Water Quality Act of 1965 permitted the creation of federal water quality standards for interstate waters and portions thereof. ("Standards" refer to how much and what kind of pollution are permissible.) For the first time, the federal government could act to prevent water pollution instead of having to wait until pollution created health or welfare problems. But the problems of standards and who sets and enforces them rage on and on.

The Water Quality Act also created the Federal Water Pollution Control Administration within the Department of Health, Education and Welfare. FWPCA was transferred to the Department of Interior in 1966, based on the concept that water like other natural resources should be under Interior's control.

Balancing a state's economic interests against a state and federal concern with water purity has kept administrators of the Act in constant turmoil.

* * * * * *

The first clearly-labeled air pollution control law was passed in 1955 and provided for research, technical aid and training. This act also set the basic policy that primary responsibility for air pollution control rests with state and local governments, with the federal government providing leadership and support. Public Law 86-493 in 1960 gave the Public Health Service the authority to study the effects of motor vehicle pollution on health and report to Congress.

The Clean Air Act of 1963 tried to increase the ability of state and local governments to handle air pollution problems by authorizing financial grants to them to develop abatement programs. Auto exhaust was recognized as an uncontrolled air pollution offender, and the Secretary of Health, Education and Welfare got authority to set emission standards for new motor vehicles.

The Clean Air Act Amendments and Solid Waste Disposal Act of 1965 brought garbage into the laps of the federal government.

"We have now reached the point," said President Johnson at the bill signing ceremony, "where our factories, our automobiles, our furnaces, and our municipal dumps are spewing more than 150 million tons of pollutants annually into the air we breathe—almost one-half million tons a day. . . . This has become a health problem that is national in scope. . . . Rachel Carson once wrote: 'In biological history, no organism has survived long if its environment became in some way unfit for it. But no organism before man has deliberately polluted its own environment'."

The 1965 Act created "A National Solid Waste Program," according to the heads of the Public Health Service agency in Washington which would administer many of its provisions. Its two basic purposes were:

1. To initiate and accelerate a national research and development program for new and improved methods of proper and economic solid-waste disposal including

studies directed toward the conservation of natural resources by reducing the amount of wastes and unsalvageable materials and by recovery and utilization of potential resources in solid wastes, and

2. To provide technical and financial assistance to state and local governments and interstate agencies in the planning, development and conduct of solid waste disposal programs.

Two leading PHS officials, Wesley Gilbertson and Ralph Black predicted: "The Congress has forged a chain binding the Federal Government to responsibility for leadership of a national solid waste program. The Solid Waste Disposal Act authorized four major links in the chain: research, demonstration, training and planning. We believe this legislation will help to raise our solid waste programs to high levels of accomplishment."

* * * * * *

The Act did set up and encourage research which had been neglected for so long. The use of wastes in reclaiming worthless land is demonstrated in a number of projects. Other grants hope to reveal how improved waste handling practices and regional solid waste management authorities can help solve problems. Up to two-thirds of the total cost of projects may be financed by Federal funds.

Other projects are specifically concerned with composting and have helped finance plants at Johnson City, Tennessee, and Gainesville, Florida. New waste utilization concepts (converting wastes into marketable chemicals and food supplements) are also being sponsored.

Unfortunately, while the Act is sponsoring a great deal of very interesting research, it does not appear to be attacking the fundamental problem. We are still unable to overcome our heritage of waste. After visiting the headquarters of the agency in Cincinnati which is responsible for assigning grants and judging their merits, I got the distinct impression that most of the men there are about as pessimistic as you can get over chances for developing successful solutions to the solid waste problem. Their depression, despite all their efforts to get the job done well, is further indication that more research is only part of the story.

Prior to passage of the Solid Waste Disposal Act of 1965, Prof. McGauhey of the University of California testified to the lawmakers:

> "Although researchers have been contributing to a clarification of the problem of solid wastes management, the public attitude of financial disinheritance of its rejected goods has been reflected in its support of research. Lack of research, however, has *not* been the critical factor in delaying the solution to solid wastes disposal problems. Research and development must go hand in hand and *development has been slow for many reasons.*
>
> "For example, the principles of composting were clarified by the University of California in 1954. Yet in spite of wide public enthusiasm for conservation of organic matter, the method is but little used today. Nor does the fault lie with developers. Their efforts were stymied by economic factors and institutional aspects which are hard to overcome.

The pertinent fact is that *after a certain point had been reached by research, there was little more that the research could do until development proceeded.*"

At an Engineering Research Conference in July, 1968, Prof. McGauhey in his role as director of the conference concluded:

> "To me, as a long time traveler along the road to composting, the session on that subject was disappointing. Not that the research reported was not in good hands and well reported, but that it revealed a considerable dedication of limited resources of men and money to a replowing of areas well plowed in past years."

* * * * * *

In other words, we know that city wastes can be composted successfully. We have demonstration plants here and abroad which show us clearly that we have solved the problem of taking hideous, unmentionable wastes and making them into something attractive and useful. No longer must we spend a great amount of time and money trying to show people that the principles of composting municipal wastes all work. *We know composting works!*

What we must solve is the problem of what to do with the compost after it has been made. And this gets us into a welter of complications that involve the rights of city governments versus the rights of private corporations, the complex problem of distributing a commodity whose price must be kept fairly low, the problem of promoting a new product and creating a demand for it.

We must get our nation's lawmakers—at the federal, state and local levels—to change their views on public wastes. Our legislators must be made to realize that top priority must be given to recycling wastes back into our environment. To accomplish recycling, we must recognize that we have to pay to deliver wastes to where they can best be used, just as we must pay what it costs to treat them. We must get our officials to stop looking at compost as if this were a new way to make money and begin to look at it as a new, less expensive, more beneficial way to treat wastes.

* * * * * *

Early in 1968, a bill was introduced into the New York State Senate which would create a public corporation for the purpose of improving the handling of solid wastes and reducing environmental pollution. Though the bill right now has been shuffled off to some committee, this type of legislation can provide the right framework for genuine waste management.

As envisioned by its sponsor, Senator Whitney North Seymour, Jr., the bill would help to eliminate most of the barriers now in the way of effective utilization of waste materials, and specifically provide for ways "to develop means for converting organic wastes into usable by-products."

Who knows, it could be the forerunner of a kind of a Tennessee Valley Authority for garbage treatment and utilization.

State Senator Seymour believes that the government should stop acting like a traffic cop in dealing with air pollution in our cities, and start taking an active role of its own in helping to clean up the air. "Not only is the penal system approach for bringing about compliance with air pollution standards a generally unsatisfactory way to engender cooperation, because of the considerable enforcement problems, it is also hypocritical—since government itself is frequently one of the major violators of the same air pollution regulations."

Here are excerpts from Sen. Seymour's bill:

"The problem of handling and disposing of solid waste materials has strained the capabilities of most units of local government. . . . The time has come for affirmative state action to develop improved techniques for handling solid waste, for reclaiming salvageable waste products, and for disposing of refuse and garbage in ways that are least harmful to the environment, in order to promote the general welfare of the state and its inhabitants.

"A public corporation is hereby created, to be known as the New York state technology and reclamation council.

"The purposes of the corporation shall be to encourage and promote research and development in respect to new and improved methods of handling, disposing of and reclaiming solid wastes, to publicize such new methods and to participate in the application of such new methods of solid waste handling, disposal and reclamation.

". . . The Corporation shall have the following powers:

"To develop procedures for reclaiming waste materials, to encourage the implementation of the same, and, where necessary, to *develop market outlets for the sale of salvaged or other reclaimed materials.*

"To provide for the removal and disposition of abandoned automobile hulks.

"To develop means for converting organic wastes into usable by-products."

* * * * * *

WE NEED NEW LAWS! The existing laws which affect what we do with our garbage are most inadequate, if not downright poor, at all government levels. The Solid Waste Disposal Act of 1965, the last time Congress made any real attempt to do something about garbage dumps in this country, is a feeble effort considering the magnitude of the problem. Even the President's own commission to look into waste management and control stressed the gaps in present laws:

"We need new public policies and institutional arrangements before we can attempt many of the technological innovations that are possible.

". . . The role of the federal government is very large. It is only by federal government initiative that the large-scale experimentation and demonstration models can be produced, that the advice for regional agencies and assistance can be generated, that critical data on which criteria and standards can be based can emerge, that research on legislative precedents and the role of the courts in pollution can be stimulated, and that research in engineering, biological, and ecological studies that must form the foundation of both systems development and controls can be supported. It is primarily by federal government initiative that all these can be started, and that industry can be stimulated to perceive the really large business potentials in reprocessing for reuse and other aspects of the management of residues."

The National Academy of Science members, who made up the presidential commission, in their report entitled, *Waste Management and Control,* continue:

"Most obvious, perhaps, are the vast heterogeneity and almost complete lack of coordination among the laws relating to pollution, with respect both to different government levels and to the several pollution environments. There is a lack of uniformity among the states in the development of legislation, in the content of legislation that has been adopted, and in the degree of enforcement. In general, local control ordinances *cannot* cope with pollution sources beyond the jurisdictions involved."

We need new laws to *prevent* pollution *before* it happens. We must have laws which give cities the ability to approach solid waste management on a community-wide or regional basis. I believe that cities can achieve this ability only with the support of the federal government. The Congress of the United States must stop allowing local garbage to make a dump of every state in the union. Indeed, we must make a federal case out of city garbage!

The difficulties in the city-federal relationships are many, complex and, many believe, hopeless. Whether republican or Democrat, some believe that Washington has taken over city rights; some believe Washington ignores city problems.

(Austin Heller, Commissioner of New York City's Department of Air Pollution Control, stresses the need for a new federal partnership with cities. "If the federal government continues to act as if the cities did not exist, failure will result in anti-pollution programs.")

I think the absolute need for federal support to solve crises in urban America has been made abundantly clear by studies in a variety of areas. To argue that the federal government cannot get involved in something as local as garbage, is just as wrong in principle as it is in practice.

The kind of legislation which I consider essential will not make the city *fearful* of using a waste treatment process that yields a residue. More positively, the new legislation will make cities *want* to treat their wastes by a process that will produce the most useful product.

Until we have legislation providing such support, no city official in his right mind is about to stake his elected position on a residue that soon becomes a mountainous headache to get rid of. That's why as great a method as composting is not being widely used.

Under the provisions of the present Solid Waste Disposal Act, the federal government supports research and pilot plants for composting. Great! We're closer to a national policy of turning "garbage into black gold."

We start out with a waste and turn it into a natural resource to improve our nation's soils.

But that is not happening now. We're letting the "black gold" become garbage again by forcing it to be sold. Mayors and public works directors know that all the compost made from their cities' garbage can't be sold, so what are they going to do with ton after ton of the stuff—no matter how good or safe it is? The result is that city officials don't even seriously consider composting.

What kind of legislation would keep this one-time garbage that looks and acts so much like peat moss in the asset category? Since I'm not a lawyer or congressman, I can speculate without handicap. Here goes:

1. The new law would provide financial incentive to a city if it could create a hygienic, stable end product from its municipal wastes. Just as it does with sewage

treatment plants, the federal government could help finance original construction. More important, there could be a federal subsidy to cities based on the number of tons of composted waste used to reclaim public land. Such a subsidy could pay for, or at least defray, transportation costs for moving compost from plant site to utilization area.

2. The new law would encourage states, counties, municipalities and designated regions to set up public corporations to arrange for the recycling of wastes. Such an agency would have the power to arrange for use of treated wastes or to stockpile the materials for use by future generations. Decisions would be based upon where wastes could be used to greatest advantage or sold most profitably.

By creating public agencies to supervise the disposal of treated wastes, we shall take a tremendous step toward recognizing residues as a by-product to be used rather than a pollutant to be dumped somewhere.

A public corporation charged with managing wastes would offer the best solution to a most delicate problem —avoiding competition between government and private industry. No elected official wants to compete with a taxpayer. If a city runs a parking lot, it tries to make sure that it doesn't underprice nearby privately-run lots. If a city was producing an unlimited supply of compost, it would not like to disrupt the marketing methods of local commercial fertilizer firms. A public corporation, charged with overall utilization of treated wastes, would be better able to iron out the problems than elected city officials themselves.

The federal government has a direct commitment to see that city wastes become part of our nation's natural resources. Every natural resource is our heritge for future generations. Once garbage is composted, it becomes a natural resource. Through financial incentives to public corporations charged with managing wastes, the federal government can make it right economically as well as morally for city garbage to be used where it does the most good.

From the ruined land in Appalachia to the eroding slopes of national parks, composted garbage could play a tremendous role in improving the national scene.

* * * * * *

A discussion of new laws which would help garbage become a national asset gets us into the critical area of federal-city government relations. Many of the major campaign themes of the 1968 presidential election revolved around this relationship:—What should the federal government do about law and order on city streets; what should the federal government do to improve city ghettos; what should the federal government do to stop cities poisoning themselves with their own wastes?

Charles Abrams, chairman of Columbia University's city planning department, analyzed the complicated federal-city enigma in an excellent book, *The City Is The Frontier:*

"If the national effort to rebuild our cities limps along aimlessly and fruitlessly, an important reason is that the national power to deal with its urban problems is checked by antiquated theories of states' rights, home rule, and local autonomy that no longer make sense in an era in which 70 per cent of the American people are concentrated in cities and their sprawling metropolitan formations."

Prof. Abrams explains that though the primary responsibility for the general welfare since the Depression has been conceded to the federal government, the state continues to check its exercise in all matters in which the welfare of urban people is involved. This handicap becomes especially evident in developing new solutions to city waste treatment problems.

Under what Abrams refers to as the "cloak of home rule and local autonomy," states have given over much of the responsibility to local governments. The consequences have been a national impotence to deal with problems of poverty and substandard education, and, along with these, the problems of waste treatment. This system means that problems of metropolitan regions with a great many minor jurisdictions must be treated individually even though the problems are common to an entire area.

When Prof. Abrams finishes his analysis of the "bizarre federalism" we have developed—a kind of bastardized luxury in an era of national prosperity—it appears we should give as much attention to the crime-*of*-the-cities as to crime-in-the-cities.

* * * * * *

The change from country-bumpkin America to slick-city America has come about too quickly for us to adjust to—mentally, philosophically, technologically, financially, physically—both as a society and as individuals making up that society.

Just 60 years ago, more than 60 per cent of the total tax revenues was levied by state and local governments. Today, nearly 70 per cent is levied by the federal government. But the power to tax has not been accompanied by the power to act.

And, time and again, the statement is made in the form of the question: "What can be more local a problem than local waste?"

To paraphrase Prof. Abrams, although the central city has become the center of the solid waste problem and therefore should invoke the federal welfare power and bring federal aid to the cities, the federal government—except for modest support in research grants—has remained isolated from the central city's waste problems. The traditional cry of states' rights has further weakened the ability of the federal government to develop breakthrough programs.

In 1961, an Advisory Commission on Intergovernmental Relations suggested the need for "moderate federal action": . . . Wholesale assumption of metropolitan area functions by the federal government is not recommended by few, if any, thoughtful people; but it will surely come to pass if the only alternative is chaos, disintegration, and bickering at the local level. To those who question the justification or the degree of increased federal responsibility recommended in this report, the Commission would point out that moderate federal action now, designed to stimulate more effective state and local action, is much to be preferred to a more unitary approach at a later date."

But years after that report, notes Prof. Abrams, no better prospect of cooperation has appeared, and more and more "thoughtful people" are concluding that the only alternative to "chaos, disintegration, and bickering" is precisely a more positive assertion of federal power. What we need, he says, "is not simply some financial aid to prevent air pollution. What is needed is a regional renewal program for the nation's metropolitan areas and the regional agencies to carry it out. . . . Regional renewal calls for the same imaginative enterprise that spawned the Tennessee Valley Authority, and in some areas it may even require a similar assumption of federal financial responsibility and leadership, including federal land acquisition, land planning, and land development if necessary. It calls for a recasting of federal aid programs so that they will meet not only the needs of cities but also of the regions as a whole."

In *The City Is The Frontier,* Abrams describes how the crisis of the Depression allowed Roosevelt's New Deal to inaugurate federal programs without resistance from the states. He writes:

"A tradition of 150 years which had made the power over environment the preserve of the states and the forbidden territory of the federal government was thus broken. The first onslaught on the formula of state sovereignty since the Civil War was met not by the states calling out their militias, but by welcoming the invader with open arms. The states rights monolith fell quiet under the impact of emergency.

"If there were some who feared the consequences of the federal intrusion, the fears subsided with the promise of local participation. When the Tennessee Valley Authority assured grass roots administration to the governments in the region, it was not long before TVA was not only welcomed but also actively supported by the states wherever cooperation was needed.

". . . Thus, up to 1936, a formula had been carved out of the necessities of the era under which the federal government had assumed the initiative and responsibility for rebuilding urban and suburban America. . . . Had the programs gone on unchallenged, . . . urban conservation might have taken its place of honor with soil conservation, and federal open space programs would have won the esteem held for federal parks. . . . This, however, did not happen. The formula of direct federal building faded from view. The old pattern of hundreds of little autonomous districts was restored. . . . The federal government became the paymaster and underwriter of private risks under a string of assorted programs without theme or aim."

* * * * * *

"There are no easy roads to the better city—and it is in fact easier to build roads than cities, which might be one explanation for the direction of federal policy. But no society can be a great society without great cities. And a nation of ever-widening suburban enclaves thrusting outward from bankrupt cores is no pathway to a great society or even a middling one.

"It is late but not too late to alter the stream of events. But it will require a change in the nation's philosophy.

"The new philosophy must acknowledge that there are values worth preserving in cities as there are in suburbs. . . . must acknowledge that the central city and suburb are an entity. . . . A suburb requires an urb. . . . The new philosophy would redefine state and federal functions in fulfilling the general welfare."

* * * * * *

If truly effective action is to arise, the new laws in the waste treatment field must be specific about the overall concept of making wastes into a national asset. There is relatively little, if any, room for private profit from using all the garbage we produce. I believe that researchers will someday fairly soon uncover procedures to make treated wastes commercially valuable—as raw materials to be used in certain manufacturing processes. But that is not the case today when we're smack in the middle of a garbage crisis. The good law we need today will not make the public good depend upon the good will of private enterprise.

"When the entrepreneurial and the general welfare are bracketed in the same legislation, it should not be surprising that the social purpose will be subordinated."

That's how Prof. Abrams describes the perversion of the Housing Act of 1949 which contained the Urban Renewal Formula. Congress had enacted this legislation to help solve the slum problem, but actually a great many more slum dwellers were evicted than housed.

"The economic motivation had been the dominant ingredient in federal housing recipes from the inception and the stated ideal of better housing for everybody had simply supplied the sweetening, the coloring, and some of the political palatability. Since the welfare of the building industry had won equal place with the people's welfare in the 1949 act, it seemed inevitable that sooner or later the interests of the lower-income families would be forgotten."

* * * * * *

Reconciling our heritage of private enterprise with our commitment to public welfare presents a tremendous legislative hurdle. The challenge is no less great whether the Congress is mulling over public housing or public wastes. Concerning wastes, so far only destruction—burying or burning—have been emphasized. The real challenge is to legislate for *waste construction*.

Thus the present situation regarding waste disposal laws boils down to this: Very, very few persons want Washington to tell the what to do with their local problems. We have had a tradition of solving local problems locally rather than depending upon help and direction from the federal government.

But our present circumstances are altogether different from any that have prevailed in the past. Constitutional guarantees that were satisfactory in the days of the covered wagon have no relevance in a time when there can be no such thing really as a *local* problem of pollution. If the solid waste disposal methods of your community destroy the drinking water of a city in a neighboring state, the federal government must come into the picture. If the air pollution of your city's outmoded incinerator befouls the air of neighboring states, any solution is likely to degenerate into mere squabbling between state legislatures unless there is some kind of federal intervention. Pollution has a way of continuing unabated while debate goes on.

What's more, the problem of pollution abatement has become so expensive that local and state governments need financial help to achieve effective solutions. The federal government alone has the monetary power to help finance model installations and help train personnel.

When it comes to recycling wastes, local governments are almost invariably too limited *physically* and *psychologically,* as well as financially, to develop the potentials of utilization.

The federal government can serve to shore up the flagging spirits of the local and state public works officials who right now are up to their necks in garbage. A few successful waste recycling plants in this country would do wonders toward overcoming the malaise we're in now.

CHAPTER XVIII

Anti-Poverty and Anti-Pollution or Add a C to the Old CCC

Waste utilization could become a constructive force in the anti-poverty drive.

Look at it this way. A city like New York sets up a public corporation to manage the composted wastes produced from its garbage. Suppose that 1,000 tons a day (only a fraction of the total daily refuse in the City) of compost would be used to reclaim lands, build park areas, improve shorelines, hauled to upstate forest lands, etc.

A public corporation, handling that much material,

would generate a great deal of wages for unskilled and semi-skilled workers. In addition to providing jobs quickly for many persons, such work would also allow on-the-job training.

To illustrate the potential in relating conservation goals to anti-poverty programs, let's go back 36 years—to another time of domestic crisis.

* * * * * *

On March 31, 1933, the Civilian Conservation Corps was signed into law. It provided for work on reforestation, road building, flood control, range improvement, and other conservation projects. The CCC was given an appropriation of $323 million, and was to provide jobs for 250,000 unemployed. During those Depression days, about 20 per cent of all unemployed were in the 20 to 24 age group.

In seven years, the CCC enrolled more than 2,250,000 young men in 1,500 camps in every state. Age limits were 17 to 25. Time limit for participation was two years.

The Departments of Agriculture and Interior laid out the work projects which were determined by community needs. The men got training in skilled and unskilled jobs, how to handle tools and equipment. There were educational facilities, where many learned to read and write.

The crisis in cities today is as great as the crisis which existed in the Thirties, if not greater. Why not link up the problems of jobs and wastes—and come up with a solution that would really benefit city dwellers?

What I propose is something that might be called the Civilian Conservation Compost Corps—the CCCC. The composted refuse produced from city wastes could be used to reclaim lands within a 50-mile radius of metropolitan areas—to create parks and greenbelts, to landscape new highway construction, to build and beautify new jetports.

The report of the National Advisory Commission on Civil Disorders has called for the creation of one million jobs in the public sector over a 3-year period. A federally-subsidized program of pollution control-conservation build-up could very definitely achieve that goal. Here's a way that a resource could not only be created out of city wastes but recycled—at the same time providing employment for many persons now in need of jobs.

One of the big problems today is that technology has changed the job market so tremendously that many people in ghettos need training to qualify for employment. And in the past there has been no incentive for Negroes to make the effort to get training, since the jobs weren't open to them anyway. Now many more jobs are open but there's going to be a time gap until training can be carried out.

The job crisis is so overwhelming that re-shaping the Conservation Corps could be an effective way to create opportunities for unskilled workers — opportunities which have become relatively scarce in the Nineteen Sixties. These opportunities in the earlier days of our history made life hopeful for the immigrants who gave birth to today's middle class. We can, in effect, use waste treatment to turn back the clock in the employment field.

Waste utilization is currently playing a role in job creation and rehabilitation. In April, 1968, the Public Health Service announced a demonstration grant of $207,000 to St. Louis for a project to train city workhouse inmates to convert logs, tree trimmings and other forestry wastes into pulpwood for making paper.

As part of a rehabilitation and training program, inmates will operate a facility on workhouse grounds to prepare waste wood, collected in city parks, for pulpwood. Wood that can't be salvaged as well as other bulky, flammable objects from municipal refuse collection will be reduced in size so that they can be handled in a near-by incinerator. A local private hauler has agreed to purchase the pulpwood for resale to paper mills in the area.

* * * * * *

"America is hung up on its cities. More than 70 per cent of all Americans live in only 2 per cent of the space, crowded together into the narrow confines of urban America," says a poverty specialist, former assistant director of the Office of Economic Opportunity and director of Vista, who is at present U.S. Ambassador to Australia.

William H. Crook writes in the *Saturday Review* for September 14, 1968, that our present policy of planning to spend hundreds of billions of dollars trying to make our cities habitable, is short-sighted and doomed to failure. As fast as we wipe out one area of poverty, we become aware of another even larger area. By the time we tear down and rebuild one section of a city, it's past time to tear down and rebuild another section.

To take care of increasing millions of urban population in the near future we are feverishly making plans to pile apartments on apartments, floor on floor and family on family. Says Mr. Crook: "Nowhere does the human spirit suffer more from the enervating effects of poverty than within the packed confines of the urban ghetto."

Let us imagine, he suggests, a world in which these poor ghetto dwellers, presently blamed for their lack of responsibility in rioting and looting, were given something to be responsible *for*. Let's suppose they were given a vested interest in America, something of their own, rather than welfare handouts, urban renewal, and blame.

He goes on to make the audacious and startling proposal that we inaugurate a new Homestead Act. There are 454 million acres of publicly-owned land presently standing barren and uninhabited. This is an area larger than all of Mexico. Why don't we, asks Mr. Crook, give this land away to low income families who are eager to leave the cities and willing to assume responsibility for making the land useful. We can sell the land to them with no down payment and with fifty-year, interest-free loans to provide for improvements needed to make it habitable and profitable.

He is not talking about giving each family "40 acres and a mule" since agriculture in modern times is not successful on such a basis. Nor can we successfully move millions of even the most hopeful and cooperative slum dwellers out to the country overnight and expect them to adapt to this kind of living, know how to deal with their land and make successful lives for themselves. It will be a prodigious and expensive job to make the changeover, but, he says, *it can be done for less money than it will take to rehabilitate our cities.* And it provides a partial answer to some problems that, at the moment, seem to have no answers.

Today six per cent of Americans are responsible for feeding 200 million of us and some 700 million people in other countries. Last year we shipped one-quarter of our wheat crop to India and prevented a famine in that country. In another ten years, the population of India will have grown to such numbers that our entire wheat crop would be needed just to save this one country from famine. As American agriculture moves out into hitherto uncultivated last under the tremendous pressure which the population explosion will bring, the price of land is bound to soar.

Why not give the land *now* to people who will farm it—refugees from blighted urban areas who, in ten or twenty years, can make this land productive and profitable thus helping to feed the burgeoning American population of the coming years, as well as the starving millions in other countries?

Nor is it necessary that all the government-owned land be used for farming, Mr. Crook suggests. "We should not just think of new farms, but of new factories and new stores, of new towns and new cities, of new communities with new needs and new productivity and new opportunities generating new demand—and new profit." Land could be sold at today's prices to industries that will hire the poor.

Mr. Crook points out that very little of the billions we plan to pour into rehabilitating cities will go to creating facilities that will become self-supporting and profit-producing. They will go instead to solve just the pressing problems of city living, chiefly transportation and housing. But if these billions were to be used for reclaiming land by compost applications, irrigation, desalinization and so forth, we will be well on our way to solving not just the problems of congested, uninhabitable cities, but also the problems of feeding a future hungry world.

What is the old Homestead Act which Mr. Crook wants us to revitalize for today's ghetto occupants? The middle of the last century found America facing many of the same problems as today, only not nearly so pressing and so immense. Large numbers of recent immigrants were crowded into Eastern cities, along with native-born Americans—poor, unemployed and, many of them, unemployable. They lived in unbelievable squalor and filth. At the same time, there were millions of acres of uninhabited land in the West and Middle West. In 1862 the Homestead Act was passed by Congress and signed into law by President Lincoln. It provided for the government to give 160 acres of publicly-owned land to anyone over 21, a citizen or an intended citizen, provided that he showed his good faith by cultivating the land and staying with it for five years.

After six months he could obtain full ownership by paying $1.25 an acre. In the 1870's further legislation made it possible for homesteaders to get additional land in unwooded and arid regions.

Within twenty years, over 50 million acres were transferred to private ownership. In 1860 there had been two million farms in the USA. Fifty years later there were more than six million. The total area of the newly homestead productive land was equal to that of Germany and France combined. During these years our output of wheat rose from 173 million bushels to nearly 700 million—one-sixth of the total crop of the world.

True, the Homestead Act did not provide instant paradise. There were abuses galore. Industries, lumber and mining interests fraudulently obtained many of the acres meant for poor farmers. Since no more than four homesteads were allowed to a square mile, the settlers were isolated and lonely, so many of the young folks went back to the cities, just as they are doing today. But, generally speaking, the Homestead Act of 1862 succeeded, at a crucial moment of our history, in providing a dignified and profitable way of life for the urban poor, while it also broadened the base of American agriculture and started us on the road to agricultural plenty.

Modern homesteaders would not encounter the same problems that made 19th century homesteading difficult. Transportation, modern schools, TV and radio almost guarantee a lively social life, even if one lives reasonably far from neighbors. Modern equipment has abolished the back-breaking labor of the last century.

Mr. Crook's emphasis is on relieving the social and economic plight of the ghetto dwellers and providing a meaningful and dignified way of life. To the city planner and engineer, his proposal may represent a partial solution to other problems that concern them specifically. It is well known that problems of waste disposal and sanitation are greater in slum areas than in other parts of the city. It is well known, too, that the enormity of all these problems plus those of transportation and housing is heightened by the very numbers of people living in cities. If large numbers of these folks move out, the situation will be eased for everyone, at least for the time being.

Congestion on grossly overcrowded highways may be sufficiently lessened so there is not the present extreme pressure on future planning. With the exodus of many people who lived in the worst slums, "urban renewal" may take on a new meaning so that far-reaching plans can be made to provide housing for the urban poor who remain.

The problems of waste disposal can be solved by the countryman far more easily than by his city cousin. The compost pile is an essential in the country, not just an experimental luxury. There is no problem of disposing of the finished product. With what agricultural experts now know of soil needs and successful farming methods, homesteaders from the cities could be supplied with a basic education in farming as a necessary part of their job training. They could work closely with government experts. Perhaps, with enough financial backing, the whole experiment could be turned into *a great new showcase for new methods of reclaiming formerly useless land.* Possibly using the compost from nearby city waste disposal plants would become an attractive proposal. One could hardly ask for a more challenging future for people whose lives, we are told, have been, up to now, mostly without direction and meaningless.

Times of great crisis demand extraordinary measures —bold, novel, daring. Perhaps this suggestion by a man well versed in solving poverty problems is just what we need.

* * * * * *

We keep hearing about how many billions of dollars pollution will cost us if we continue our present disposal methods. And we as a nation realize how many billions must be spent to provide jobs for people who are at present unemployable.

Is it too dreamlike to seek a simultaneous solution to pollution and poverty? I believe not, if we are to be optimistic about our ability to solve either problem!

CHAPTER XIX

How to Beautify the "Ugly American"

The Ugly American, a novel of several years back, described repeated attempts to foist U.S.-style solutions on developing countries. A 4-lane super-highway was the answer to economic woes for a young nation without a transportation industry. Diesels were the answer when bikes would have been closer to the truth. Time and again, the would-be aider would ignore the open sewage pits and garbage-strewn streets and concentrate on "aiding" with the latest technological marvel.

I believe we often make the same mistake today when we try to aid emerging nations with their waste disposal problems. We make the mistake by recommending and helping to pay for equipment that requires skilled supervisors. We hire consulting engineering firms to make recommendations and their expertise is mainly with U.S.-oriented waste treatment methods. Some of these engineers have visited me to obtain information about municipal composting. And unfortunately they seem to worry about whether the compost made from garbage and human wastes can be sold.

In today's world, where hunger and poverty shape the course of events, no one can deny the vital role of using land to its fullest potential. Producing food, treating wastes to eliminate health hazards, and reclaiming land are tremendous problems.

That's why compost already plays an important role in many countries overseas. And that's why our foreign aid program would benefit by helping those countries build compost plants.

In such countries as India, Taiwan, Israel, Indonesia and Nigeria, compost from refuse makes an important contribution toward improving soils for crop growth, as well as treating wastes to prevent health hazards. Why would it not be valid for our foreign aid programs to encourage composting in many other nations? Composting could be easily adapted so it needs little if any large-scale mechanized equipment or structures. It could and would help solve both the hygienic and hunger problems facing those peoples.

* * * * * *

The idea of including composting in our aid programs has made some progress. At least, members of the Peace Corps seem to recognize its potential. In late 1962, a Peace Corps Volunteer in the Philippine Islands wrote me of his interest in getting more residents there to make and use compost. After an exchange of letters, we came up with the idea that other volunteers might also want to encourage the use of composting to solve agricultural and sanitary problems in the areas they serve. A series of announcements in Peace Corps publications that Compost Science would be sent free to any volunteer asking for it has since resulted in over 3,000 requests.

I really do not know how much (if any) more compost is being made as a result of the Volunteers reading Compost Science. But I do know from the letters many Volunteers have written me that there is a need for the kind of solution compost offers. Here are excerpts from some of those letters:

Bogota, Colombia—"As to soil improvement practices in my area, they are virtually nil. People here cannot afford tractors. The farmers cultivate every inch of land available including steep hillsides. They are not able to fertilize the burned-over fields with chemical fertilizers because they are too expensive. They do manage, when it is available, to fertilize with manures from cattle, horses, mules or whatever."

Addis Ababa, Ethiopia—"There is little organization for systematic disposal of wastes. Cesspools are found by every house and large containers about the size of a gravel truck box are set along the roads in the residential districts and emptied periodically. Some material picked up in this manner is used as fertilizer. Dysentery is also prevalent in some areas since human waste (not composted) is used as fertilizer."

Mato Grosso, Brazil—"The idea of compost piles is practically unheard of also. I am explaining this idea and have an experiment started showing the 'Hows and Whys of Composting.' On sewage disposal, the people here use privies, but a recent survey showed that only about half enough privies for the number of houses in the village exist. People not having one use their neighbor's or their back yard. We hope to do something about this before we leave."

Gabon, Africa—"As far as human waste materials go (and go they must), there is little use for this material in agricultural production. I have seen in some of the villages collections of goat dung mixed with black topsoil which is found in thin layers in the jungle. Weeds and other grass are placed on top of this until a considerable amount is collected to use for a garden."

Cholutcia, Honduras, Central America—"Animal wastes are not used as fertilizers. As of this date, there is still a great lack of knowledge in investigation of compost science. Perhaps it will be possible some day, after enough research is done, to begin to educate the people here concerning the usable wastes."

Chunya, Tanzania—"As far as I can determine about two-thirds of the farmers (peasants) use manure from their cows (not human) as fertilizer. Their results are noticeable better than those who use none whatsoever. The crops are harvested from March to June. The dead plants are left where they are and (I believe) are gathered together before recultivating the ground in December and burned.

"As for human organic waste, each house has its own outdoor lavatory which is a hole about 8 feet deep with a grass fence around it for privacy."

Guatemala, Central America—"On the use of waste materials in our pueblo, I can only reply that there are none. Everything is utilized in one way or the other. These people have very few material possessions. They mainly utilize natural resources. All of the houses are either made out of adobe or tree limbs. As an example of their resourcefulness, they make a crude sugar candy out of sugar cane. They utilize every part of the cane. The leaves they use for wrapping the finished candy, while the crushed cane fibers are put into the sun to dry and used in their cooking kilns. It is really an efficient operation."

Ghana—"In the Northern Regions of Ghana, on the granitic soils, the farmers practice a settled system of farming in which the land around the compound (area around the house) is cropped continuously.

"Non-compound land beyond is also rested to restore fertility. The compound land receives all the household refuse and also farm yard manure of cow dung, from the cattle which are kept in Kraals. This method of producing farm yard manure is taught by the Extension Service of the Ministry of Agriculture whereby farmers are encouraged to bed their cattle on straw and thus accumulate as much of the nutrients from the excreta as possible during the night. By day the cattle are driven to the field beyond to graze, so that in effect the compound land is being sustained by imports of nutrients.

"The mixture of grass bedding and animal droppings are periodically applied to cereal crops grown around the compound."

* * * * * *

In India, where available cattle manure is so very inadequate for soil needs, "using urban wastes is of the utmost importance," says Dr. Amrik Singh Cheema, Agricultural Commissioner of New Delhi. "These wastes are especially useful for growing vegetables, fodders and short-duration crops for which there is large and rapidly growing demands in the rural areas."

The Indian Ministry of Food and Agriculture has been stressing the appropriate use of urban wastes for increasing agricultural production. Under the Five-Year-Plan in operation now, some 3,000 urban centers hope to achieve a total production of 4.4 million tons of urban compost.

You can see why I believe we should support composting programs wherever we can.

Here's another example, this one from Indonesia,

reported by Ralph E. Carlyle of the United Nations' Food and Agricultural Organization:

"Experiences in Bogor City, Indonesia, indicate that many of the smaller towns and cities in the tropics and sub-tropics could convert organic wastes into useful compost. Admittedly, many cannot afford modern incinerators or elaborate composters. However, almost all towns could have a compost farm piling the organic wastes in orderly rows. The metallic and non-decomposable debris has to be removed, but where labor is cheap this material can be raked out by hand as the waste is unloaded."

Compost has even been a factor in the survival of Nationalist China on the island of Taiwan. Y. S. Tsiang, Commissioner, Sino-American Joint Commission on Rural Reconstruction in Taipei has written: "The use of compost on the Nationalist Chinese island of Taiwan is a key reason why the yield per plot of ground is the highest in the world. It is estimated that about 3½ millon tons of compost can be produced each year on the island."

It just might be that we'd make many more friends throughout the world if our foreign aid program would show how to use garbage and other wastes to grow more food. We could do a lot worse than convincing our allies we're interested in their garbage as well as their politics.

CHAPTER XX

Technology vs. Man (Can Man Win?)

Today it is impossible to analyze any major national problem—from Viet Nam to the urban crisis to waste management—without considering the power of technology and the power of man to control that technology. Our ability to develop controls holds the key to a better world. This does not mean we need *more* technology to develop controls; it does mean we need more understanding and more aggressive action to take control of existing technology.

Every technological step taken results in changes in our environment, changes in the life of human beings. Mostly the steps are taken with little or no thought as to what the ultimate consequences may be. Very often damage done cannot be undone. Yet apparently there is no possibility of slowing down this rapid advance of technology.

In the field of waste disposal, any new invention poses a new problem, for, no matter what is manufactured, there is bound to be some waste in the process which must be disposed of. And as soon as the object has been used and discarded, it presents a new problem of disposal. Will it burn? How soon will it disintegrate if it is buried? Will it pollute waters if it is cast into the sea? Will it rust? Can it be crushed or compacted? Usually we do not have the answers to these questions until many years have been passed. By then it may be too late.

It is a complex and completely new kind of challenge. New rules must be made to deal with it. The future of our world may be at stake. Warnings come from all sides. Here are some of the strong voices raised to predict what kind of a future we may have (if any future at all) if we do not immediately begin to exert some mighty controls on what technology is rushing us into. The men quoted here are recognized authorities in their fields. All are desperately concerned with what we are doing to the world we live in—*the only world we have.*

* * * * * *

"The Most Important Task on the American Agenda"

The regulation of technology is the most important intellectual and political task on the American agenda. Technology is subtracting as much or more from the sum of human welfare as it is adding. We are substituting a technological environment for a natural environment.

All the trouble begins in the American confidence that technology is ultimately the medicine for all ills. It has become the American theology, promising salvation by material works. There is a growing list of things we *can* do that we *must not* do. My view is that toxic and tonic potentialities are mingled in technology and that our most challenging task is to sort them out.

What is needed is a firm grasp on the technology itself, and an equally clear conviction of the primacy of men, women and children in all the calculations.

I am convinced only that political institutions and theory developed in other times for other conditions offer little hope.

The basic way to get at it (our technical society) would be through a revision of the Constitution of the United States. Up to now the attitude has been to keep hands off technological development until its effects are plainly menacing. Public authority usually has stepped in only after damage almost beyond repair has been done: in the form of ruined lakes, gummed-up rivers, spoilt cities and countrysides, armless and legless babies, psychic and physical damage to human beings beyond estimate. The measures that seem to me urgently needed to deal with the swiftly expanding repertoire of toxic technology go much further than I believe would be regarded as Constitutional.

What is required is not merely extensive police power to inhibit the technically disastrous, but legislative and administrative authority to *direct* technology in positive ways: the power to encourage as well as forbid, to slow down as well as speed up, to plan and initiate as well as to oversee developments that are now mainly determined by private forces for private advantage.

—Wilbur H. Ferry,
Vice President,
Fund for the Republic, Inc.

(Excerpted from "Must We Rewrite the Constitution to "Control Technology," Saturday Review, March 2, 1968)

Power to Control . . . or Pollute

The increasing power of technology to destroy and disrupt the natural environment and the ecological systems upon which life depends has reduced man's margin for error in the use of this power.

In the past, pollution of the environment has tended to be slow. But now, the increased power of technology to damage nature, the loss of insulating space, and the compression of time—all reduce our margin for error and the cushion which space and time once provided for our mistakes.

That time and space cushion for our mistakes no longer exists.

The adverse effect on nature of our technological power was not inevitable. It could have been partially avoided with some foresight and thought—with some attention to applying our technological power to preventing pollution, to cleaning it up when it occurred, and to restoring the environment which pollution destroyed.

We are only now starting to restore the balance in the use of our technological power by not only promoting economic growth, but also using our technology, our legal and institutional power to prevent pollution, to limit ecological damage when pollution occurs and to restore what pollution destroys.

We still have a long way to go before our power to prevent and control pollution is equal to our power to pollute—a power which is great, which is increasing, and which has taken many generations of neglect and growth of population, industry, of cities and technology to achieve.

Economic growth and environmental quality go

together. In the long run, economic growth and life itself depend on environmental quality. And, in the short run, of what use is economic growth if, in promoting economic growth, we create an environment in which it is not a joy to live?

—Jacob I. Bregman
Deputy Assistant Secretary
of the Interior for Water
Pollution Control

(Excerpt from speech at the Annual Meeting of the American Institute of Mining Engineers, New York, February 27, 1968)

"Ways We Do Not Understand"

There has been a fundamental change in the way we affect our environment and how it, in turn, recoils upon us. Before the Scientific Revolution, the population explosion, and the concentration of mankind in cities, our external environment was what scientists call in an equation a given or constant. The balance between animal and plant life, the weather, the energy of the sun—i.e., the ecological balance—had gone unchanged for centuries. When a man felled trees to build a house, or shot a deer, or laid rails across the plains, he knew the result of what he was doing because he was slightly changing a constant—the vast pattern of nature. There were not enough of him, nor was the power of his science great enough, to upset the ecological balance in which man himself lived.

Now that has changed. When we alter our urban

environment, we are not changing a system that has been stable for a great period of time. We are changing an extension of ourselves that is in the process of rapid complex growth, and about which we have a limited understanding. We spray pesticide and the birds vanish; we build a superhighway and start a riot; we extend charity and break up families; we pass a law to renew our cities and spill slums over the land. Each action we take alters the total system in ways we do not understand and makes the next crash program more immediately necessary and more ultimately hazardous.

—Lt. Gen. James M. Gavin
USA (Ret.)
Chairman of the Board
Arthur D. Little, Inc.

(Excerpted from "The Crisis of the Cities," Saturday Review, February 24, 1968)

We Need "A Humanistic Technology"

(These comments have been excerpted from a speech by Admiral H. G. Rickover before the British Association for the Advancement of Science.)

There is a tendency in contemporary thinking to regard technology as an irresistible force rather than a tool.

. . . Why should the ease and affluence made possible by technology affect precepts that have guided Western man for centuries? This may brand me as old-fashioned but I have not yet found occasion to discard a single principle that was accepted in the America of my youth. Why should anyone feel in need of a new ethical code because he is healthier or has more possessions or more leisure?

. . . It disturbs me to be told that technology "demands" an action the speaker favors, that "you can't stop progress." It troubles me that we are so easily pressured by purveyors of technology into permitting so-called "progress" to alter our lives, without attempting to control it—as if technology were an irrepressible force of nature to which we must meekly submit. If we reflected, we might discover that not everything hailed as progress contributes to happiness; that the new is not always better nor the old always outdated.

. . . Technology cannot claim the authority of science. It is properly a subject of debate, not alone by experts but by the public as well. It has proved anything but *infallibly* beneficial. Much harm has been done to man and nature because technologies have been used with no thought for the possible consequences of their interaction with nature. A certain ruthlessness has been encouraged by the mistaken belief that to disregard human considerations is as necessary in technology as it is in science. The analogy is false.

. . . Science, being pure *thought,* harms no one; therefore it need not be humanistic. But technology is *action,* and often potentially dangerous. Unless it is made to adapt itself to human interests, needs, values, and principles, more harm will be done than good. Never before, in all his long life on earth, has man possessed such enoromus power to injure his human fellows and his society as has been put into his hands by modern technology.

This is why it is important to maintain a humanistic attitude toward technology; to recognize clearly that, since it is a product of human effort, technology can have no *legitimate* purpose but to serve man—man in general, not merely some men; future generations, not merely those who currently wish to gain advantage for themselves; man in the totality of his humanity, encompassing all his manifold interests and needs, not merely some one particular concern of his. Humanistically viewed, technology is not an end in itself but a means to an end, the end being determined by man himself in accordance with the laws prevailing in his society.

. . . Irretrievable damage has been done by those who use technology without giving thought to its effect on our environment. Waste products, carelessly emitted, create a massive problem of soil, water and air pollution—we may be damaging the atmosphere permanently by changing its chemical composition.

. . . It takes firm commitment to a humanistic technology to push through needed legislation as well as a thorough understanding of the filibustering tactics of opponents, and great skill in combating these tactics. No wonder public opinion and the law have nowhere fully caught up with those who misuse technology.

. . . As citizens of a free, self-governing society, we are individually and severally responsible for the *quality* of our society. The values making for civilized life are neither created nor preserved without continuous effort. In a democracy it is the people themselves who must make this effort.

. . . In an oversimplified way, one might say that in a free society citizens have private liberties and public responsibilities; they safeguard their private liberties by faithfully discharging their public duties.

. . . How we use technology affects profoundly the shape of our society. In the brief span of time —a century or so—that we have had a science-based technology, what use have we made of it? We have multiplied inordinately, wasted irreplaceable fuels and minerals and perpetrated incalculable and irreversible ecological harm. I have thought much about this, and I can find no evidence that man contributes anything to the balance of nature—anything at all. On the strength of his knowledge of nature, he sets himself above nature; he presumes to change the natural environment for *all* the living creatures on this earth. Do we, who are transients and not overly wise, really believe we have the right to upset the order of nature, an order established by a power higher than man?

These are complicated matters for ordinary citizens to evaluate and decide. How in future to make wiser use of technology is perhaps the paramount public issue facing electorates in all industrial democracies; a problem difficult enough in itself but rendered still more so by the strategies of those who wish to continue using harmful technologies.

. . . It cannot be said too often that government has as much a duty to protect the land, the air, the water, the natural environment of man against such damage, as it has to protect the country against foreign enemies and the individual against criminals; conversely, that every citizen is duty bound to make an effort to understand how technology operates and what are its possibilities and limitations.

—Vice Admiral H. G. Rickover
(Copyright 1965, H. G. Rickover)

* * * * * *

Man As Custodian

The material condition of men's lives can profoundly affect the spiritual. Science and technology must be the servant of man, not the master. Therefore, the forces operating to change the agricultural picture must not be permitted to develop without guidance.

Farmers alone cannot assume the total responsibility for conservation. Responsibility must rest on all people if the public wishes to achieve conservation of land and water. Conservation must be supported as a basic national policy just as defense, public health, education, roads, and other measures for the benefit of all.

The basic importance of land and water to the material and moral welfare of man makes conservation a "must." Soil and water are God-given heritages to all people. Man is merely the custodian and good stewardship is a sacred responsibility. The amount of conservation of soil, water and related renewable natural resources that we achieve will have a direct bearing on the future standards of living of the nation and the world.

—Elmer L. Sauer
President
Soil Conservation Society of America

* * * * * *

Environmental Needs in An Affluent Society

Part of the reason why we have not really been able to get ahead of the erosion problem lies in our attitude toward land. The person responsible for erosion is not penalized for allowing his land to deteriorate nor is he charged for the damage to the waterways of the nation. We reward and acclaim, not those who care for land but, those who can make a fortune through land speculation and exploitation. . . .

A problem that is just as important as soil erosion is the decline of soil fertility. We attempt to arrest this with application of fertilizer. . . . Yet

Barry Commoner (noted biologist) has pointed out some of the undesirable side-effects of this heavy reliance upon simple chemical fertilizers. Not only is there some evidence of declining soil structure, but a very serious problem of stream and lake pollution. . . . The community at large is expected to pay the costs of misuse of fertilizer.

Pollution, or the accumulation of substances harmful to living things, in the air, the waters, and the soils of the biosphere, has become a major conservation problem of our affluent society. Yet pollution is by no means a new problem. Even in the absence of man, substances toxic to life accumulated in excessive quantities in some areas. We see them in the salt flats and alkaline pans of our desert states and in the soils formed from copper, lead, zinc or cobalt deposits. Even in the absence of man organic wastes accumulated in excessive quantities, faster than they could be broken down and recycled in the ecosystem. Without such a process there would be no peat, no coal, petroleum, nor natural gas. . . . In the long run, however, if populations continue to grow, we will not be able to afford to dump significant amounts of any pollutant into the common environment. The long range solution to pollution lies in the recycling and reclamation of waste products.

. . . We use pesticides because today's crisis in food and agriculture appears much more important than the more major crisis we may be building for tomorrow through their continued use.

Those who recommend the use of deadly agricultural chemicals growl: "What's more important a dying eagle or a starving child?" But we could have a world with room for eagles in which children did not starve, if we applied our knowledge, our technology, and our affluence in a more sensitive way.

In a society that is not affluent, in a society where people live balanced on the edge of survival, it may be necessary to take chances with environmental health to risk future dangers that we may survive today. But we are living in an affluent society. *We need not endanger the health of our environment in order to survive.* We need not risk the future in order to get past today. We can afford to keep the environment healthy. If we have any concern for our children and grandchildren, we cannot afford to have it any other way.

—Raymond F. Dasmann
The Conservation Foundation
Washington, D.C.

What Price—A Quality Environment

Recently a staff study within the Federal Government made a first rough estimate of the likely cost of abating some pollutants. When the study started, the participants, myself included, almost unanimously felt that their findings would reveal that even limited abatement of some air and water pollutants would cost society many billions of dollars annually. One guess was $20 billion. Moreover, it was felt that industry may have a heavy burden with air and water pollution abatement expenditure; certainly heavier than routine real wage changes each year. You can imagine the study members' surprise when the results of the study indicated that society's and industry's abatement costs may be *much lower than anyone dare guess.*

"For example, the additional *annual* cost of reducing human exposure to sulphur oxides and particulates by 60 to 75 per cent in all metropolitan areas (containing 70 per cent of the population) was estimated to about $3.4 billion. . . . In contrast, another industry cost, labor costs, rose by 5 per cent or 30 times as much during 1967.

". . . Maintaining the quality of the environment is primarily a problem for society because those who benefit are not necessarily those who pay and environmental quality is a public good."

—Jack W. Carlson
Director, Program Evaluation Staff
Bureau of the Budget

Fidelity to An Ideal

Beyond all plans and programs, true conservation is ultimately something of the mind—an ideal of men who cherish their past and believe in their future. Our civilization will be measured by its fidelity to this ideal as surely as by its art and poetry and system of justice. In our perpetual search for abundance, beauty, and order we manifest both our love for the land and our sense of responsibility toward future generations.

A century ago, we were a land-conscious, outdoor people: the American face was weather-beaten, our skills were muscular, and each family drew sustenance directly from the land. Now marvelous machines make our lives easier, but we are falling prey to the weakness of an indoor nation and the flabbiness of sedentary society.

We can have abundance and an unspoiled environment if we are willing to pay the price. We must develop a land conscience that will inspire these daily acts of stewardship which will make America a more pleasant and more productive land. If enough people care enough about their continent to join in the fight for a balanced conservation program, this generation can proudly put its signature on the land. But this signature will not be meaningful unless we develop a *land ethic*.

Only an ever-widening concept and higher ideal of conservation will enlist our finest impulses and move us to make the earth a better home both for ourselves and for those as yet unborn.

—Stewart Udall
Secretary of the Interior
The Quiet Crisis

CHAPTER XXI

Creative Conservation and the Compost Philosophy

"Conservation can be defined as the wise use of our natural environment: It is, in the final analysis, the highest form of national thrift—the prevention of waste and despoilment while preserving, improving and renewing the quality and usefulness of all our resources."

President John F. Kennedy spoke those words in his Conservation Message to Congress in 1962.

Conservation must be basic to our national policy

and actions. In the field of waste prevention (which directly relates to the quality of our conservation efforts), composting can serve a most valuable purpose. I hope these pages have made this quality of the compost process clear to you.

President Kennedy's attitudes on conservation and using all our resources are, luckily for us, shared by others. And it remains our challenge to translate moral commitments to something as ineloquent as garbage.

Time and again, we have heard our leaders and ourselves declare that it is within our power to determine that our nation be beautiful as well as healthy. We know we must recognize and legislate for our aesthetic and spiritual values as well as our physical and monetary goals.

To continue on our present course with garbage and other wastes is to accept the certainty of final defeat.

As a responsible citizen, you cannot ignore the fate of garbage. If you care about the quality of our land, water and air, you've got to make sure we don't ruin them with the wastes produced today.

I have tried to show you that something better *can* be done with garbage. To improve, though, we've got to *think* differently and we've got to *act* differently. We must do what is necessary to provide a favorable environment for life on this earth, and we can start right now with our garbage.

Don't be satisfied with a city garbage disposal plan that creates a dump 20, 50, 100 or 1,000 miles away from where you live. When your garbage creates a dump, sooner or later, it will make a dump of where you live.

We must create resources not dumps. Then and only then will we achieve genuine creative conservation. Then and only then will we achieve the goal of improving the world around us.

Composting as a waste treatment process has a key role in taking us to that goal. Instead of burning or burying, we can turn our garbage into a resource that can help build America. It makes sense morally, technically and economically.

Of all alternative methods for treating our wastes, composting stands alone in its ability to create a potentially-valuable resource. Compost *conserves* what may be useful, just as other wastes *destroy* what may be useful. That is why composting deserves to be called creative conservation.

It's going to take a lot of talking with public officials, experts and neighbors if you are going to help your city solve its garbage problem. But like other critical problems we face today, if we don't make the effort with our voice and our vote, we'll just keep sliding further and further away from effective solutions.

I hope this book has convinced you why we must do something now and why we can be optimistic if the something we do is composting.

APPENDIX

Technical and Mechanical Aspects of Composting

The following section offers a brief summary of composting technology used in United States composting plants, (some of which are no longer in operation), and is included for those readers directly concerned with the engineering aspects of waste treatment.

Over the years, quite a number of machines and methods have been developed to compost refuse. Some methods have made direct use of experiences and technology developed in Europe, while others have been strictly American-bred.

Without attempting to get you bogged down in detail, I would like to give you some highlights of some of the methods used in United States composting plants and some briefer descriptions of methods in use overseas. For most of these descriptions, I am leaning heavily on reports published in *Compost Science* and also upon such authorities in this country as John Wiley of the Public Health Service and Dr. Charles Harding of the consulting engineering firm of Reynolds, Smith & Hills. A report issued by the Johnson & Anderson engineering company of Pontiac, Michigan and The Ducker Research Company of Birmingham, Michigan has also been excerpted.

Johnson City, Tennessee—Windrow Composting

Under the combined efforts of the Public Health Service, the Tennessee Valley Authority and municipal officials of Johnson City, Tennessee, a 60-ton-per-day capacity windrow type plant began operation in 1967. Refuse is brought into the plant, manually sorted and ground in either a Williams (St. Louis, Mo.) hammermill or a Dorr Oliver (Stamford, Conn.) rasp. The material is then moistened and conveyed to an outside area where windrow piles are made. The windrows are turned five to ten times with a Cobey (Crestline, Ohio) turning unit during about five weeks of composting. Windrow composting of this nature has been done successfully in many European plants and is primarily suited for smaller cities with adequate land sites available. It is estimated that about 30 acres would be required for a windrow plant to serve a city of about 100,000 population.

Altoona, Pennsylvania—Fairfield-Hardy System

Since 1951, Altoona has been composting garbage. In December 1963, Fairfield Engineering Company (Marion, Ohio) set up a Hardy digester. The pilot plant processes about 25 tons—a Williams hammermill is used as a primary grinder, and a secondary grinding is done in a wet pulper. After going through a dewatering press, the material is fed into the circular digester. Stirring is done by augers suspended from a rotating bridge in the circular tank. Air is provided by a blower and air pipes are embedded in the floor of the tank. The material is composted for about five days and operating temperatures are about 140-160° Fahrenheit. Successful experiments have been run in which digested sewage sludge was added to the ground refuse.

Altoona Fam, Inc., reputedly has a contract with the City of Altoona (population 67,000) which gives the company a flat fee of $32,000 annually. Projecting the plant's average input of 25 tons per day, over one year (computed on the basis of a 5-day week), the plant now processes 6,500 tons of waste per year making the per ton income from the city about $4.92.

Altoona Fam makes 9 tons of compost each day from its 25-ton input, putting yearly production in excess of 2,000 tons from the plant. The company had no difficulty in selling its 1966-67 production of compost. The selling price was $16.50 per bulk ton F.O.B. the plant, or $42.50 per ton in 40-lb. bags of compost (the bags retail at $1.95 under the trade name "Sun Soil"). Most of the compost is sold in bulk to fertilizer blending plants in Pennsylvania and in Ohio. For the 1967-68 season, the compost was priced at $20.50 per bulk ton. According to a Company official, fertilizer blending plants had to be "rationed"; there was not enough Altoona Fam compost to meet the demand. He said, ". . . we know from our marketing experience that we could easily handle a plant here in Ohio that would produce 30,000 tons of compost per year."

All of Altoona Fam Corporation's compost output is pelletized and screened.

St. Petersburg, Florida—International Disposal Corporation System

Incoming refuse is sorted, then run through a magnetic separator at the 105-ton-per day International Disposal Corporation plant in St. Petersburg. The material is then fed into a rotary mixer called a pulverator, after which it enters a grinder. The digester is housed in a vertical building with horizontal, moving belts on which the ground refuse becomes compost. Air is blown into the pile, and temperatures are kept in the thermophilic (160°-180° F.) range. The digester consists of five nine-foot wide steel apron conveyors operating in individual, insulated, heat-retaining cells stacked one above the other. The first two are 165 feet long; the rest, 150 feet long. Each conveyor travels its own length in 12 hours and remains stationary for an equal period.

The material is reground after two days, and at the end of five days, the material is removed and passed through a screen. Westinghouse Electric (Pittsburgh, Pa.) designed much of the St. Petersburg plant.

The International Disposal Corporation (Shawnee, Okla.), previously known as the Naturizer Company, had operated plants in San Fernando, California, 70 tons per day, and Norman, Oklahoma, 35 tons per day. The Norman plant featured a specially-designed swing hammermill as well as the digester with a movable floor in each story.

Houston, Texas—Metropolitan Waste

The Houston plant, as described elsewhere, is a 350-ton-per day design with the incoming refuse hand sorted, then ground in either a hammermill or a Joy Centriblast unit, providing inertial separation. The material is then passed through a magnetic separator, a secondary grinder and is moistened with sewage solids or nitrogen solution prior to composting. The batch digesters are horizontal tanks with perforated bottoms. The ground refuse is kept in the tanks for 4 to 6 days, depending upon plant operating conditions. Air can be blown through the bottom on a controlled basis. A special agitator-unloader is used to mix the material or to unload it when composting is completed.

The Metropolitan Waste Company has also set up a 150-ton-per-day plant in Gainesville, Florida.

Mobile, Alabama—Windrowing

At the 300-ton-per-day Mobile plant, after separation, the material goes through a primary and secondary grinding. A Gruendler (St. Louis, Mo.) unit is used.

Reduced refuse is made into windrows which are

periodically turned. The plant which had been constructed with private capital, was taken over by the city and is operated to meet the demand for the finished compost, which is marketed by the City of Mobile under the brandname, *Mobile-Aid*.

Mobile plans to sell bagged compost in units of 25 and 50 pounds. Prices would be about $1.10 and $1.85. The present selling price per bulk ton is reported to be $8.00 (the plant makes fibrous, not pellitized, compost). Of the 20,000 tons produced since start-up, a reported 5,000 tons was sold (Auburn University bought 2,000 tons; Battleship Park took 200 tons, as examples), 10,000 tons has been given away to the Mobile citizenry in a public relations program, and the remainder is in storage at the plant.

In terms of formal marketing information, the City does have a preliminary survey, which is said to have concluded that the market for compost in Mobile is unlimited because the humus content of the sandy soil is very low.

Phoenix, Arizona—The Dano System

The Phoenix, Arizona, composting plant was the second Dano-type plant in the United States, the first one being in Sacramento, California. The Phoenix plant, with a 300-ton-per-day capacity, originally had two Dano biostabilizers, and an additional two units were added in March, 1963, shortly before the plant closed down. The plant provided sorting, magnetic separation, salvage, hammermill grinding, composting in the biostabilizers for 24 to 30 hours, screening, regrinding, deglassing and further composting in windrows. A Pennsylvania Crusher (Broomall, Pa.) was used to shred garbage.

Dano plants are in use in a number of countries such as Israel, Switzerland, Germany and Holland.

Holland—Dorr-Oliver Rasping Unit

In just about 20 minutes, raw refuse is separated, ground and piled in neat windrows at the Arnhem, Netherlands, composting plant. This 125-ton-per-day plant serves a population of 160,000. Total cost of the plant was estimated at about $830,000. After original sorting, material is conveyed under a magnet and then discharged onto a third conveyor and then discharged into the rasping machine, made by the Dorr-Oliver Company.

The rasping machine consists of a cylindrical tank (5.5 meters diameter) with double floors. The upper floor consists of segments which are perforated with holes of 22 mm. The upstanding side is fitted with pins. Eight heavy steel arms rotate at a clearance of 10 cm. above the plates, and are also free to move in a vertical direction. These arms press on the refuse and carry it around by their weight. The reduced refuse falls through to the lower floor, where four arms rotate in the same manner, driven from the same shaft.

The discharged material is then conveyed to a hammermill which further reduces the size of the material and also serves as a ballistic separator.

The reduced refuse is carried on to an electrically-driven belt conveyor. The material is dropped from a slowly rotating "arm," which discharges it into a semicircular heap. After setting four weeks, the heaps are turned over for the first time with a front-end loader. After another four weeks, the heaps are turned over for a second time. The compost is piled on a brick pavement with built-in drains and does not attract flies or rodents.

The rasping system developed in the Netherlands is in widespread use.

Other European Methods

There are many examples of high speed shredders in use at composting plants in Europe. These include Hazemag, Gondard, Buhler, Tollemache, and British-Jeffrey-Diamond. Composting systems in use include those developed by Earp-Thomas, with plants in Italy and Greece; the "Fermascreen"—a refuse reduction unit developed by the John Thompson Company.

At Jersey in the British Isles, a 60-ton-per-day plant has been processing sludge as well as garbage. After sorting, refuse is delivered into either of two pulverizing machines. The refuse including sludge is loaded by a crane into the top floor of the fermentation cells. The cells contain six floors, one above the other, each arranged to open and discharge the contents of the one below (or into vehicles in the case of the bottom one), thoroughly turning and aerating the material in the process. This operation is carried out each day—discharging the material from the bottom floor and emptying the top floor to receive the new day's supply; therefore the compost is in the cells for a week.

The plant in Bangkok, Thailand, works on basically the same principle as the Jersey plant. The Jersey plant started in 1955 and has been operating continually since. The Bangkok plant began operating in 1960.

At the town of Cheadle in England, the Simon plant uses a 3-floor totally enclosed digester with the refuse being slowly swept from floor to floor by radial arms mounted on a central shaft. Retention period is three days.

Moscow has built a 600-ton-per-day plant using a Triga "Hygienizator", made by a Paris firm. This is a type of digester divided vertically, which offers a continuous "closed system" with the material recirculated through the digester during a four-day fermentation.

In 1964, G. J. Kupchick, under a World Health Organization Fellowship, undertook a comprehensive study of composting costs in Europe and Israel. His analysis of 14 composting plants, serving a total population of over 3,136,000, indicated an average composting cost of $4.55 per short ton (2,000 pounds) of raw refuse received. Although the sale price of the completed compost averaged $2.73 per short ton, with a maximum price of $5.45 per short ton, it averaged only $0.90 income per ton of raw refuse received by the composting plant. (Salable compost weight was between 16 and 70 percent of the weight of the incoming raw refuse, the loss being due to removal of noncompostables, moisture loss, and the organic degradation loss that converts refuse to humus.) The net cost of composting (with allowance for iron, rag, and paper salvage, if conducted) amounted to $3.38 per short ton of raw refuse accepted.

The Public Health Service publication on "Composting—European Activity and American Potential" includes a survey of those 14 European composting plants. The plants described are as follows:

Bad Kreuznach, Germany, (population served, 45,000). Type of plant, Dano.

Blaubeuren, Germany, (population served, 20,000). Type of plant, windrowing of shredded refuse.

Duisburg-Huckingen, Germany, (population served, 90,000). Type of plant, Dano.

Heidelberg, Germany, (Population served, 30,000). Type of plant, Multi-Bactor compost tower.

Schweinfurt, Germany, (population served, 85,000). Type of plant, Caspari-Brikollare.

St. Georgen, Germany, (population served, 14,000). Type of plant, windrow of ground refuse.

Stuttgart (the suburb of Mohringen), Germany, (population served, 75,000). Type of plant, windrow of ground refuse.

Versailles, France, (population served, 82,000). Type of plant, Triga, silo type.

Buchs, Switzerland, (population served, 40,000). Type of plant, windrow of ground refuse.

Hinwill, Switzerland, (population served, 100,000, in 23 surrounding communities). Type of plant, Dano.

La Chaux-de-Fonds, Switzerland, (population served, 15,000 [estimated]). Type of plant, original Dano (built in 1953).

Turgi, Switzerland, (population served, 70,000, in 10 communities; Turgi population 5,000). Type of plant, Multi-Bacto compost tower.

Arnhem, Holland, (population served, 134,000). Type of plant, Windrow, of ground refuse.

Wijster, Holland, (population served, 1 million plus: Den Haag and other communities). Type of plant, Van Maanen system.

Landfill and Grinding—Prelude to Composting

Recently, much research has been done on the advantages of grinding refuse prior to landfill. In this process, the word "stabilization" is used, although the process itself is a form of composting.

Briefly stated, the purposes of shredding are to extend the life of a landfill, eliminate the need for a soil cover of the refuse, and allow for a controlled stabilization (or composting) to take place.

In this country, a pilot project at Madison, Wisconsin, supported by a PHS grant, is making an economic study of the Gondard Process (a French machine distributed in this country by the Heil Company) of milling solid wastes. A visit to the project area in July, 1968, indicated that the milling is working successfully, eliminating the need for a soil cover. Refuse, which was ground 12 months earlier, had compost-like qualities.

In a study entitled, "New Waste Process Concept to Extend Disposal Area Use Life" by Harold G. Mason of the URS Corporation of Burlingame, California, statistics are given to show that a combination of grinding-landfill-composting could make landfill areas in the

San Francisco Bay area effective for another 25 years at least. Another advantage mentioned for this procedure is that, while buildings cannot be erected on a sanitary landfill, construction can be done on top of refuse that has been shredded and composted.

Economic Feasibility Study

During the autumn of 1968, a report was issued by the Johnson and Anderson consulting engineering company of Pontiac, Michigan, and the Ducker Research Company of Birmingham, Michigan. This excellent analysis, supported in part by a grant from the Public Health Service, is entitled "Economic Feasibility Study for Refuse and Sludge Composting Plant."

The report analyzes costs of the compost method, as well as costs for disposing of excess compost in landfill. Total cost for mechanized composting is slightly less than incineration. The big economic advantage for composting is the income that can be expected from salvaging and from sale of compost. This is estimated at $6.74 per ton of refuse, and the authors predict an additional income of at least $1.00 per ton of refuse from the sale of compost after 2 or 3 years of successful plant operations.

Observes the report: "The future of composting municipal wastes is dependent upon federal subsidy and control. Clearly, the conversion of refuse into compost is a process that would have died through lack of success were it not for the acute problem of disposing of municipal wastes that the nation faces today. Air pollution has caused incineration of refuse to fall from favor; its costs have risen too. Landfill sites near many large cities are becoming more difficult to find. Turning waste substances into a semi-nutrient product that acts as a good soil conditioner as well seems, suddenly, to be an answer, partially at least, to the question of what to do next.

"It is expensive to build compost plants. Lack of past success may serve as a deterrent to private investors. . . . If much is to be accomplished in proving the concept of composting municipal wastes, Federal funds will be needed to provide the research and the facilities."

The market analysis for the Riverview, Michigan, area establishes a market for 25,000 tons of compost per year. This amounts to approximately 100 tons of compost per day on a 5-day-per-week basis. A 150-ton plant will produce approximately 70 tons of compost per day and a 300-ton plant will have an output of between 100 to 170 tons per day, depending on the proportion of fibrous to pelletized output, degree of salvaging, etc.

Disposal of non-compostable, non-salvageable materials will require landfilling operations. It is estimated that 10 per cent by weight of incoming refuse i.e., 30-tons-per-day, will be disposed of by land filling at a cost of $2.00 per ton.

Since the market analysis for the sale of compost indicates that a market exists for only a portion of the total compost production of the plant, an additional operational expense occurs in disposing of the non-saleable compost. Until such time as the market for compost increases to the point of erasing this extra cost, and indeed turns it into additional income, the disposal of the extra compost is a very real part of the cost picture. For the 300-ton-per-day plant under consideration, 70 tons of compost per day are estimated as nonsaleable for as long as the market for compost remains at 25,000 tons per year.

The cost of disposal to a nearby landfill is estimated at $2.00 per ton of compost, thus arriving at a cost per day of $140.00 and a cost per ton of refuse of 140/300 = $0.47.

Interviews by the marketing consultants held with the two major companies now operating compost plants have yielded the following price range for salvageable materials:

Cardboard and paper—$10 to $15 per ton
If exceptionally clean, it may sell for $22 per ton
Metal and Cans—$8 to $12 per ton
Glass—$8 to $10 per ton

Experience of compost operating companies shows the existence of ready markets for salvaged metals and glass. One compost company reports there is a demand from glass companies for salvaged glass separated by colors, as well as for mixed colors.

Following are additional excerpts from the "Economic Feasibility Study for Refuse and Sludge Composting Plant" prepared by Johnson and Anderson and the Ducker Research Company:

Most of the compost plants started in the United States have shut down, often because the fees paid by municipalities to plant operators for each ton of refuse processed ranged from $1 to $5, while the cost to process a ton of refuse is between $5 and $10. Running a deficit, the operators looked belatedly into the marketing of compost. Still believing that there was a wide requirement for the commodity, they tried to sell it to the people they thought had the greatest need for humus—farmers. They were, except in Florida, usually disappointed. The market was not there.

Another reason for failure has been the choice of a poor location for the plant, with the result that neighbors strenuously objected.

Even now, after the lengthy list of shut-down plants has become common knowledge, refuse disposal contracts are negotiated with municipalities at rates that are below the cost of plant operation. The same causes are at work . . . eagerness to set up a deal and get a plant built, plus the competitive effect of other disposal methods such as landfilling and incineration. Therefore, there is a continued need for the additional income of end-product sales in order to make a profit.

Home Market for Organic Soil Conditioners

The home garden market has experienced a rapid, steady growth over the past ten years. In 1965, retail sales alone are estimated to have been $370,000,000.

In a recent survey conducted by Aaron L. Mehring, former Fertilizer Specialist with the U.S. Department of Agriculture, consumption of non-farm fertilizers was synthesized, using approved research techniques (no direct statistics are available). Data developed for the year 1960 was employed in this report and then projected through 1965 using a 6 per cent per annum growth rate which experts regard as entirely realistic.

Total non-farm fertilizer consumption is expressed in tons by geographical region and by state within regions, for the years 1960 and 1965. Total consumption in 1960 of 2,650,500 tons is shown increasing to 3,546,-500 tons in 1965.

In the East, North Central and West North Central regions, 50 per cent of the households use fertilizers, but unit consumption is low; about 32 pounds per annum. National usage is similar; only 50.9 per cent use fertilizer. Based upon the Mehring survey, there are indications that 70 per cent of higher income households use fertilizers while as few as 2.5 per cent of low income home owners use fertilizers. Thus, there appears to be a large sales potential for a lower-priced, good-quality fertilizer.

Where Compost Has Value and Economics Still Play

In a recently-published PHS report, *Solid Wastes Management / Composting European Activity and American Potential,* author Samuel Hart discusses use of compost to reclaim despoiled land such as that found in strip mine areas. (There are some 800,000 acres of such land in the U.S.) Prof. Hart also envisions a way in which composting can aid landfilling operations by using composted garbage to produce cover material for burying the rest of the refuse. "In this way a rubbish mountain, covered with trees and grass growing on compost, would benefit future inhabitants of the city."

Prof. Hart goes into great detail on the use of the

composting process when it is "nonbeneficial but economic." Here are some excerpts:

> It is a middle ground, between using compost at 15 to 50 tons per acre-year for crop production or land improvement, and using the land as a refuse burial site. (About 800 tons of refuse will raise the elevation of 1 acre of landfill site by 1 foot.) Conceive of the application of 100, 200, or 300 tons of rough-quality compost per acre per year, year after year. The land would be an acceptor, degrader, and stabilizer of the waste; crop production would probably not occur; but neither would the land be lost to a future use.
>
> . . . Put bluntly, it would be insulting Mother Earth without quite violating her. Each year, as much rough compost—or perhaps just ground refuse—would be incorporated into the soil as could be assimilated. When the land is needed for another purpose, the compost application would be stopped, and the land would recover. This concept of compost utilization might be likened to a biological incinerator. The applied organic matter would be burned and consumed by the soil microbiological activity. The operating cost of this biological incinerator will probably be higher than running a thermal one, but the capital cost, and thus the amortized cost, should be much lower. For many cities this scheme might well be preferred over conventional incineration. As a hypothetical example, if a soil could take 100 tons of rough compost per acre per year, produced from

200 tons of domestic refuse, this would be a land requirement of 1 acre per 500 people. This is even less land than is required for sewage lagoons.

. . . At this stage, it is impossible to estimate what percentage of the nation's refuse might be so managed. It would appear that this system might have merit for communities with populations between 10,000 and 100,000. There are 1,760 such communities in the United States, with a total population of nearly 50 million, or 25 percent of the nation's population. If this method of refuse disposal were appropriate for even one-third of these communities, it would be a substantial avenue of waste disposal.

Acknowledgement

My special thanks go to my wife for having suggested I write this book in the first place, and to Ruth Adams who reviewed the several versions of the manuscript and whose additions appear throughout.